VIDEO NOTEBOOK

COLLEGE ALGEBRA
WITH INTERMEDIATE ALGEBRA
A BLENDED COURSE

Judith A. Beecher

Judith A. Penna

Barbara L. Johnson
Ivy Tech Community College of Indiana

Marvin L. Bittinger
Indiana University Purdue University Indianapolis

PEARSON

Boston   Columbus   Indianapolis   New York   San Francisco
Amsterdam   Cape Town   Dubai   London   Madrid   Milan   Munich   Paris   Montreal   Toronto
Delhi   Mexico City   Sao Paulo   Sydney   Hong Kong   Seoul   Singapore   Taipei   Tokyo

Copyright © 2017 Pearson Education, Inc.
Publishing as Pearson, 501 Boylston Street, Boston, MA 02116.

ISBN-13: 978-0-13-455590-4
ISBN-10: 0-13-455590-2

1   16

www.pearsonhighered.com

CONTENTS

Set Notation and the Set of Real Numbers

ESSENTIALS

A **set** is a collection of objects. A **subset** is a set contained within another set. **Roster notation** is a way of designating a set by enclosing a list of its elements in braces.

Natural numbers (or **counting numbers**) are those numbers used for counting:
$\{1, 2, 3, \ldots\}$.

Whole numbers are the set of natural numbers with 0 included: $\{0, 1, 2, 3, \ldots\}$.

Integers are the set of all whole numbers and their opposites:
$\{\ldots, -4, -3, -2, -1, 0, 1, 2, 3, 4, \ldots\}$.

The **opposite** of a number is found by reflecting it across the number 0 on the number line. The natural numbers are called **positive integers**. The opposites of the natural numbers are called **negative integers**.

Set-builder notation is a way of designating a set that specifies conditions under which a number is in a set.

A **rational number** can be expressed as an integer divided by a nonzero integer. The set of rational numbers is

$$\left\{ \frac{p}{q} \,\middle|\, p \text{ is an integer, } q \text{ is an integer, and } q \neq 0 \right\}.$$

Rational numbers are numbers whose decimal representation either terminates or has a repeating block of digits. **Irrational numbers** are numbers whose decimal representation neither terminates nor has a repeating block of digits. They cannot be represented as the quotient of two integers. The set of all rational numbers, combined with the set of all irrational numbers, gives us the set of **real numbers**. The set of real numbers is
$\{x \mid x \text{ is a rational number } or \ x \text{ is an irrational number}\}$.

Examples

- Represent the set consisting of the first 6 odd natural numbers.
  Roster notation: $\{1, 3, 5, 7, 9, 11\}$

  Set-builder notation: $\{n \mid n \text{ is an odd number between 0 and 13}\}$

- List the set(s) of numbers to which 3.5 belongs.
  Since 3.5 is a terminating decimal, it is a rational number, and it is also a real number.

- π and $\sqrt{13}$ are examples of irrational numbers.

GUIDED LEARNING   (1) **Textbook**   (👤) **Instructor**   (▶) **Video**

EXAMPLE 1	YOUR TURN 1
Find the opposite of 4. When 4 is reflected across 0 on the number line, we find the opposite. The opposite is the number ☐ .	Find the opposite of –6.
EXAMPLE 2	YOUR TURN 2
Use roster notation to write the set of natural numbers greater than 20. $\{21, 22, \boxed{}, 24, \ldots\}$	Use roster notation to write the set of whole numbers between 6 and 15.
EXAMPLE 3	YOUR TURN 3
Use set-builder notation to write the set $\{-5, -4, -3, -2, -1, 0, 1\}$. $\{x \mid x$ is an integer greater than $\boxed{}$ and less than $\boxed{}\}$	Use set-builder notation to write the set $\{7, 8, 9, 10\}$.
EXAMPLE 4	YOUR TURN 4
Which numbers in the list are (a) natural numbers? (b) whole numbers? (c) integers? (d) rational numbers? (e) irrational numbers? (f) real numbers? $5, -\dfrac{1}{3}, -4, \pi, 3.6, 0, \sqrt{2}$ a) Natural number: 5 b) Whole numbers: 5, ☐ c) Integers: 5, ☐ , ☐ d) Rational numbers: 5, ☐ , –4, ☐ , 0 e) Irrational numbers: ☐ , $\sqrt{2}$ f) Real numbers: $5, -\dfrac{1}{3}, -4, \boxed{}, \boxed{}, 0, \sqrt{2}$	Which numbers in the list are (a) natural numbers? (b) whole numbers? (c) integers? (d) rational numbers? (e) irrational numbers? (f) real numbers? $0.2, 0, -16, -\dfrac{1}{9}, \sqrt{7}, 125, 1$

YOUR NOTES   Write your questions and additional notes.

Order for the Real Numbers

ESSENTIALS

Inequality Symbols

< means "is less than."

> means "is greater than."

≤ means "is less than or equal to."

≥ means "is greater than or equal to."

If x is a positive real number, then $x > 0$.

If x is a negative real number, then $x < 0$.

$a < b$ also has the meaning $b > a$.

Sentences containing $<, >, ≤,$ or $≥$ are called **inequalities**.

Examples

- $-6 > -9$ because -6 is to the right of -9 on the number line.
- The inequalities $x > 8$ and $8 < x$ have the same meaning.
- Write the meaning of the inequality and determine whether it is a true statement: $-4 ≥ -2\dfrac{1}{4}$.

    "-4 is greater than or equal to $-2\dfrac{1}{4}$" is false because -4 is to the left of $-2\dfrac{1}{4}$ on the number line.

GUIDED LEARNING	📖 **Textbook** 👤 **Instructor** ▶ **Video**
EXAMPLE 1	**YOUR TURN 1**
Use either < or > for ☐ to write a true sentence. 7 ☐ 2 Since 7 is to the right of 2 on the number line, 7 is greater than 2, so $7 > 2$.	Use either < or > for ☐ to write a true sentence. -11 ☐ -6
EXAMPLE 2	**YOUR TURN 2**
Use either < or > for ☐ to write a true sentence. -8 ☐ 3 Since -8 is to left of 3 on the number line, -8 is less than 3, so $-8 < 3$.	Use either < or > for ☐ to write a true sentence. 1 ☐ -15

EXAMPLE 3	YOUR TURN 3
Use either $<$ or $>$ for ☐ to write a true sentence. 6.24 ☐ 6.56 Since 6.24 is to left of 6.56 on the number line, 6.24 is less than 6.56, so $6.24 < 6.56$.	Use either $<$ or $>$ for ☐ to write a true sentence. $\dfrac{2}{5}$ ☐ $\dfrac{4}{7}$
EXAMPLE 4	YOUR TURN 4
Write a different inequality with the same meaning as $3 \geq x$. x ☐ 3	Write a different inequality with the same meaning as $y < 10$.
EXAMPLE 5	YOUR TURN 5
Determine whether the following is true or false. $-6 \geq -1.8$ The statement is _____ since neither $-6 > -1.8$ nor $-6 = -1.8$ is true.	Determine whether the following is true or false. $3 \geq -7\dfrac{3}{4}$

YOUR NOTES   Write your questions and additional notes.

Graphing Inequalities on the Number Line

ESSENTIALS

A replacement that makes an inequality true is called a **solution**. The set of all solutions is called the **solution set**. A **graph** of an inequality is a drawing that represents its solution set.

Examples

- Graph $x \le -2$.

  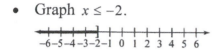

- Graph $x > 3$.

  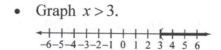

	📖 **Textbook**	👤 **Instructor**	▶ **Video**

GUIDED LEARNING

EXAMPLE 1	YOUR TURN 1
Graph: $x > 1$.	Graph: $x < 2$.
The solutions consist of all real numbers _____ 1. less than / greater than	←―――――――――――→ –6–5–4–3–2–1 0 1 2 3 4 5 6
They are shown on the number line by shading all numbers to the _____ of 1. left / right	
The parenthesis at 1 indicates that 1 _____ a is / is not solution.	
←―――――――――――→ –6–5–4–3–2–1 0 1 2 3 4 5 6	
EXAMPLE 2	**YOUR TURN 2**
Graph: $x \le -3$.	Graph: $x \le 4$.
The graph consists of ☐ as well as the numbers less than -3.	←―――――――――――→ –6–5–4–3–2–1 0 1 2 3 4 5 6
They are shown on the number line by shading all the numbers to the _____ of -3. left / right	
The bracket at -3 indicates that -3 _____ a is / is not solution.	
←―――――――――――→ –6–5–4–3–2–1 0 1 2 3 4 5 6	

YOUR NOTES   Write your questions and additional notes.

Absolute Value

ESSENTIALS

The **absolute value** of a number is its distance from 0 on the number line. We use the symbol $|x|$ to represent the absolute value of a number x.

Examples

- $|-9| = 9$ The distance of -9 from 0 is 9.

- $|20| = 20$ The distance of 20 from 0 is 20.

- $|0| = 0$ The distance of 0 from 0 is 0.

	📖 **Textbook**	👤 **Instructor**	▶ **Video**
GUIDED LEARNING			

EXAMPLE 1	YOUR TURN 1
Find the absolute value: $|14|$. The distance of 14 from 0 is ☐ , so $|14| = $ ☐ .	Find the absolute value: $|-39|$.
EXAMPLE 2	YOUR TURN 2
Find the absolute value: $|-8.1|$. The distance of -8.1 from 0 is ☐ , so $|-8.1| = $ ☐ .	Find the absolute value: $\left\|\dfrac{7}{8}\right\|$.

YOUR NOTES   Write your questions and additional notes.

Practice Exercises

Readiness Check

Determine if each statement is true or false.

1. _____ The number 2.61324189 . . . is a rational number.

2. _____ The number 0 is a natural number.

3. _____ All rational numbers and irrational numbers are real numbers.

4. _____ Absolute value is never negative.

Set Notation and the Set of Real Numbers

Which numbers in the list provided are (a) natural numbers? (b) whole numbers? (c) integers? (d) rational numbers? (e) irrational numbers? (f) real numbers?

5. $-3.1, \ -1, \ 0, \ \sqrt{10}, \ \dfrac{75}{2}, \ 4$

6. $-\dfrac{3}{8}, \ \dfrac{1}{2}, \ \sqrt{12}, \ \dfrac{43}{2}, \ 51$

Use roster notation to write each set.

7. The set of letters in the word "history"

8. The set of all natural numbers that are multiples of 5

Use set-builder notation to write each set.

9. The set of all multiples of 10 between 20 and 100.

10. $\{72, 74, 76, 78, 80\}$

Order for the Real Numbers

Use either < or > for ☐ *to write a true sentence.*

11. 1 ☐ −15

12. −42 ☐ −21

13. −13.2 ☐ −15.1

14. Write a second inequality with the same meaning as $x \geq -7$.

Write true or false.

15. $0 \leq 2$

16. $6 \geq 6$

17. $-2 < -10\dfrac{1}{3}$

Graphing Inequalities on the Number Line

Graph on the number line.

18. $x < -4$

19. $x \geq 1$

Absolute Value

Find the absolute value.

20. $|5|$

21. $|-3.8|$

22. $\left|\dfrac{0}{5}\right|$

Addition

ESSENTIALS

To find $a + b$ using the number line, we start at 0, move to a, and then move according to b.

- If b is positive, move to the right.
- If b is negative, move to the left.
- If b is 0, stay at a.

Rules for Addition of Real Numbers

1. *Positive numbers*: Add the numbers. The result is positive.

2. *Negative numbers*: Add absolute values. Make the answer negative.

3. *A positive number and a negative number*:
   - If the numbers have the same absolute value, the answer is 0.
   - If the numbers have different absolute values, subtract the smaller absolute value from the larger. Then:
     a) If the positive number has the greater absolute value, make the answer positive.
     b) If the negative number has the greater absolute value, make the answer negative.

4. *One number is zero*: The sum is the other number. This rule is known as the **identity property of 0**. It says that for any real number a, $a + 0 = a$.

Examples

- To add $5 + (-8)$ on the number line, we start at 0 and move 5 units right. Since -8 is negative, we then move 8 units left. $5 + (-8) = -3$

   $$-6\ -5\ -4\ -3\ -2\ -1\ \ 0\ \ 1\ \ 2\ \ 3\ \ 4\ \ 5\ \ 6$$

- Add: $-7 + (-5)$.

  Both numbers are negative. Add absolute values, 7 and 5, and make the answer negative: $-7 + (-5) = -12$.

- Add: $12 + (-12)$.

  One number is positive and one is negative. The numbers have the same absolute value. The answer is 0: $12 + (-12) = 0$.

GUIDED LEARNING

📖 **Textbook**   👤 **Instructor**   ▶ **Video**

EXAMPLE 1	YOUR TURN 1
Add: $-3 + 7$.	Add: $-2 + 5$.
Start at 0 and move ☐ units left. Then move ☐ units _____.	$-6\ -5\ -4\ -3\ -2\ -1\ \ 0\ \ 1\ \ 2\ \ 3\ \ 4\ \ 5\ \ 6$
$-6\ -5\ -4\ -3\ -2\ -1\ \ 0\ \ 1\ \ 2\ \ 3\ \ 4\ \ 5\ \ 6$	
$-3 + 7 = $ ☐	

EXAMPLE 2	YOUR TURN 2
Add: $-10+(-28)$. Both numbers are _____. _____ absolute values, 10 and 28, and make the answer negative. $-10+(-28)=\boxed{}$	Add: $-6+(-11)$.

EXAMPLE 3	YOUR TURN 3
Add: $-9+0$. One number is zero. The sum is the other number. $-9+0=\boxed{}$	Add: $-13+13$.

EXAMPLE 4	YOUR TURN 4
Add: $-2.5+10.3$. One number is positive and one is _____. _____ the smaller absolute value from the greater absolute value: $10.3-2.5=7.8$. Since $\lvert10.3\rvert>\lvert-2.5\rvert$, the answer is _____. $-2.5+10.3=\boxed{}$	Add: $-0.5+8.5$.

EXAMPLE 5	YOUR TURN 5
Add: $\dfrac{1}{4}+\left(-\dfrac{2}{3}\right)$. One number is _____ and one is negative. _____ the smaller absolute value from the greater absolute value: $\dfrac{2}{3}-\dfrac{1}{4}=\dfrac{8}{12}-\dfrac{3}{12}=\dfrac{5}{12}$. Since $\left\lvert-\dfrac{2}{3}\right\rvert>\left\lvert\dfrac{1}{4}\right\rvert$, the answer is _____. $\dfrac{1}{4}+\left(-\dfrac{2}{3}\right)=\boxed{}$	Add: $\dfrac{3}{10}+\left(-\dfrac{1}{5}\right)$.

YOUR NOTES   Write your questions and additional notes.

Opposites, or Additive Inverses

ESSENTIALS

Two numbers whose sum is 0 are called **opposites**, or **additive inverses**, of each other. For any real number a, the opposite, or additive inverse, of a, which is denoted $-a$, is such that

$$a + (-a) = 0.$$

A negative number is sometimes said to have a "negative sign." A positive number is said to have a "positive sign." When we replace a number with its opposite, or additive inverse, we can say that we have "*changed the sign.*"

For any real number a, the absolute value of a, denoted $|a|$, is given by

$$|a| = \begin{cases} a, & \text{if } a \geq 0, \\ -a, & \text{if } a < 0. \end{cases}$$ For example, $|6| = 6$ and $|0| = 0$.
For example, $|-9| = -(-9) = 9$.

(The absolute value of a is a if a is nonnegative. The absolute value of a is the opposite of a if a is negative.)

Examples

* The opposite of 27 is -27 because $27 + (-27) = 0$.

* The opposite of $-\dfrac{3}{5}$ is $\dfrac{3}{5}$ because $-\dfrac{3}{5} + \dfrac{3}{5} = 0$.

* The opposite of 0 is 0 because $0 + 0 = 0$.

	🔲 **Textbook**	👤 **Instructor**	▶ **Video**
GUIDED LEARNING			

EXAMPLE 1	YOUR TURN 1
Find the opposite, or additive inverse, of 11. $-(11) = \boxed{}$ The opposite of 11 is $\boxed{}$.	Find the opposite, or additive inverse, of -15.
EXAMPLE 2	YOUR TURN 2
Evaluate $-x$ and $-(-x)$ when $x = -12$. If $x = -12$, then $-x = -(-12) = \boxed{}$. If $x = -12$, then $-(-x) = -(-(-12)) = \boxed{}$.	Evaluate $-x$ and $-(-x)$ when $x = 7$.

EXAMPLE 3	YOUR TURN 3
Change the sign of $-\dfrac{3}{14}$.	Change the sign of $\dfrac{2}{25}$.
(Find the opposite, or additive inverse.)	(Find the opposite, or additive inverse.)
The opposite of $-\dfrac{3}{14}$ is $\boxed{}$.	

YOUR NOTES   Write your questions and additional notes.

Subtraction

ESSENTIALS

The difference $a - b$ is the number c for which $a = b + c$.

For any real numbers a and b, $a - b = a + (-b)$. (We can subtract by adding the opposite, or additive inverse, of the number being subtracted.)

Examples

- $8 - 9 = 8 + (-9) = -1$ Changing the sign of 9 and adding

- $-3 - (-7) = -3 + 7 = 4$ Changing the sign of -7 and adding

GUIDED LEARNING	📖 **Textbook**	👤 **Instructor**	▶ **Video**

EXAMPLE 1	YOUR TURN 1
Subtract: $5 - 9$. Change the sign of 9 and add. $5 - 9 = 5 + \left(\boxed{} \right) = \boxed{}$	Subtract: $1 - 6$.
EXAMPLE 2	YOUR TURN 2
Subtract: $-2.6 - (-9.1)$. Change the sign of -9.1 and add. $-2.6 - (-9.1) = -2.6 + \boxed{} = \boxed{}$	Subtract: $\dfrac{1}{4} - \left(-\dfrac{1}{2} \right)$.

YOUR NOTES   Write your questions and additional notes.

Multiplication

ESSENTIALS

To multiply a positive number and a negative number, multiply their absolute values. Then make the answer negative.

To multiply two negative numbers, multiply their absolute values. The answer is positive.

Examples

- $2(-5) = -10$
- $(-4)(-5) = 20$
- $-2(-1.8) = 3.6$

GUIDED LEARNING	📖 **Textbook**	👤 **Instructor**	▶ **Video**

EXAMPLE 1	YOUR TURN 1
Multiply: $(-3)(20)$. Since one number is negative and one is positive, the answer is _____. <u>positive / negative</u> $(-3)(20) = \boxed{}$	Multiply: $-8(11)$.
EXAMPLE 2	YOUR TURN 2
Multiply: $\left(-\dfrac{2}{7}\right)\left(-\dfrac{3}{7}\right)$. Since both numbers are negative, the answer is _____. <u>positive / negative</u> $\left(-\dfrac{2}{7}\right)\left(-\dfrac{3}{7}\right) = \boxed{}$	Multiply: $\left(-\dfrac{4}{5}\right)\left(-\dfrac{1}{3}\right)$.

YOUR NOTES   Write your questions and additional notes.

Division

ESSENTIALS

The quotient $a \div b$, or $\dfrac{a}{b}$, where $b \neq 0$, is that unique real number c for which $a = b \cdot c$.

To multiply or divide two real numbers:
1. Multiply or divide the absolute values.
2. If the signs are the same, then the answer is positive.
3. If the signs are different, then the answer is negative.

Division by 0 is not defined and is not possible.

Two numbers whose product is 1 are **reciprocals** (or **multiplicative inverses**). Every nonzero real number a has a **reciprocal** (or **multiplicative inverse**) $1/a$. The reciprocal of a positive number is positive. The reciprocal of a negative number is negative.

For any real numbers a and b, $b \neq 0$,

$$a \div b = \frac{a}{b} = a \cdot \frac{1}{b}.$$

(To divide, we can multiply by the reciprocal of the divisor.)

For any numbers a and b, $b \neq 0$,

$$\frac{-a}{b} = \frac{a}{-b} = -\frac{a}{b} \quad \text{and} \quad \frac{-a}{-b} = \frac{a}{b}.$$

Examples

- $\dfrac{8}{-2} = -4$ $\qquad$ $\dfrac{-8}{2} = -4$ $\qquad$ $-8 \div (-2) = 4$

- $\dfrac{2}{0}$ is not defined. $\qquad$ $0 \div 2 = 0$ $\qquad$ $\dfrac{2}{x-x}$ is not defined.

- The reciprocal of $-\dfrac{3}{4}$ is $-\dfrac{4}{3}$ because $\left(-\dfrac{3}{4}\right)\left(-\dfrac{4}{3}\right) = 1.$

- The reciprocal of 6 is $\dfrac{1}{6}$ because $6\left(\dfrac{1}{6}\right) = 1.$

- $\dfrac{2}{5} \div \left(-\dfrac{3}{10}\right) = \dfrac{2}{5} \cdot \left(-\dfrac{10}{3}\right) = -\dfrac{20}{15}$, or $-\dfrac{4}{3}$

GUIDED LEARNING   📖 **Textbook**   👤 **Instructor**   ▶ **Video**

EXAMPLE 1	YOUR TURN 1
Divide: $4.2 \div (-3)$.	Divide: $-14.8 \div (-2)$.
Do the long division $3\overline{)4.2}$ with 1.4 above.	
The answer is negative.	
$4.2 \div (-3) = \boxed{}$	

EXAMPLE 2	YOUR TURN 2
Divide, if possible: $\dfrac{-8}{0}$.	Divide, if possible: $\dfrac{0}{4}$.
$\dfrac{a}{0}$ is not defined for any real number a, so $\dfrac{-8}{0}$ is _____.	

EXAMPLE 3	YOUR TURN 3
Find the reciprocal of $\dfrac{-1}{6}$.	Find the reciprocal of $-\dfrac{4}{7}$.
The reciprocal of $\dfrac{-1}{6}$ is $\dfrac{\boxed{}}{\boxed{}}$, or $\boxed{}$.	

EXAMPLE 4	YOUR TURN 4
Divide by multiplying by the reciprocal of the divisor: $\left(-\dfrac{3}{8}\right) \div \left(-\dfrac{1}{2}\right)$.	Divide by multiplying by the reciprocal of the divisor: $\dfrac{1}{3} \div \left(-\dfrac{5}{9}\right)$.
$\left(-\dfrac{3}{8}\right) \div \left(-\dfrac{1}{2}\right) = \left(-\dfrac{3}{8}\right) \cdot \left(-\dfrac{\boxed{}}{\boxed{}}\right) = \dfrac{6}{8}$ $= \dfrac{3 \cdot \boxed{}}{4 \cdot \boxed{}} = \dfrac{3}{4} \cdot \dfrac{\boxed{}}{\boxed{}} = \dfrac{\boxed{}}{\boxed{}}$	

YOUR NOTES   Write your questions and additional notes.

Practice Exercises

Readiness Check

Choose the word or phrase below the blank that will make the statement true.

1. To add two negative numbers, we _____ their absolute values and make the answer negative.
   add / subtract

2. To multiply two negative numbers, multiply their absolute values. The answer is _____.
   positive / negative

3. Division by zero is _____.
   defined to be 0 / not defined

4. To divide by a fraction, _____ by its reciprocal.
   divide / multiply

Addition

Add.

5. $-11 + (-12)$

6. $11.6 + (-31.8)$

7. $-\dfrac{3}{7} + \dfrac{9}{14}$

Opposites, or Additive Inverses

Evaluate $-a$ for each of the following.

8. $a = 13.6$

9. $a = -\dfrac{7}{2}$

10. $a = 0$

11. Evaluate $-(-x)$ when $x = 14$.

12. Find the opposite of -2.5. (Change the sign.)

Subtraction

Subtract.

13. $0-(-32)$ **14.** $-90.6-(-43.2)$ **15.** $\dfrac{2}{3}-\dfrac{4}{5}$

Multiplication

Multiply.

16. $-5\cdot 3$ **17.** $-8\cdot(-6.1)$ **18.** $\left(\dfrac{3}{4}\right)\cdot\left(-\dfrac{1}{2}\right)$

Division

Divide, if possible.

19. $-12\div(-6)$ **20.** $\dfrac{-77}{11}$ **21.** $5\div 0$

Find the reciprocal of each number, if it exists.

22. $-\dfrac{7}{5}$ **23.** -8

Divide.

24. $-\dfrac{4}{3}\div\left(-\dfrac{5}{6}\right)$ **25.** $\dfrac{3}{4}\div\left(-\dfrac{3}{16}\right)$

Exponential Notation

ESSENTIALS

Using **exponential notation**, we can write the product $2 \cdot 2 \cdot 2 \cdot 2 \cdot 2$ as 2^5.

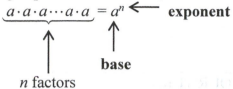

$$\underbrace{a \cdot a \cdot a \cdots a \cdot a}_{n \text{ factors}} = a^n \longleftarrow \textbf{ exponent}$$

base

We read a^n as "a to the nth power," or simply "a to the nth." We can read a^2 as "a-squared" and a^3 as "a-cubed."

For any number a, we agree that a^1 means a. For any nonzero number a, we agree that a^0 means 1.

Examples

- $4^3 = 4 \cdot 4 \cdot 4 = 64$
- $4^1 = 4$
- $3^0 = 1$

GUIDED LEARNING   📖 **Textbook**   👤 **Instructor**   ▶ **Video**

EXAMPLE 1	YOUR TURN 1
Write exponential notation: $x \cdot x \cdot x \cdot x$. There are 4 factors of x. $x \cdot x \cdot x \cdot x = x^{\square}$	Write exponential notation: $m \cdot m \cdot m \cdot m \cdot m \cdot m$.
EXAMPLE 2	YOUR TURN 2
Evaluate: $(-6)^2$. $(-6)^2 = \boxed{} \cdot \boxed{} = \boxed{}$	Evaluate: $(-7)^2$.

EXAMPLE 3	YOUR TURN 3
Evaluate: $\left(\dfrac{3}{5}\right)^3$. $\left(\dfrac{3}{5}\right)^3 = \dfrac{3}{5} \cdot \boxed{} \cdot \boxed{} = \dfrac{27}{125}$	Evaluate: $\left(\dfrac{5}{8}\right)^2$.

EXAMPLE 4	YOUR TURN 4
Evaluate: $(4.6)^3$. $(4.6)^3 = \boxed{} \cdot \boxed{} \cdot \boxed{} = \boxed{}$	Evaluate: $(10.1)^2$.

EXAMPLE 5	YOUR TURN 5
Evaluate: $-(2)^4$. $-(2)^4 = -\boxed{} \cdot \boxed{} \cdot \boxed{} \cdot \boxed{} = -16$	Evaluate: $(-2)^4$.

YOUR NOTES   Write your questions and additional notes.

Negative Integers as Exponents

ESSENTIALS

For any real number a that is nonzero and any integer n,

$$a^{-n} = \frac{1}{a^n}.$$

A negative exponent does not necessarily indicate that an answer is negative!

Examples

- $n^{-3} = \dfrac{1}{n^3}$

- $\dfrac{1}{x^{-6}} = x^6$

GUIDED LEARNING	📖 **Textbook** 👤 **Instructor** ▶ **Video**	

EXAMPLE 1	YOUR TURN 1
Rewrite using a positive exponent: b^{-4}. $$b^{-4} = \frac{1}{\boxed{}}.$$	Rewrite using a positive exponent: t^{-9}.

EXAMPLE 2	YOUR TURN 2
Rewrite using a positive exponent: $(-6)^{-3}$. Evaluate, if possible. $$(-6)^{-3} = \frac{1}{(-6)^{\square}} = \frac{1}{(-6)(-6)(-6)} = -\frac{1}{\boxed{}}$$	Rewrite using a positive exponent: $(5)^{-2}$. Evaluate, if possible.

EXAMPLE 3	YOUR TURN 3
Rewrite using a positive exponent: $\left(\dfrac{2}{3}\right)^{-2}$. Evaluate, if possible. $$\left(\frac{2}{3}\right)^{-2} = \frac{1}{\left(\frac{2}{3}\right)^2} = \frac{1}{\boxed{}} = 1 \cdot \frac{9}{4} = \frac{\boxed{}}{\boxed{}}$$	Rewrite using a positive exponent: $\left(\dfrac{3}{5}\right)^{-3}$. Evaluate, if possible.

EXAMPLE 4	YOUR TURN 4
Rewrite using a negative exponent: $\dfrac{1}{y^7}$. $\dfrac{1}{y^7} = y^{\square}$	Rewrite using a negative exponent: $\dfrac{1}{(-9)^2}$.

YOUR NOTES   Write your questions and additional notes.

Order of Operations

ESSENTIALS

Rules for Order of Operations

1. Do all the calculations within grouping symbols, like parentheses, before operations outside.
2. Evaluate all exponential expressions.
3. Do all multiplications and divisions in order from left to right.
4. Do all additions and subtractions in order from left to right.

When parentheses occur within parentheses, **computations in the innermost ones are to be done first**.

In addition to parentheses, brackets, and braces, a fraction bar and absolute-value signs can act as grouping symbols.

Example

$$
\begin{aligned}
7 - 6^2 \div 2(-3) - (1 - 5) &= 7 - 6^2 \div 2(-3) - (-4) \\
&= 7 - 36 \div 2(-3) - (-4) \\
&= 7 - 18(-3) - (-4) \\
&= 7 + 54 - (-4) \\
&= 61 - (-4) \\
&= 65
\end{aligned}
$$

GUIDED LEARNING   🔖 **Textbook**   👤 **Instructor**   ▶ **Video**

EXAMPLE 1	YOUR TURN 1
Simplify: $4 - 15 \div 3 - 7$.	Simplify: $3 - 4 \cdot 5 + 9$.

There are no grouping symbols or exponential expressions, so we begin with the division.

$$
\begin{aligned}
4 - 15 \div 3 - 7 &= 4 - \boxed{} - 7 \\
&= \boxed{} - 7 \\
&= \boxed{}
\end{aligned}
$$

EXAMPLE 2	YOUR TURN 2
Simplify: $18 \div 2 - \left\{ 7 - \left[(1-4)^2 + 3 \right] \right\}$.	Simplify: $-5 - \left\{ \left[(9+7) \div 2^3 \right] + 6 \right\} + 4$.

We do the calculations in the innermost grouping symbols first.

$$18 \div 2 - \left\{ 7 - \left[(1-4)^2 + 3 \right] \right\}$$

$$= 18 \div 2 - \left\{ 7 - \left[(\boxed{})^2 + 3 \right] \right\}$$

$$= 18 \div 2 - \left\{ 7 - \left[\boxed{} + 3 \right] \right\}$$

$$= 18 \div 2 - \left\{ 7 - \boxed{} \right\}$$

$$= 18 \div 2 - \left\{ \boxed{} \right\}$$

$$= 18 \div 2 + \boxed{}$$

$$= \boxed{} + 5 = \boxed{}$$

EXAMPLE 3	YOUR TURN 3
Simplify: $\dfrac{2 - 4\left[1 - (5-7)^3 \right]}{6^2 - 5^2}$.	Simplify: $\dfrac{12 - \left[5 - (2^2 - 3^2) \right]}{7 - (3-4)^5}$.

We simplify the numerator and the denominator separately.

$$\frac{2 - 4\left[1 - (5-7)^3 \right]}{6^2 - 5^2} = \frac{2 - 4\left[1 - (\boxed{})^3 \right]}{\boxed{} - \boxed{}}$$

$$= \frac{2 - 4\left[1 - (\boxed{}) \right]}{\boxed{}}$$

$$= \frac{2 - 4\left[\boxed{} \right]}{11}$$

$$= \frac{2 - \boxed{}}{11} = \frac{\boxed{}}{11} = \boxed{}$$

YOUR NOTES   Write your questions and additional notes.

Practice Exercises

Readiness Check

In each of Exercises 1-6, name the operation that should be performed first. Do not calculate.

1. $2(12 \div 3 + 7 - 9)$

2. $14 - 5 \cdot 3 + 8$

3. $3\left[4(24 \div 12) - 15\right]$

4. $(9 - 5) \cdot 4 + 16$

5. $4 \cdot 3 - 9 \cdot 5 \div 3$

6. $12 - 3(7 + 4)$

Exponential Notation

Write exponential notation.

7. $t \cdot t \cdot t$

8. $\dfrac{5}{11} \cdot \dfrac{5}{11} \cdot \dfrac{5}{11} \cdot \dfrac{5}{11} \cdot \dfrac{5}{11}$

9. $(-54.9)(-54.9)(-54.9)(-54.9)$

Evaluate.

10. 2^4

11. $(-5)^3$

12. 1.2^1

13. $(-6)^0$

Negative Integers as Exponents
Rewrite using a positive exponent. Evaluate, if possible.

14. $\left(\dfrac{3}{2}\right)^{-3}$

15. $\dfrac{1}{y^{-5}}$

16. -3^{-3}

Rewrite using a negative exponent.

17. $\dfrac{1}{5^3}$

18. $\dfrac{1}{x^5}$

19. $\dfrac{1}{(-6)^3}$

Order of Operations

Simplify.

20. $14 - 7 \cdot 3 + 8$

21. $6 - 2(4-9)^2 + 1$

22. $72 \div 3(1-5) - 2^2 - 8$

23. $\left[4 - (-11-3)\right] \div 3^2 - 9$

24. $\dfrac{(-2)^3 + 3^2}{4 \cdot 5 - 6^2 + 2 \cdot 7}$

25. $\dfrac{-22 + 4^3}{6 - (3-4)^5}$

Translating to Algebraic Expressions

ESSENTIALS

When a letter is used to represent various numbers, it is called a **variable**. If a letter represents one particular number, it is called a **constant**.

An **algebraic expression** consists of variables, numbers, and operation signs, such as $+, -, \cdot, \div$. When an equals sign, $=$, is placed between two expressions, an **equation** is formed.

Key Words

Addition	Subtraction	Multiplication	Division
add	subtract	multiply	divide
sum	difference	product	quotient
plus	minus	times	divided by
total	decreased by	twice	ratio
increased by	less than	of	per
more than			

Examples

Translate to an algebraic expression.

- Nine less than some number: $w - 9$

- Ten more than three times a number: $3y + 10$, or $10 + 3y$

- Four less than twenty-three percent of some number: $23\%x - 4$, or $0.23x - 4$

- The sum of a number and 83: $b + 83$, or $83 + b$

- Some number divided by 0.7: $\dfrac{x}{0.7}$, or $x \div 0.7$

GUIDED LEARNING	🔖 **Textbook**	👤 **Instructor**	▶ **Video**

EXAMPLE 1	YOUR TURN 1
Translate the phrase to an algebraic expression.	Translate the phrase to an algebraic expression.
The sum of four and a number	Five less than some number
Let $n =$ the number.	Let $s =$ the number.
Expression: _____	Expression: _____

EXAMPLE 2	YOUR TURN 2
Translate the phrase to an algebraic expression.	Translate the phrase to an algebraic expression.
Eight more than three times some number	One-third of a number minus four
Let $x =$ the number.	Let $n =$ the number.
	(Think of this as four less than one third of a number.)
Expression: _____	Expression: _____
EXAMPLE 3	YOUR TURN 3
Translate the phrase to an algebraic expression.	Translate the phrase to an algebraic expression.
Fifteen percent of some number	Forty-two percent of a number
Let $y =$ the number.	Let $w =$ the number.
Expression: _____	Expression: _____
EXAMPLE 4	YOUR TURN 4
Translate the phrase to an algebraic expression.	Translate the phrase to an algebraic expression.
Five less than the product of two numbers	Two more than the quotient of two numbers
Let x and $y =$ the numbers.	Let a and $b =$ the numbers.
Expression: _____	Expression: _____

YOUR NOTES   Write your questions and additional notes.

Evaluating Algebraic Expressions

ESSENTIALS

When we replace a variable with a number, we say that we are **substituting** for the variable. Carrying out the resulting calculation is called **evaluating the expression**. The result is called the **value** of the expression.

Example

- Evaluate $7x + y^2$ for $x = 5$ and $y = 4$.

$$7x + y^2 = 7(5) + 4^2 = 7(5) + 16 = 35 + 16 = 51$$

GUIDED LEARNING	📖 **Textbook**		👤 **Instructor**	▶ **Video**	

EXAMPLE 1	YOUR TURN 1
Evaluate $2x - 5y$ for $x = 12$ and $y = -3$. $$2x - 5y = 2\left(\boxed{}\right) - 5(-3)$$ $$= 24 - \left(\boxed{}\right)$$ $$= 24 + \boxed{}$$ $$= \boxed{}$$	Evaluate $11rs - 2t$ for $r = 4$, $s = -2$, and $t = 21$.
EXAMPLE 2	**YOUR TURN 2**
The area of a triangle with base of length b and height of length h is given by the formula $A = \dfrac{1}{2}bh$. Find the area when $b = 2.1$ m and $h = 5$ m. $$A = \frac{1}{2}bh$$ $$= \frac{1}{2}(2.1)(5)$$ $$= 1.05(5)$$ $$= \boxed{} \text{ sq m, or } \boxed{} \text{ m}^2$$	The area of a triangle with base of length b and height of length h is given by the formula $A = \dfrac{1}{2}bh$. Find the area when $b = 3.6$ ft and $h = 7$ ft.

EXAMPLE 3	YOUR TURN 3
Evaluate $12 - x^5 + 14 \div 2y^3$ for $x = 1$ and $y = 2$.	Evaluate $18 \div 3y^2 + 15 - xy$ for $x = 4$ and $y = 3$.

$$12 - x^5 + 14 \div 2y^3 = 12 - (1)^5 + 14 \div 2(2)^3$$
$$= 12 - 1 + 14 \div 2(\boxed{})$$
$$= 12 - 1 + \boxed{}(8)$$
$$= 12 - 1 + 56$$
$$= 11 + \boxed{}$$
$$= \boxed{}$$

YOUR NOTES   Write your questions and additional notes.

Practice Exercises

Readiness Check

Match each key word(s) to the corresponding symbol.

1. product   **a)**   +

2. increased by   **b)**   −

3. less than   **c)**   ×

4. quotient   **d)**   ÷

Translating to Algebraic Expressions

Translate each phrase to an algebraic expression.

5. Three times d

6. Five more than twice x

7. Six more than the difference of a and b

8. Seven less than twenty percent of y

9. 10 increased by m

10. The product of $\dfrac{1}{5}$ and three times q

11. 1 less than 25 times y

12. 34 subtracted from r

13. What is the price of a refrigerator after a 15% reduction if the price before the reduction was P?

14. Jen's part-time job pays $8.65 per hour. How much does she earn for working h hours?

Evaluating Algebraic Expressions

Evaluate.

15. $5(x-8)+1$, for $x=9$

16. $(n-12)^2 -10$, for $n=20$

17. $12a-(b+a^2)$, for $a=2$ and $b=-6$

18. $x^3 +8-y$, for $x=-3$ and $y=14$

19. $3a^3b - 4b$, for $a=2$ and $b=-1$

20. $7x \div (15-y+2)$, for $x=6$ and $y=14$

21. Find the area of a triangular fireplace with a base of 2.1 ft and a height of 9 ft. Use $A=\dfrac{1}{2}bh.$

Equivalent Expressions

ESSENTIALS

Two expressions that have the same value for all *allowable* replacements are called **equivalent expressions**.

Example

- Complete the following table by evaluating each of the expressions for the given values. Then look for expressions that appear to be equivalent.

VALUE	$4x - 2x$	$2x$	$3x + x$
$x = -3$			
$x = 2$			
$x = 0$			

We substitute and find the value of each expression. For example, for $x = -3$,

$$4x - 2x = 4(-3) - 2(-3) = -12 + 6 = -6,$$
$$2x = 2(-3) = -6, \quad \text{and}$$
$$3x + x = 3(-3) + (-3) = -9 - 3 = -12.$$

VALUE	$4x - 2x$	$2x$	$3x + x$
$x = -3$	-6	-6	-12
$x = 2$	4	4	8
$x = 0$	0	0	0

Because the values of $4x - 2x$ and $2x$ are the same for all given values of x, and are in fact the same for any allowable real-number replacement of x, the expressions $4x - 2x$ and $2x$ are **equivalent**. The expressions $4x - 2x$ and $3x + x$ are not equivalent, and the expressions $2x$ and $3x + x$ are not equivalent, since values are not the same for *all x*.

GUIDED LEARNING   **Textbook**   **Instructor**   **Video**

EXAMPLE 1	YOUR TURN 1

EXAMPLE 1

Complete the table by evaluating each expression for the given values. Then look for expressions that may be equivalent.

VALUE	x^3	$3x$	$2x + x$
$x = -5$			
$x = 2$			
$x = 2.1$			

For $x = -5$,

$$x^3 = (-5)^3 = -125,$$
$$3x = 3(-5) = \boxed{}, \quad \text{and}$$
$$2x + x = 2(-5) + (-5) = \boxed{} - 5 = -15.$$

Also evaluate the expressions for $x = 2$ and for $x = 2.1$.

VALUE	x^3	$3x$	$2x + x$
$x = -5$	-125	-15	-15
$x = 2$	8	$\boxed{}$	6
$x = 2.1$	$\boxed{}$	$\boxed{}$	6.3

The expressions $\boxed{}$ and $2x + x$ are equivalent.

YOUR TURN 1

Complete the table by evaluating each expression for the given values. Then tell if the expressions are equivalent.

VALUE	$x^2 + 3$	$5x$
$x = -4$		
$x = 1$		
$x = 3.4$		

YOUR NOTES   Write your questions and additional notes.

Equivalent Fraction Expressions

ESSENTIALS

The Identity Property of 1

For any real number a,

$$a \cdot 1 = 1 \cdot a = a.$$

(The number 1 is the **multiplicative identity**.)

Examples

- Use multiplying by 1 to find an expression equivalent to $\dfrac{5}{8}$ with a denominator of $16a$.

  Because $16a = 8 \cdot 2a$, we multiply by 1, using $\dfrac{2a}{2a}$ as a name for 1:

  $$\frac{5}{8} = \frac{5}{8} \cdot 1 = \frac{5}{8} \cdot \frac{2a}{2a} = \frac{10a}{16a}.$$

- Simplify: $-\dfrac{27z}{18z}$.

  $$-\frac{27z}{18z} = -\frac{3 \cdot 9z}{2 \cdot 9z} = -\frac{3}{2} \cdot \frac{9z}{9z} = -\frac{3}{2} \cdot 1 = -\frac{3}{2}$$

GUIDED LEARNING	📘 **Textbook** 👤 **Instructor** ▶ **Video**
EXAMPLE 1	**YOUR TURN 1**
Write a fraction expression equivalent to $\dfrac{2}{3}$ with a denominator of $15y$. $$\frac{2}{3} = \frac{2}{3} \cdot 1 = \frac{2}{3} \cdot \frac{\boxed{}}{\boxed{}} = \frac{\boxed{}}{15y}$$	Write a fraction expression equivalent to $\dfrac{4}{7t}$ with a denominator of $21t$.
EXAMPLE 2	**YOUR TURN 2**
Simplify: $\dfrac{13s}{52s}$. $$\frac{13s}{52s} = \frac{\boxed{} \cdot 1}{\boxed{} \cdot 4} = \frac{\boxed{}}{\boxed{}} \cdot \frac{1}{4}$$ $$= 1 \cdot \frac{1}{4} = \frac{\boxed{}}{\boxed{}}$$	Simplify: $\dfrac{11ab}{55a}$.

EXAMPLE 3	YOUR TURN 3
Simplify: $-\dfrac{36b}{8b}$.	Simplify: $-\dfrac{63x}{18x}$.

$$-\frac{36b}{8b} = -\frac{9 \cdot \boxed{}}{2 \cdot \boxed{}} = -\frac{9}{2} \cdot \frac{\boxed{}}{\boxed{}}$$

$$= -\frac{9}{2} \cdot 1 = -\frac{\boxed{}}{\boxed{}}$$

YOUR NOTES   Write your questions and additional notes.

The Commutative Laws and the Associative Laws

ESSENTIALS

The Commutative Laws

Addition. For any numbers a and b,

$a + b = b + a.$

(We can change the order when adding without affecting the answer.)

Multiplication. For any numbers a and b,

$ab = ba.$

(We can change the order when multiplying without affecting the answer.)

The Associative Laws

Addition. For any numbers a, b, and c,

$a + (b + c) = (a + b) + c.$

(Numbers can be grouped in any manner for addition.)

Multiplication. For any numbers a, b, and c,

$a \cdot (b \cdot c) = (a \cdot b) \cdot c.$

(Numbers can be grouped in any manner for multiplication.)

Examples

- Evaluate $m + n$ and $n + m$ when $m = 12$ and $n = 4$.

  $m + n = 12 + 4 = 16;$ $n + m = 4 + 12 = 16$

- Evaluate mn and nm when $m = 8$ and $n = 3$.

  $mn = 8 \cdot 3 = 24;$ $nm = 3 \cdot 8 = 24$

- Evaluate $r + (s + t)$ and $(r + s) + t$ when $r = 5$, $s = 7$, and $t = 2$.

$$r + (s + t) = 5 + (7 + 2) \qquad (r + s) + t = (5 + 7) + 2$$
$$= 5 + 9 \qquad\qquad\qquad = 12 + 2$$
$$= 14; \qquad\qquad\qquad\quad = 14$$

- Evaluate $r \cdot (s \cdot t)$ and $(r \cdot s) \cdot t$ when $r = 3$, $s = 5$, and $t = 4$.

$$r \cdot (s \cdot t) = 3 \cdot (5 \cdot 4) \qquad (r \cdot s) \cdot t = (3 \cdot 5) \cdot 4$$
$$= 3 \cdot 20 \qquad\qquad\qquad = 15 \cdot 4$$
$$= 60; \qquad\qquad\qquad\quad = 60$$

GUIDED LEARNING   🔖 **Textbook**   👤 **Instructor**   ▶ **Video**

EXAMPLE 1	YOUR TURN 1
Evaluate $x+y$ and $y+x$ when $x=-2$ and $y=6$. $x+y=-2+\boxed{}=\boxed{}$; $y+x=6+(-2)=\boxed{}$	Evaluate xy and yx when $x=10$ and $y=-7$.

EXAMPLE 2	YOUR TURN 2
Evaluate $a\cdot(b\cdot c)$ and $(a\cdot b)\cdot c$ when $a=9$, $b=3$, and $c=2$. $a\cdot(b\cdot c)=\boxed{}\cdot(3\cdot2)$ $\qquad\quad =9\cdot\boxed{}$ $\qquad\quad =54$; $(a\cdot b)\cdot c=(9\cdot3)\cdot2$ $\qquad\quad =\boxed{}\cdot2$ $\qquad\quad =\boxed{}$	Evaluate $a+(b+c)$ and $(a+b)+c$ when $a=2$, $b=6$, and $c=7$.

EXAMPLE 3	YOUR TURN 3
Use the commutative laws and the associative laws to write at least three expressions equivalent to $(c+1)+d$. Answers may vary. $(c+1)+d=c+\left(\boxed{}+d\right)$ $(c+1)+d=\boxed{}+(c+1)$ $(c+1)+d=\left(1+\boxed{}\right)+d$	Use the commutative laws and the associative laws to write at least three expressions equivalent to $x\cdot(3\cdot y)$. Answers may vary.

YOUR NOTES   Write your questions and additional notes.

The Distributive Laws

ESSENTIALS

The Distributive Law of Multiplication over Addition

For any numbers a, b, and c,

$$a(b+c) = ab + ac, \quad \text{or} \quad (b+c)a = ba + ca.$$

(We can add and then multiply, or we can multiply and then add.)

The Distributive Law of Multiplication over Subtraction

For any real numbers a, b, and c,

$$a(b-c) = ab - ac, \quad \text{or} \quad (b-c)a = ba - ca.$$

(We can subtract and then multiply, or we can multiply and then subtract.)

The reverse of multiplying is called **factoring**. Factoring an expression involves factoring its terms. **Terms** of algebraic expressions are the parts separated by addition signs.

To **factor** an expression is to find an equivalent expression that is a product. If $N = a \cdot b$, then a and b are **factors** of N.

Examples

- Evaluate $6(x-y)$ and $6x - 6y$ when $x = 8$ and $y = 1$.

$$
\begin{aligned}
6(x-y) &= 6(8-1) & 6x - 6y &= 6 \cdot 8 - 6 \cdot 1 \\
&= 6(7) & &= 48 - 6 \\
&= 42; & &= 42
\end{aligned}
$$

- $4(x-2) = 4 \cdot x - 4 \cdot 2 = 4x - 8$

- The terms of $-5x - 2y + 3z = -5x + (-2y) + 3z$ are $-5x$, $-2y$, and $3z$.

- The factorization of $3a + 3y$ is $3(a+y)$.

GUIDED LEARNING	📍 **Textbook**	📍 **Instructor**	📍 **Video**

EXAMPLE 1	YOUR TURN 1
Multiply: $4(x-3)$.	Multiply: $5(x+2)$.
$4(x-3)$ $= 4 \cdot \boxed{} - 4 \cdot \boxed{}$ Using the distributive law of multiplication over subtraction $= 4x - \boxed{}$	

EXAMPLE 2	YOUR TURN 2
Multiply: $-8(3x+4y-z)$.	Multiply: $-2(5m-7n+9p)$.
$-8(3x+4y-z)$ $$= -8 \cdot (\boxed{}) + (-8) \cdot (\boxed{}) - (-8) \cdot \boxed{}$$ $$= -24x - \boxed{} + \boxed{}$$	

EXAMPLE 3	YOUR TURN 3
Factor: $12x - 15y + 6$.	Factor: $8a - 12b + 20$.
$12x - 15y + 6$ $$= \boxed{} \cdot 4x - \boxed{} \cdot 5y + \boxed{} \cdot 2$$ $$= \boxed{}(4x - 5y + 2)$$	

EXAMPLE 4	YOUR TURN 4
Factor: $ax + bx - cx$.	Factor: $mx + my + mz$.
$ax + bx - cx$ $$= x \cdot \boxed{} + x \cdot b - x \cdot \boxed{}$$ $$= x\left(a + \boxed{} - \boxed{}\right)$$	

YOUR NOTES   Write your questions and additional notes.

Practice Exercises

Readiness Check

Choose from the column on the right an equation that illustrates the law.

1. Commutative law of multiplication
2. Commutative law of addition
3. Associative law of multiplication
4. Associative law of addition
5. Distributive law of multiplication over subtraction

a) $5(6-2)=5\cdot 6-5\cdot 2$

b) $9\cdot\left(\dfrac{1}{3}\cdot 7\right)=\left(9\cdot\dfrac{1}{3}\right)\cdot 7$

c) $4\cdot 7=7\cdot 4$

d) $2+10=10+2$

e) $2(7+4)=2\cdot 7+2\cdot 4$

f) $3+\left(1+\dfrac{1}{4}\right)=(3+1)+\dfrac{1}{4}$

Equivalent Expressions

Complete the table by evaluating each expression for the given values. Then look for expressions that are equivalent.

6.

VALUE	$4(x+5)$	$4x+5$	$4x+20$
$x=-2$			
$x=1.8$			
$x=0$			

Equivalent Fraction Expressions

Use multiplying by 1 to find an equivalent expression with the given denominator.

7. $\dfrac{3}{7}$; $7y$

8. $\dfrac{2}{5}$; $15a$

Simplify.

9. $-\dfrac{64xy}{24xy}$

10. $\dfrac{45ab}{20b}$

The Commutative Laws and the Associative Laws

Use a commutative law to find an equivalent expression.

11. $6+n$

12. xy

Use an associative law to find an equivalent expression.

13. $(u+v)+3$

14. $(5 \cdot c) \cdot p$

Use the commutative laws and the associative laws to write three equivalent expressions.

15. $5+(y+z)$

16. $(x \cdot y) \cdot 4$

The Distributive Laws

Multiply.

17. $3(x+10)$

18. $-3(7x-2y-z)$

List the terms of each expression.

19. $5x-3y$

20. $-3a-b+18c$

Factoring

Factor.

21. $7a+7b$

22. $4x-2$

23. $6m+12n-15$

24. $2 \cdot \pi \cdot r \cdot h + 2 \cdot \pi \cdot r \cdot r$

Collecting Like Terms

ESSENTIALS

If two terms have the same letter or letters, we say that they are **like terms**, or **similar terms**. (If powers, or exponents, are involved, then like terms must have the same letters raised to the same powers.) If two terms are simply numbers with no letters, they are also like terms.

We can simplify by **collecting**, or **combining**, **like terms**.

Examples

Collect like terms.

- $6x - 2x = (6 - 2)x = 4x$

- $5x + 4y + x - 8y$

$$= 5 \cdot x + 4 \cdot y + 1 \cdot x - 8 \cdot y$$
$$= 5 \cdot x + 1 \cdot x + 4 \cdot y + (-8 \cdot y)$$
$$= (5 + 1)x + (4 - 8)y$$
$$= 6x - 4y$$

GUIDED LEARNING　　　　🔲 **Textbook**　　　👤 **Instructor**　　　▶ **Video**

EXAMPLE 1	YOUR TURN 1
Collect like terms: $9x + 13x$. $$9x + 13x = \left(\boxed{} + \boxed{}\right)x$$ $$= \boxed{}\,x$$	Collect like terms: $14x - x$.
EXAMPLE 2	YOUR TURN 2
Collect like terms: $11.2x - 6.3y - 9.8x + 2.7y$. $$11.2x - 6.3y - 9.8x + 2.7y$$ $$= 11.2x + (-6.3y) + (-9.8x) + 2.7y$$ $$= \left(\boxed{} - 9.8\right)x + \left(-6.3 + \boxed{}\right)y$$ $$= \boxed{}\,x - 3.6y$$	Collect like terms: $7x + 1.3 - 2x + 5.6$.

EXAMPLE 3	YOUR TURN 3
Collect like terms: $\dfrac{3}{4}x + \dfrac{5}{6}y + \dfrac{1}{3}x - \dfrac{2}{3}y$.	Collect like terms: $\dfrac{4}{7}x - \dfrac{7}{12}y - \dfrac{1}{2}x + \dfrac{3}{8}y$.

$\dfrac{3}{4}x + \dfrac{5}{6}y + \dfrac{1}{3}x - \dfrac{2}{3}y$

$= \left(\dfrac{3}{4} + \dfrac{\boxed{}}{\boxed{}} \right)x + \left(\dfrac{5}{6} - \dfrac{2}{3} \right)y$

$= \left(\dfrac{9}{12} + \dfrac{\boxed{}}{12} \right)x + \left(\dfrac{\boxed{}}{\boxed{}} - \dfrac{4}{6} \right)y$

$= \dfrac{\boxed{}}{\boxed{}}x + \dfrac{1}{6}y$

YOUR NOTES   Write your questions and additional notes.

Multiplying by –1 and Removing Parentheses

ESSENTIALS

The Property of –1

For any number a,

$$-1 \cdot a = -a.$$

(Negative 1 times a is the opposite of a. In other words, changing the sign is the same as multiplying by -1.)

The Opposite of a Difference

For any real numbers a and b,

$$-(a-b) = b-a.$$

(The opposite of $a-b$ is $b-a$.)

Examples

- $-(6x) = -1(6x)$
$$= (-1 \cdot 6)x$$
$$= -6x$$

- $-(-14y) = -1(-14y)$
$$= \left[-1(-14)\right]y$$
$$= 14y$$

- $-(2x-4y+9) = -1(2x-4y+9)$
$$= -2x+4y-9$$

- $3y+(4x+5)-(10y-2) = 3y+4x+5-10y+2$
$$= -7y+4x+7$$

	⦿ **Textbook**	⦿ **Instructor**	⦿ **Video**
GUIDED LEARNING			

EXAMPLE 1	YOUR TURN 1
Find an equivalent expression without parentheses: $-(21a-16b+28)$.	Find an equivalent expression without parentheses: $-(-19a-8b-12)$.
$-(21a-16b+28)$	
$= -21a \boxed{} -28$	

EXAMPLE 2	YOUR TURN 2
Remove parentheses and simplify: $2x-(4x+6)$. $2x-(4x+6)=2x-4x\ \boxed{}$ $\qquad\qquad\ =\boxed{}-6$	Remove parentheses and simplify: $7x-(5y-8)+(3x-14)$.

EXAMPLE 3	YOUR TURN 3
Simplify: $8c-\{3[6(2c-1)-4(2c+4)+5]-7\}$. $8c-\{3[6(2c-1)-4(2c+4)+5]-7\}$ $=8c-\{3[12c-6-8c-\boxed{}+5]-7\}$ $=8c-\{3[4c-17]-7\}$ $=8c-\{12c-\boxed{}-7\}$ $=8c-\{12c-58\}$ $=8c-12c+\boxed{}$ $=\boxed{}c+58$	Simplify: $4d-\{9d-[(d+8)-(d-2)]\}$.

YOUR NOTES   Write your questions and additional notes.

Practice Exercises

Readiness Check

Determine whether each set of expressions are equal or opposite.

1. $-(-4x)$, $4x$ _____

2. $x-5$, $5-x$ _____

3. $2x-4y+15$, $-2x+4y-15$ _____

4. $7y-(-2y)$, $9y$ _____

Collecting Like Terms

Collect like terms.

5. $2m+7m$

6. $8n-11n$

7. $14p-p$

8. $10a+13b-4a$

9. $4.3s-1.1t-6.5t+s$

10. $\dfrac{1}{3}x-\dfrac{4}{5}x-\dfrac{2}{5}y+\dfrac{1}{2}y$

Multiplying by −1 and Removing Parentheses

Find an equivalent expression without parentheses.

11. $-(-5a)$

12. $-(d+14)$

13. $-(3x+2y-8)$

14. $-(-3f + 4g + 2h - j)$

15. $-\left(-x - \dfrac{1}{5}w - 4.7y + 15.4z\right)$

Simplify by removing parentheses and collecting like terms.

16. $b + (8b - 7)$

17. $9x - (x + 23)$

18. $6y - (3y + 9) + 4(2y - 12)$

19. $\dfrac{1}{3}(9a - 6) - \dfrac{1}{4}(16a + 20) - 5$

Simplify.

20. $3\{-2 - 6[4x + 1 - 3(5x + 7)]\}$

21. $[16 - 4(8y + 3)] - [2(y + 1) - 4]$

Multiplication and Division

ESSENTIALS

The Product Rule

For any number a and any integers m and n,

$$a^m \cdot a^n = a^{m+n}.$$

(When multiplying with exponential notation, add the exponents if the bases are the same.)

The Quotient Rule

For any nonzero number a and any integers m and n,

$$\frac{a^m}{a^n} = a^{m-n}.$$

(When dividing with exponential notation, subtract the exponent of the denominator from the exponent of the numerator if the bases are the same.)

Examples

- Multiply and simplify: $\left(2a^2b^{-3}\right)\left(3ab\right)$.

$$\left(2a^2b^{-3}\right)\left(3ab\right) = 2 \cdot 3 \cdot a^2 \cdot a \cdot b^{-3} \cdot b$$

$$= 6a^{2+1}b^{-3+1} = 6a^3b^{-2} = \frac{6a^3}{b^2}$$

- Divide and simplify: $\dfrac{-8x^{-7}y^5}{2x^2y^3}$.

$$\frac{-8x^{-7}y^5}{2x^2y^3} = -\frac{8}{2} \cdot x^{-7-2} \cdot y^{5-3} = -4x^{-9}y^2 = -\frac{4y^2}{x^9}$$

GUIDED LEARNING   🛈 **Textbook**   👤 **Instructor**   ▶ **Video**

EXAMPLE 1	YOUR TURN 1
Multiply and simplify: $\left(7y^{2n}\right)\left(3y^n\right)$.	Multiply and simplify: $\left(-3x^{4n}\right)\left(8x^{2n}\right)$.
$\left(7y^{2n}\right)\left(3y^n\right) = 7 \cdot \boxed{} \cdot y^{2n+\boxed{}}$	
$\qquad = 21y^{\boxed{}}$	

EXAMPLE 2	YOUR TURN 2
Multiply and simplify: $\left(-5x^{-5}y^{10}\right)\left(4x^3y^7\right)$. $$\left(-5x^{-5}y^{10}\right)\left(4x^3y^7\right) = -5 \cdot 4 \cdot x^{-5+3}y^{\square+\square}$$ $$= \boxed{}$$	Multiply and simplify: $\left(-7a^{10}b^{-2}\right)\left(2ab^3\right)$.
EXAMPLE 3	YOUR TURN 3
Divide and simplify: $\dfrac{24x^{4n}}{6x^{5n}}$. $$\frac{24x^{4n}}{6x^{5n}} = \frac{\boxed{}}{6} \cdot \frac{x^{4n}}{x^{5n}}$$ $$= 4x^{\square-5n} = 4x^{\square} = \frac{4}{x^n}$$	Divide and simplify: $\dfrac{-27y^{8n}}{9y^{12n}}$.
EXAMPLE 4	YOUR TURN 4
Divide and simplify: $\dfrac{-24m^6n^8}{-8m^5n^2}$. $$\frac{-24m^6n^8}{-8m^5n^2} = \frac{-24}{-8}m^{\square-\square}n^{8-2}$$ $$= \boxed{}$$	Divide and simplify: $\dfrac{-18m^{10}n^2}{2m^7n}$.

YOUR NOTES   Write your questions and additional notes.

Raising Powers to Powers and Products and Quotients to Powers

ESSENTIALS

The Power Rule

For any real number a and any integers m and n,

$$\left(a^m\right)^n = a^{mn}.$$

(To raise a power to a power, multiply the exponents.)

Raising a Product to a Power

For any real numbers a and b and any integer n,

$$\left(ab\right)^n = a^n b^n.$$

(To raise a product to the nth power, raise each factor to the nth power.)

Raising a Quotient to a Power

For any real numbers a and b and any integer n,

$$\left(\frac{a}{b}\right)^n = \frac{a^n}{b^n}, b \neq 0; \quad \text{and} \quad \left(\frac{a}{b}\right)^{-n} = \left(\frac{b}{a}\right)^n = \frac{b^n}{a^n}, a \neq 0, b \neq 0.$$

(To raise a quotient to the nth power, raise the numerator to the nth power and divide by the denominator to the nth power.)

Examples

- $\left(y^2\right)^6 = y^{2 \cdot 6} = y^{12}$

- $\left(-3x^4 y^{-2}\right)^4 = (-3)^4 \left(x^4\right)^4 \left(y^{-2}\right)^4 = 81x^{16} y^{-8} = \dfrac{81x^{16}}{y^8}$

- $\left(\dfrac{3x^5}{y^2}\right)^{-2} = \left(\dfrac{y^2}{3x^5}\right)^2 = \dfrac{\left(y^2\right)^2}{\left(3x^5\right)^2} = \dfrac{y^4}{9x^{10}}$

GUIDED LEARNING	📘 **Textbook** 👤 **Instructor** ▶ **Video**
EXAMPLE 1	YOUR TURN 1
Simplify: $\left(x^3\right)^{-7s}$.	Simplify: $\left(x^{-6}\right)^{4p}$.
$\left(x^3\right)^{-7s} = x^{3(-7s)} = x^{\boxed{}}$	

EXAMPLE 2	YOUR TURN 2
Simplify: $\left(-2x^6y^{-2}z\right)^{-3}$.	Simplify: $\left(-3x^{-7}y^{-2}z^5\right)^{-4}$.

For Example 2:

$$\left(-2x^6y^{-2}z\right)^{-3} = (-2)^{-3} \cdot \left(x^6\right)^{\Box} \cdot \left(y^{-2}\right)^{\Box} \cdot z^{-3}$$

$$= \frac{1}{(-2)^3} \cdot x^{-18} \cdot y^6 \cdot z^{-3}$$

$$= \boxed{} \cdot \frac{1}{x^{18}} \cdot y^6 \cdot \frac{1}{z^3}$$

$$= -\frac{y^6}{8x^{18}z^{\Box}}$$

EXAMPLE 3	YOUR TURN 3
Simplify: $\left(\dfrac{3x^5}{y^2}\right)^{-2}$.	Simplify: $\left(\dfrac{x}{2y^{-3}}\right)^{-4}$.

For Example 3:

$$\left(\frac{3x^5}{y^2}\right)^{-2} = \left(\frac{y^2}{3x^5}\right)^2$$

$$= \frac{\left(y^2\right)^{\Box}}{\left(3x^5\right)^{\Box}}$$

$$= \frac{y^4}{\Box x^{\Box}}$$

YOUR NOTES   Write your questions and additional notes.

Scientific Notation

ESSENTIALS

Scientific notation for a number is an expression of the type

$$M \times 10^n,$$

where n is an integer, M is greater than or equal to 1 and less than 10 ($1 \leq M < 10$), and M is expressed in decimal notation. 10^n is also considered to be scientific notation when $M = 1$.

A positive exponent in scientific notation indicates a large number (greater than or equal to 10) and a negative exponent indicates a small number (between 0 and 1).

Examples

- Convert 129,040,000 to scientific notation.

  $$129{,}040{,}000.$$ $$129{,}040{,}000 = 1.2904 \times 10^8$$

  We must move the decimal point 8 places. The number is large so the exponent is positive.

- Convert 1.903×10^{-4} to decimal notation.

  $$0.0001.903$$ $$1.903 \times 10^{-4} = 0.0001903$$

  The exponent is negative, so the answer is a small number. Move the decimal point 4 places.

- Multiply: $(2.1 \times 10^9)(5.4 \times 10^7)$

  $$(2.1 \times 10^9)(5.4 \times 10^7) = (2.1 \times 5.4)(10^9 \times 10^7)$$
  $$= 11.34 \times 10^{16}$$
  $$= (1.134 \times 10^1) \times 10^{16}$$
  $$= 1.134 \times 10^{17}$$

- Divide: $\dfrac{1.8 \times 10^{-11}}{1.5 \times 10^{-17}}$.

  $$\frac{1.8 \times 10^{-11}}{1.5 \times 10^{-17}} = \frac{1.8}{1.5} \times \frac{10^{-11}}{10^{-17}}$$
  $$= 1.2 \times 10^6$$

	🔖 **Textbook**	👤 **Instructor**	▶ **Video**
GUIDED LEARNING			

EXAMPLE 1	YOUR TURN 1
Convert 0.000046 to scientific notation. 0.000046 (continued)	Convert 70,100,000,000,000 to scientific notation.

We move the decimal point ☐ places.

The number is small so the exponent is

_____.

$0.000046 = \boxed{} \times 10^{\boxed{}}$

EXAMPLE 2	YOUR TURN 2
Convert 5.47×10^6 to decimal notation.	Convert 4.078×10^{-3} to decimal notation.

The exponent is _____, so the answer is a large number.

Move the decimal point ☐ places.

5.470000.

$5.47 \times 10^6 = \boxed{}$

EXAMPLE 3	YOUR TURN 3
Multiply: $(3.3 \times 10^8)(2.5 \times 10^{-13})$. Write scientific notation for the result.	Multiply: $(3.6 \times 10^{-12})(2.2 \times 10^{15})$. Write scientific notation for the result.

$(3.3 \times 10^8)(2.5 \times 10^{-13}) = 3.3 \times 2.5 \times 10^8 \times 10^{-13}$

$= \boxed{} \times 10^{\boxed{}}$

EXAMPLE 4	YOUR TURN 4
Divide: $\dfrac{3.6 \times 10^{30}}{4.8 \times 10^{50}}$. Write scientific notation for the result.	Divide: $\dfrac{1.2 \times 10^{-7}}{1.5 \times 10^{-18}}$. Write scientific notation for the result.

$\dfrac{3.6 \times 10^{30}}{4.8 \times 10^{50}} = \dfrac{3.6}{4.8} \times \dfrac{10^{30}}{10^{50}}$

$= \boxed{} \times 10^{-20}$

$= \left(\boxed{} \times 10^{\boxed{}} \right) \times 10^{-20}$

$= 7.5 \times 10^{\boxed{}}$

YOUR NOTES Write your questions and additional notes.

Practice Exercises

Readiness Check

Choose the word from the list below that will make each statement true. Not all words will be used.

negative   positive   multiply   add   scientific

1. When multiplying with exponential notation, _____ the exponents if the bases are the same.

2. To raise a power to a power, _____ the exponents and leave the base unchanged.

3. The number 3.1×10^{-4} is written in _____ notation.

4. In scientific notation, _____ exponents are used to represent small numbers between 0 and 1.

Multiplication and Division

Multiply and simplify.

5. $3^9 \cdot 3^2$

6. $8x^0 \cdot 4x^4$

7. $\left(-2a^{-5}\right)\left(9a^3\right)$

Divide and simplify.

8. $\dfrac{x^{10}}{x^3}$

9. $\dfrac{20x^{-12}}{4x^8}$

10. $\dfrac{-45x^9 y^5}{5x^6 y}$

Raising Powers to Powers and Products and Quotients to Powers

Simplify.

11. $\left(-6xy^2\right)^2$

12. $\left(2x^{-2}y^3\right)^{-2}$

13. $\left(\dfrac{3x^4y^7}{4xy^8}\right)^3$ **14.** $\left(\dfrac{2x^4y^{-3}}{3x^{-7}y^{-9}}\right)^{-4}$

Scientific Notation

Convert each number to scientific notation.

15. 0.047   **16.** 501,790,000,000

17. Before the Dalles Dam was built, Celilo Falls was the largest waterfall on the Columbia River. Its average discharge was 190,000 cubic feet of water per second. Express this number in scientific notation.

Convert each number to decimal notation.

18. 8.13×10^7 **19.** 2.04×10^{-3}

20. There are about 2.4×10^4 grains of rice in a pound. Express this number in decimal notation.

Multiply or divide and write scientific notation for the result.

21. $\left(2\times10^5\right)\left(3\times10^4\right)$ **22.** $\left(1.5\times10^3\right)\left(8.7\times10^{-5}\right)$

23. $\dfrac{9.2\times10^{-3}}{2.3\times10^{17}}$ **24.** $\dfrac{2.4\times10^6}{9.6\times10^{-10}}$

Equations and Solutions

ESSENTIALS

An **equation** is a number sentence that says that the expressions on either side of the equals sign, =, represent the same number.

The replacements for the variable that make an equation true are called the **solutions** of the equation. The set of all solutions is called the **solution set** of the equation. When we find all the solutions, we say that we have **solved** the equation.

Equations with the same solutions are called **equivalent equations**.

Examples

- $2 + 5 = 7$ The equation is *true*.

- $9 - 3 = 3$ The equation is *false*.

- $x - 8 = 11$ The equation is *neither* true nor false, because we do not know what number x represents.

- Determine whether 3 is a solution of $2x + 2 = 9$.

$$2x + 2 = 9 \qquad \text{Writing the equation}$$
$$2 \cdot 3 + 2 \;?\; 9 \qquad \text{Substituting 3 for } x$$
$$6 + 2$$
$$8 \quad | \quad \text{FALSE}$$

The number 3 is not a solution of the equation.

GUIDED LEARNING   📖 **Textbook**   👤 **Instructor**   ▶ **Video**

EXAMPLE 1	YOUR TURN 1
Determine whether the equation is true, false, or neither.	Determine whether the equation is true, false, or neither.
$4 - 6 = 2$	$5 - 9 = -4$
The equation is _____. true / false / neither	
EXAMPLE 2	**YOUR TURN 2**
Determine whether the equation is true, false, or neither.	Determine whether the equation is true, false, or neither.
$13 + 7 = 5 + 15$	$12 + 4 = 7 + 7$
The equation is _____. true / false / neither	

EXAMPLE 3	YOUR TURN 3
Determine whether the equation is true, false, or neither. $x + 5 = 14$ The equation is _____. true / false / neither	Determine whether the equation is true, false, or neither. $7 + 3 = x$
EXAMPLE 4	**YOUR TURN 4**

Determine whether -6 is a solution of $10 - y = 16$.
 	
$\underline{10 - y = 16}$ Writing the equation
 	
$10 - \left(\boxed{}\right) \overset{?}{=} 16$ Substituting -6 for y
 	
$\boxed{}$ TRUE
 	
The statement $16 = 16$ is _____.
              true / false
 	
-6 _____ a solution of $10 - y = 16$.
   is / is not
	Determine whether -22 is a solution of $x + 2 = 20$.

EXAMPLE 5	**YOUR TURN 5**
Determine whether $a = 13$ and $2a + 11 = 39$ are equivalent. $2a + 11 = 39$ $2(13) + 11 = 39$ Substituting 13 for a $26 + 11 = 39$ $37 = 39$ The statement $37 = 39$ is _____. true / false The equations $a = 13$ and $2a + 11 = 39$ _____ equivalent. are / are not	Determine whether $x = -4$ and $-3x - 9 = 3$ are equivalent.

YOUR NOTES   Write your questions and additional notes.

The Addition Principle

ESSENTIALS

The Addition Principle

For any real numbers a, b, and c,

$a = b$ is equivalent to $a + c = b + c$.

Examples

- Solve: $w - 9 = 2$.

$$w - 9 = 2$$
$$w - 9 + 9 = 2 + 9$$
$$w + 0 = 11$$
$$w = 11$$

Check: $\dfrac{w - 9 = 2}{11 - 9 \,?\, 2}$

$2 \,|\,$ TRUE

- Solve: $x + 3 = -7$.

$$x + 3 = -7$$
$$x + 3 + (-3) = -7 + (-3)$$
$$x + 0 = -10$$
$$x = -10$$

Check: $\dfrac{x + 3 = -7}{-10 + 3 \,?\, -7}$

$-7 \,|\,$ TRUE

GUIDED LEARNING

📖 **Textbook**   👤 **Instructor**   ▶ **Video**

EXAMPLE 1	YOUR TURN 1
Solve: $y + 13 = -2$.	Solve: $x + 9 = -15$.

$$y + 13 = -2$$

$$y + 13 + (\boxed{}) = -2 + (\boxed{}) \quad \text{Adding } -13 \text{ on both sides}$$

$$y + 0 = -15$$

$$y = \boxed{} \quad \text{Identity property of 0}$$

Check: $\dfrac{y + 13 = -2}{-15 + 13 \,?\, -2}$ Substituting -15 for y

$-2 \,|\,$ TRUE

The solution is -15.

EXAMPLE 2	YOUR TURN 2
Solve: $-5.3 = b - 2.4$.	Solve: $-7.6 = x - 4.2$.

EXAMPLE 2 content:

$$-5.3 = b - 2.4$$

$$-5.3 + \boxed{} = b - 2.4 + \boxed{}$$

$$\boxed{} = b$$

Check: $-5.3 = b - 2.4$

$$-5.3 \,?\, -2.9 - 2.4$$

$$\Big|\, -5.3 \qquad \text{TRUE}$$

The solution is $\boxed{}$.

EXAMPLE 3	YOUR TURN 3
Solve: $\dfrac{1}{2} + x = -\dfrac{2}{5}$.	Solve: $\dfrac{2}{3} + w = -\dfrac{3}{4}$.

EXAMPLE 3 content:

$$\frac{1}{2} + x = -\frac{2}{5}$$

$$\frac{1}{2} - \boxed{} + x = -\frac{2}{5} - \boxed{}$$

$$x = -\frac{2}{5} \cdot \frac{\boxed{}}{2} - \frac{1}{2} \cdot \frac{5}{\boxed{}}$$

$$x = -\frac{4}{10} - \frac{5}{10}$$

$$x = -\frac{\boxed{}}{\boxed{}}$$

The number $-\dfrac{9}{10}$ checks.

The solution is $\boxed{}$.

YOUR NOTES   Write your questions and additional notes.

The Multiplication Principle

ESSENTIALS

The Multiplication Principle

For any real numbers a, b, and c, $c \neq 0$,

$a = b$ is equivalent to $a \cdot c = b \cdot c$.

In a product like $4x$, the number in front of the variable is called the **coefficient**.

Examples

- Solve: $\dfrac{2}{5}x = 84$.

$$\frac{2}{5}x = 84 \qquad \text{Check:} \qquad \frac{2}{5}x = 84$$

$$\frac{5}{2} \cdot \frac{2}{5}x = \frac{5}{2} \cdot 84 \qquad \qquad \frac{2}{5} \cdot 210 \,?\, 84$$

$$1 \cdot x = 210 \qquad \qquad \qquad \qquad 84 \,\Big|\, \text{TRUE}$$

$$x = 210$$

- Solve: $-7x = 42$.

$$-7x = 42 \qquad \text{Check:} \qquad -7x = 42$$

$$\frac{-7x}{-7} = \frac{42}{-7} \qquad \qquad -7(-6) \,?\, 42$$

$$1 \cdot x = -6 \qquad \qquad \qquad 42 \,\Big|\, \text{TRUE}$$

$$x = -6$$

GUIDED LEARNING 📖 **Textbook**   👤 **Instructor**   ▶ **Video**

EXAMPLE 1	YOUR TURN 1
Solve: $\dfrac{5}{9} = -\dfrac{4}{3}x$.	Solve: $\dfrac{3}{8}y = -\dfrac{9}{4}$.

$$\frac{5}{9} = -\frac{4}{3}x$$

$$-\frac{\square}{\square} \cdot \frac{5}{9} = -\frac{\square}{\square} \cdot \left(-\frac{4}{3}x\right)$$

$$-\frac{5}{12} = 1 \cdot x$$

$$-\frac{5}{12} = x$$

The number $-\dfrac{5}{12}$ checks and is the solution.

EXAMPLE 2	YOUR TURN 2
Solve: $6x = 84$.	Solve: $7y = 91$.

$$6x = 84$$

$$\frac{6x}{\boxed{}} = \frac{84}{\boxed{}}$$

$$1 \cdot x = 14$$

$$x = \boxed{}$$

The number 14 checks and is the solution.

EXAMPLE 3	YOUR TURN 3
Solve: $2.43y = 9963$.	Solve: $1.97x = 4334$.

$$2.43y = 9963$$

$$\frac{2.43y}{\boxed{}} = \frac{9963}{\boxed{}}$$

$$1 \cdot y = 4100$$

$$y = \boxed{}$$

The number 4100 checks and is the solution.

EXAMPLE 4	YOUR TURN 4
Solve: $-x = 15$.	Solve: $-a = -32$.

$$-x = 15$$

$$\frac{-x}{\boxed{}} = \frac{15}{\boxed{}}$$

$$1 \cdot x = -15$$

$$x = \boxed{}$$

The solution is $\boxed{}$.

YOUR NOTES   Write your questions and additional notes.

Using the Principles Together

ESSENTIALS

Example

- Solve: $-2x + 4 = 58$.

$$-2x + 4 = 58$$

$$-2x + 4 - 4 = 58 - 4 \qquad \text{Using the addition principle: subtracting 4}$$

$$-2x = 54 \qquad \text{Simplifying}$$

$$\frac{-2x}{-2} = \frac{54}{-2} \qquad \text{Using the multiplication principle: dividing by } -2$$

$$x = -27 \qquad \text{Simplifying}$$

The solution is -27.

Check:

$$-2x + 4 = 58$$

$$-2(-27) + 4 \overset{?}{=} 58$$

$$54 + 4$$

$$58 \qquad \text{TRUE}$$

GUIDED LEARNING 🔖 **Textbook** 👤 **Instructor** ▶ **Video**

EXAMPLE 1	YOUR TURN 1
Solve: $\dfrac{3}{4}x - \dfrac{1}{3} + \dfrac{1}{2}x = 2x - \dfrac{4}{3}$.	Solve: $3x - \dfrac{4}{5}x = \dfrac{2}{5} + 2x + \dfrac{1}{2}$.

The number $\boxed{}$ is the LCM of all the denominators.

$$\boxed{}\left(\frac{3}{4}x - \frac{1}{3} + \frac{1}{2}x\right) = \boxed{}\left(2x - \frac{4}{3}\right) \qquad \text{Multiplying by 12 on both sides}$$

$$12 \cdot \frac{3}{4}x - 12 \cdot \frac{1}{3} + 12 \cdot \frac{1}{2}x = 12 \cdot 2x - 12 \cdot \frac{4}{3}$$

$$9x - 4 + 6x = 24x - 16$$

$$\boxed{} - 4 = 24x - 16 \qquad \text{Collecting like terms}$$

$$15x - 4 - \boxed{} = 24x - 16 - \boxed{} \qquad \text{Subtracting } 15x$$

$$-4 = 9x - 16$$

$$-4 + \boxed{} = 9x - 16 + \boxed{} \qquad \text{Adding 16}$$

$$12 = 9x$$

$$\frac{12}{\boxed{}} = \frac{9x}{\boxed{}} \qquad \text{Dividing by 9}$$

$$\boxed{} = x$$

The number $\dfrac{4}{3}$ checks. The solution is $\boxed{}$.

EXAMPLE 2	YOUR TURN 2
Solve: $-15.3 + 4.2b = -3.96$.	Solve: $17.2 - 5.4x = -4.94$.

We multiply on both sides by 10^2, or 100, because the greatest number of decimal places is $\boxed{}$.

$$-15.3 + 4.2b = -3.96$$

$$\boxed{}(-15.3 + 4.2b) = \boxed{}(-3.96)$$

$$100(-15.3) + 100(4.2b) = -396 \qquad \text{Using the distributive law}$$

$$-1530 + 420b = -396$$

$$-1530 + 420b + 1530 = -396 + 1530 \qquad \text{Adding 1530}$$

$$420b = \boxed{}$$

$$\frac{420b}{\boxed{}} = \frac{1134}{\boxed{}} \qquad \text{Dividing by 420}$$

$$b = \frac{1134}{420}, \text{ or } \frac{27}{10}$$

The solution is $\frac{27}{10}$.

EXAMPLE 3	YOUR TURN 3
Solve: $7 - 3x = 6 - 3(x+1)$.	Solve: $14 + 5x = 7 + 5(x-9)$.

$$7 - 3x = 6 - 3(x+1)$$

$$7 - 3x = 6 - \boxed{} - \boxed{} \qquad \text{Using the distributive law}$$

$$7 - 3x = 3 - 3x$$

$$7 - 3x + \boxed{} = 3 - 3x + \boxed{} \qquad \text{Adding } 3x$$

$$7 = 3 \qquad \text{False}$$

There is *no* solution.

EXAMPLE 4	YOUR TURN 4
Solve: $6x + 5 = 17 + 6(x-2)$.	Solve: $-8x + 7 = 15 - 8(x+1)$.

$$6x + 5 = 17 + 6(x-2)$$

$$6x + 5 = 17 + \boxed{} - 12 \qquad \text{Using the distributive law}$$

$$6x + 5 = 6x + 5$$

$$\boxed{} + 6x + 5 = \boxed{} + 6x + 5 \qquad \text{Adding } -6x$$

$$5 = 5 \qquad \text{True for all real numbers}$$

The equation has *infinitely* many solutions. All real numbers are solutions.

YOUR NOTES   Write your questions and additional notes.

Practice Exercises

Readiness Check

Choose from the column on the right the most appropriate first step in solving each equation.

1. $-4x = 20$

2. $\dfrac{1}{4}x = 20$

3. $20x = -4$

4. $4 = -\dfrac{1}{20}x$

a) Multiply by 4 on both sides.

b) Multiply by 20 on both sides.

c) Divide by 4 on both sides.

d) Divide by 20 on both sides.

e) Multiply by -20 on both sides.

f) Divide by -4 on both sides.

Equations and Solutions

Determine whether the given number is a solution of the given equation.

5. $14; \ x + 19 = 43$

6. $24; \ \dfrac{x}{6} = 4$

7. $-4; \ 6(y - 3) = 42$

The Addition Principle

Solve using the addition principle. Don't forget to check.

8. $x + 7 = 9$

9. $-10 + y = 3$

10. $6.2 = -4.7 + y$

11. $-\dfrac{3}{4} + x = -\dfrac{5}{6}$

The Multiplication Principle

Solve using the multiplication principle. Don't forget to check.

12. $9x = 72$

13. $-x = 17$

14. $-1.6x = -1.44$

15. $-\dfrac{3}{4}c = \dfrac{9}{8}$

Using the Principles Together

Solve using the principles together. Don't forget to check.

16. $3x - 7 = 11$

17. $\dfrac{3}{5}x - \dfrac{3}{2} = -\dfrac{4}{5} + 2x$

18. $8.2x - 5.9x = 17.48$

19. $4x - 9 = 3 + 4(x - 3)$

20. $3x - 11 = 7x - 4(x - 11)$

21. $7(x + 5) - 3 = 2\left[9 + 2(x - 1)\right]$

Evaluating and Solving Formulas

ESSENTIALS

A **formula** is an equation that represents or models a relationship between two or more quantities.

Example

- Solve the formula $C = \frac{5}{9}(F - 32)$ for F.

$$C = \frac{5}{9}(F - 32)$$

$$\frac{9}{5} \cdot C = \frac{9}{5} \cdot \frac{5}{9}(F - 32) \quad \text{Multiplying on both sides by } \frac{9}{5}$$

$$\frac{9}{5}C = F - 32$$

$$\frac{9}{5}C + 32 = F - 32 + 32 \quad \text{Adding 32 on both sides}$$

$$\frac{9}{5}C + 32 = F$$

GUIDED LEARNING	📖 **Textbook**	👤 **Instructor**	▶ **Video**

EXAMPLE 1	YOUR TURN 1
The area of a triangle with base b and height h is given by $A = \frac{1}{2}bh$. Find the area of a triangle with base 4 in. and height 9 in. $A = \frac{1}{2} \cdot \boxed{} \cdot \boxed{}$ Substituting $A = \boxed{}$ Multiplying The area of the triangle is $\boxed{}$.	The distance d that a car will travel at a rate, or speed, r in time t is given by $d = rt$. A car travels at 65 miles per hour (mph) for 3.5 hr. How far does it travel?

EXAMPLE 2	YOUR TURN 2
Solve for s: $W = \dfrac{s-t}{3}$. $W = \dfrac{s-t}{3}$ $W \cdot \boxed{} = \dfrac{s-t}{3} \cdot \boxed{}$ Multiplying by 3 to clear the fraction $3W = s - t$ $3W + \boxed{} = s - t + \boxed{}$ Adding t $\boxed{} = s$	Solve for y: $A = \dfrac{x+y}{2}$. This is a formula for the average of two numbers.

EXAMPLE 3	YOUR TURN 3
Solve $B = 3m + 4n$ for m. $B = 3m + 4n$ $B - \boxed{} = 3m + 4n - 4n$ Subtracting $4n$ $B - 4n = 3m$ $\dfrac{B - 4n}{3} = \dfrac{3m}{3}$ Dividing by 3 $\dfrac{B - 4n}{3} = \boxed{}$	Solve $B = 3m + 4n$ for n.

EXAMPLE 4	YOUR TURN 4
Solve $pr + pq = m$ for p. $pr + pq = m$ $\boxed{}(r + q) = m$ Factoring out p (collecting like terms) $\dfrac{p(r + q)}{r + q} = \dfrac{m}{r + q}$ Dividing both sides by $r + q$ $\boxed{} = \dfrac{m}{r + q}$	Solve $abc + b = 2a$ for b.

YOUR NOTES   Write your questions and additional notes.

Practice Exercises

Readiness Check

Determine whether each statement is true or false.

1. To solve the formula $V = lwh$ for w you would subtract l and h on both sides of the equation.

2. When two or more terms on the same side of a formula contain the letter for which we are solving, we can factor so that the letter is only written once.

3. A formula is an equation that uses letters to represent a relationship between two or more quantities.

4. The first step in solving the equation $T = 0.3(I - 12{,}000)$ for I is to multiply on both sides by 0.3.

Evaluating and Solving Formulas

5. The wavelength w, in meters per cycle, of a musical note is given by $w = \dfrac{r}{f}$, where r is the speed of the sound, in meters per second, and f is the frequency, in cycles per second. The speed of sound in air is 344 m/sec. What is the wavelength of a note whose frequency in air is 32 cycles per second?

Solve for the given letter.

6. $d = 60t$, for t

7. $y = 26 - x$, for x

8. $T = \dfrac{a}{b}$, for a

9. $3y = 2x$, for x

10. $y = mx + b,$ for x

11. $A = \dfrac{p + q + r}{3},$ for q

12. *Perimeter of a rectangle:*

$P = 2l + 2w,$ for w

13. *Surface area of a sphere:*

$S = 4\pi r^2,$ for r^2

14. $\dfrac{d}{r} = t,$ for r

15. $x = ya^2 + yz,$ for y

16. The number of calories, K, needed each day by a moderately active woman who weighs w pounds, is h inches tall, and is a years old can be estimated by the formula

$K = 917 + 6(w + h - a).$

Allyson is moderately active, weighs 110 lb, and is 28 years old. If Allyson needs 1781 calories per day to maintain her weight, how tall is she?

17. A garden is constructed in the shape of a triangle. Using the formula for the area of a triangle, $A = \dfrac{1}{2}bh,$ find the height of the garden if the base of the triangle is 9 ft and the garden will have an area of 54 ft².

Five Steps for Problem Solving

ESSENTIALS

Five Steps for Problem Solving
1. *Familiarize* yourself with the problem situation.
2. *Translate* the problem to an equation.
3. *Solve* the equation.
4. *Check* the answer in the original problem.
5. *State* the answer to the problem clearly.

Consecutive integers like 11, 12, 13, 14 or $-25, -24, -23, -22$ can be represented in the form $x, x+1, x+2, x+3,$ and so on. **Consecutive even integers** like 12, 14, 16, 18 or $-24, -22, -20, -18$ can be represented in the form $x, x+2, x+4, x+6,$ and so on, as can **consecutive odd integers** like 11, 13, 15, 17 or $-25, -23, -21, -19$.

GUIDED LEARNING   **Textbook**   **Instructor**   **Video**

EXAMPLE 1	YOUR TURN 1
Of the top ten rice-producing countries, Brazil is the only country not in Asia. In 2012, the Philippines produced 11,000 thousand metric tons of rice. This was 4640 thousand metric tons less than two times as much as Brazil. Find the amount of rice produced in Brazil.	In 2012, India produced 99,000 thousand metric tons of rice. This was 17,000 thousand metric tons more than four times as much as Thailand. Find the amount of rice produced in Thailand.

1. **Familiarize.** We let $b =$ the amount of rice produced in Brazil.

2. **Translate.** We translate as follows.

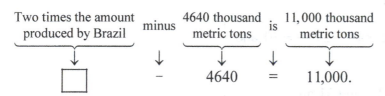

3. **Solve.** We solve the equation as follows.

$$2b - 4640 = 11,000$$
$$2b - 4640 + 4640 = 11,000 + 4640$$
$$2b = \boxed{}$$
$$\frac{2b}{\boxed{}} = \frac{15,640}{\boxed{}}$$
$$b = 7820$$

(continued)

4. **Check.** If Brazil produced 7820 thousand metric tons of rice, then the Philippines produced $2(7820) - 4640$, or 11,000, thousand metric tons of rice. The amount checks.

5. **State.** In 2012, Brazil produced 7820 thousand metric tons of rice.

EXAMPLE 2	YOUR TURN 2
The sum of three consecutive integers is 90. Find the integers.	The sum of three consecutive integers is −36. Find the integers.

1. **Familiarize.** Let x, $x+1$, and $x+2$ represent the consecutive integers.

2. **Translate.**

$$\underbrace{\text{The sum of three consecutive integers}} \quad \text{is} \quad 90$$
$$\downarrow \qquad\qquad\qquad \downarrow \quad \downarrow$$
$$x + (x+1) + (x+2) \quad = \quad 90.$$

3. **Solve.** Solve the equation.
$$x + (x+1) + (x+2) = 90$$
$$\boxed{} + 3 = 90$$
$$3x = \boxed{}$$
$$x = \boxed{}$$

The other two numbers are $x + 1 = 29 + 1 = 30$ and $x + 2 = 29 + 2 = \boxed{}$.

4. **Check.** The numbers 29, 30, and 31 are consecutive integers. Their sum is $29 + 30 + 31 = 90$. The answer checks.

5. **State.** The numbers are $\boxed{}$ $\boxed{}$, and $\boxed{}$.

YOUR NOTES   Write your questions and additional notes.

Basic Motion Problems

ESSENTIALS

The Motion Formula

Distance = Rate (or speed) · Time

$$d = rt$$

Example

- An airplane traveling 415 km/h in still air encounters a 25 km/h headwind. How long will it take the plane to travel 650 km into the wind?

  1. **Familiarize.** Because the airplane is traveling into the headwind, the speed of the wind can be subtracted from the speed of the airplane to determine the airplane's speed with the headwind. Let t represent the time, in hours, required for the plane to travel 650 km into the wind.

Distance to be traveled	650 km
Airplane's speed in still air	415 km/h
Speed of the headwind	25 km/h
Airplane's speed with the headwind	390 km/h
Time required	t

  2. **Translate.** Using the motion formula, $d = rt$, we have

     $$d = rt$$
     $$650 = 390 \cdot t.$$

  3. **Solve.** We solve the equation:

     $$650 = 390 \cdot t$$
     $$\frac{650}{390} = \frac{390 \cdot t}{390}$$
     $$\frac{5}{3} = t.$$

  4. **Check.** At a speed of 390 km/h and in a time of 5/3, or $1\frac{2}{3}$, hours, the airplane would travel $d = 390 \cdot \frac{5}{3} = 650$ km. This answer checks.

  5. **State.** The airplane will travel the distance of 650 km in 5/3, or $1\frac{2}{3}$, hours.

GUIDED LEARNING   **Textbook**   **Instructor**   ▶ **Video**

EXAMPLE 1	YOUR TURN 1
Jill rides her bicycle 15 miles to the shore at a speed of 22 mph on a still day. If riding with a 14 mph tailwind, how long will it take to bike the 15 mi?	Jinger rides her bicycle 22 mi to the shore at a speed of 25 mph. If she is biking with an 8 mph tailwind, how long will the ride take?

1. **Familiarize.** We let t = the time, in hours, it will take to bike to the shore. Because Jill is biking in the same direction as the tailwind, the two speeds can be added for total speed.

Distance to be traveled	15 mi
Jill's speed on a still day	22 mph
Speed of the tailwind	14 mph
Jill's speed with the tailwind	36 mph
Time required	t

2. **Translate.** Using the motion formula,

$$d = rt$$

$$\boxed{} = 36 \cdot t.$$

3. **Solve.** We solve the equation.

$$15 = 36 \cdot t$$

$$\frac{15}{\boxed{}} = \frac{36 \cdot t}{\boxed{}}$$

$$\frac{15}{36}, \text{ or } \frac{5}{\boxed{}} = t$$

4. **Check.** At a speed of 36 mph and a time of 5/12 hours, Jill would travel $d = 36 \cdot \dfrac{5}{12} = 15$ mi. This answer checks.

5. **State.** Jill will travel the 15 mi in 5/12 hr, or 25 min.

YOUR NOTES   Write your questions and additional notes.

Practice Exercises

Readiness Check

Fill in the steps in the five steps for problem solving in the correct order. All steps are listed below.

| Check | Translate | Familiarize | State | Solve |

1. _____

2. _____

3. _____

4. _____

5. _____

Five Steps for Problem Solving

Solve.

6. A gas station offers a 5% discount to anyone who pays cash for gas. Sue paid $65.25 in cash for gas. What would she have paid if she did not pay with cash?

7. The sum of two integers is fourteen. The larger integer is four less than the twice the smaller integer. Find the integers.

8. A worker on a production line is paid a base salary of $240 per week plus $0.81 for each unit produced. One week the worker earned $414.15. How many units were produced?

9. The sum of the page numbers on the facing pages of a book is 85. What are the page numbers?

10. In a triangle, the measure of the second angle is 5 times the measure of the first angle, and the measure of the third angle is 12° less than 10 times the measure of the first angle. Find the measure of each angle.

Basic Motion Problems

Solve.

11. A moving walkway in an airport is 880 ft long and it moves at a rate of 5 ft/sec. If Mike walks at a rate of 3 ft/sec, how long will it take him to walk its entire length?

Inequalities

ESSENTIALS

An **inequality** is a sentence containing $<$, $>$, $\leq$, $\geq$, or $\neq$.

The replacement or value for the variable that makes an inequality true is called a **solution** of the inequality. The set of all solutions is called the **solution set**. When all the solutions of an inequality have been found, we say that we have **solved** the inequality.

Examples

- 1 is a solution of the inequality $x+5<7$ because $1+5<7$, or $6<7$, is a *true* sentence.

- -4 is a solution of the inequality $x+9 \leq 5$ because $-4+9 \leq 5$, or $5 \leq 5$, is a *true* sentence.

- 0 is not a solution of the inequality $2x>x+8$ because $2(0)>0+8$, or $0>8$, is a *false* sentence.

GUIDED LEARNING   📍 **Textbook**   👤 **Instructor**   ▶ **Video**

EXAMPLE 1	YOUR TURN 1
Determine whether the given number is a solution of the inequality: $x-7<9$; 12. $\boxed{}-7<9$ $\boxed{}<9$ 5 is less than 9 is a *true* sentence so 12 is a solution of the inequality.	Determine whether the given number is a solution of the inequality: $x+3 \geq 4$; -1.
EXAMPLE 2	YOUR TURN 2
Determine whether the given number is a solution of the inequality: $2x+3>6$; -3. $2\left(\boxed{}\right)+3>6$ $\boxed{}>6$ -3 is greater than 6 is a *false* sentence, so -3 is not a solution of the inequality.	Determine whether the given number is a solution of the inequality: $3x+1>4$; -5.

EXAMPLE 3	YOUR TURN 3
Determine whether the given number is a solution of the inequality: $2x - 2 \geq x + 5;\ 8.$ $2\left(\boxed{}\right) - 2 \geq \boxed{} + 5$ $\boxed{} \geq \boxed{}$ 14 is greater than or equal to 13 is a *true* sentence, so 8 is a solution of the inequality.	Determine whether the given number is a solution of the inequality: $3x - 9 \leq x + 1;\ 5.$

YOUR NOTES   Write your questions and additional notes.

Inequalities and Interval Notation

ESSENTIALS

There are three ways to represent the solutions of an inequality such as $a < x < b$:

- A **graph**, or drawing, represents the solutions.

- **Set-builder notation** describes the solution set of an inequality.

$$\{x \mid a < x < b\}$$

This is read as "the set of all x such that x is both greater than a and less than b."

- **Interval notation** uses parentheses ( ) and brackets [ ].

$$(a, b)$$

The **endpoints** of this interval are a and b. The parentheses indicate a and b are *not* included in the interval. Brackets are used when endpoints *are* included. When an interval extends without bound in one or both directions, we use the symbols ∞ and $-\infty$, read "infinity" and "negative infinity." Parentheses are always used with ∞ and $-\infty$.

Examples

Interval Notation	Set Notation	Graph
$[3, \infty)$	$\{x \mid x \geq 3\}$	
$(-\infty, -5)$	$\{x \mid x < -5\}$	
$(-1, 4]$	$\{x \mid -1 < x \leq 4\}$	
$(-\infty, \infty)$	$\{x \mid x \text{ is a real number}\}$	

GUIDED LEARNING	📖 **Textbook** 👤 **Instructor** ▶ **Video**
EXAMPLE 1	**YOUR TURN 1**
Write interval notation for the set: $\{x \mid 0 \le x < 7\}$. Interval notation: $[\boxed{}, \boxed{})$	Write interval notation for the set: $\{x \mid -4 < x \le 3\}$.
EXAMPLE 2	**YOUR TURN 2**
Write interval notation for the set: $\{x \mid x < 4\}$. Interval notation: $\boxed{} -\infty, 4 \boxed{}$	Write interval notation for the set: $\{x \mid x < -7\}$.
EXAMPLE 3	**YOUR TURN 3**
Write interval notation for the graph. $-5\ -4\ -3\ -2\ -1\ \ 0\ \ 1\ \ 2\ \ 3\ \ 4\ \ 5\ \ 6$ Interval notation: $[\boxed{}, \boxed{})$	Write interval notation for the graph. $3\ \ 4\ \ 5\ \ 6\ 7\ 8\ \ 9\ 10\ 11\ 12\ 13\ 14$
EXAMPLE 4	**YOUR TURN 4**
Write interval notation for the graph. $-6\ -5\ -4\ -3\ -2\ -1\ \ 0\ \ 1\ \ 2\ \ 3\ \ 4\ \ 5$ Interval notation: $\boxed{} -1, \infty \boxed{}$	Write interval notation for the graph. $3\ \ 4\ \ 5\ 6\ 7\ \ 8\ \ 9\ 10\ 11\ 12\ 13\ 14$

YOUR NOTES   Write your questions and additional notes.

Solving Inequalities

ESSENTIALS

Two inequalities are **equivalent** if they have the same solution set.

The Addition Principle for Inequalities

For any real numbers a, b, and c:

$a < b$ is equivalent to $a + c < b + c$;

$a > b$ is equivalent to $a + c > b + c$.

Similar statements hold for $\le$ and $\ge$.

The Multiplication Principle for Inequalities

For any real numbers a and b, and any *positive* number c:

$a < b$ is equivalent to $ac < bc$;

$a > b$ is equivalent to $ac > bc$.

For any real numbers a and b, and any *negative* number c:

$a < b$ is equivalent to $ac > bc$;

$a > b$ is equivalent to $ac < bc$.

Similar statements hold for $\le$ and $\ge$.

Examples

- Solve $x - 3 \ge 5$.

$$x - 3 \ge 5$$
$$x - 3 + 3 \ge 5 + 3$$
$$x \ge 8$$

The solution set is $\{x \mid x \ge 8\}$, or $[8, \infty)$.

- Solve $\dfrac{1}{2} y \le 24$.

$$\frac{1}{2} y \le 24$$
$$2 \cdot \frac{1}{2} y \le 2 \cdot 24$$
$$y \le 48$$

The solution is $\{y \mid y \le 48\}$, or $(-\infty, 48]$.

- Solve $-2x > 24$.

$$-2x > 24$$
$$\frac{-2x}{-2} < \frac{24}{-2}$$
$$x < -12$$

The solution is $\{x \mid x < -12\}$, or $(-\infty, -12)$.

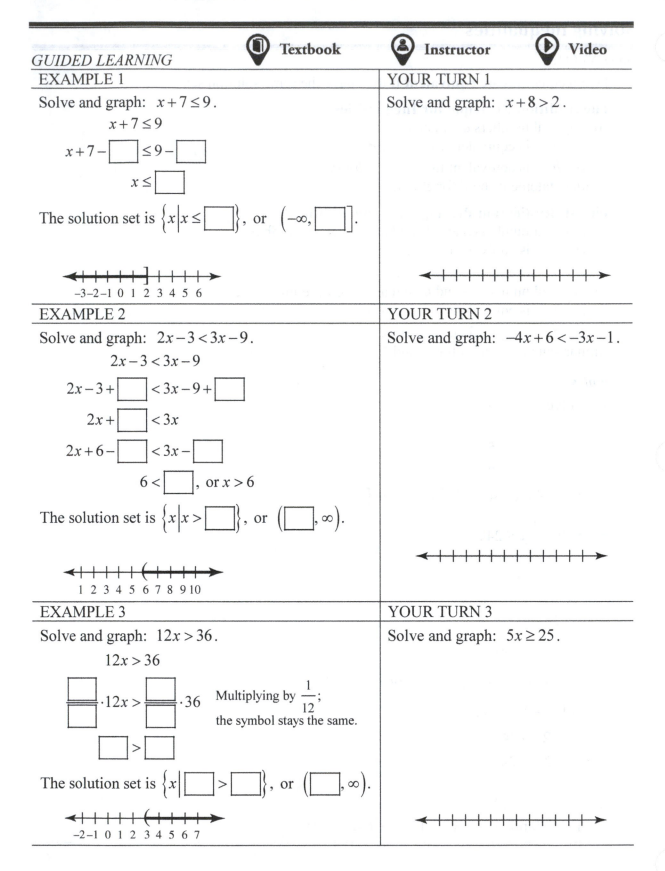

GUIDED LEARNING   📖 **Textbook**   👤 **Instructor**   ▶ **Video**

EXAMPLE 1	YOUR TURN 1

Solve and graph: $x + 7 \leq 9$.

$$x + 7 \leq 9$$

$$x + 7 - \boxed{} \leq 9 - \boxed{}$$

$$x \leq \boxed{}$$

The solution set is $\left\{ x \mid x \leq \boxed{} \right\}$, or $\left(-\infty, \boxed{} \right]$.

Solve and graph: $x + 8 > 2$.

EXAMPLE 2	YOUR TURN 2

Solve and graph: $2x - 3 < 3x - 9$.

$$2x - 3 < 3x - 9$$

$$2x - 3 + \boxed{} < 3x - 9 + \boxed{}$$

$$2x + \boxed{} < 3x$$

$$2x + 6 - \boxed{} < 3x - \boxed{}$$

$$6 < \boxed{}, \text{ or } x > 6$$

The solution set is $\left\{ x \mid x > \boxed{} \right\}$, or $\left(\boxed{}, \infty \right)$.

Solve and graph: $-4x + 6 < -3x - 1$.

EXAMPLE 3	YOUR TURN 3

Solve and graph: $12x > 36$.

$$12x > 36$$

$$\frac{\boxed{}}{\boxed{}} \cdot 12x > \frac{\boxed{}}{\boxed{}} \cdot 36 \qquad \text{Multiplying by } \frac{1}{12}; \text{ the symbol stays the same.}$$

$$\boxed{} > \boxed{}$$

The solution set is $\left\{ x \mid \boxed{} > \boxed{} \right\}$, or $\left(\boxed{}, \infty \right)$.

Solve and graph: $5x \geq 25$.

EXAMPLE 4	YOUR TURN 4
Solve and graph: $-2x > 8$.	Solve and graph: $-3x \geq -27$.

EXAMPLE 4

Solve and graph: $-2x > 8$.

$$-2x > 8$$

$$\frac{-2x}{\boxed{}} < \frac{8}{\boxed{}} \qquad \text{Dividing by } -2 \text{ and reversing the inequality symbol}$$

$$x < \boxed{}$$

The solution set is $\left\{ x \mid x < \boxed{} \right\}$, or $\left(\boxed{}, \boxed{} \right)$.

$-9\ -8\ -7\ -6\ -5\ -4\ -3\ -2\ -1\ \ 0$

YOUR TURN 4

Solve and graph: $-3x \geq -27$.

EXAMPLE 5

Solve and graph: $4x - 12 \leq 18 + 10x$.

$$4x - 12 \leq 18 + 10x$$

$$4x - 12 + \boxed{} \leq 18 + 10x + \boxed{}$$

$$4x \leq \boxed{} + 10x$$

$$4x - \boxed{} \leq 30 + 10x - \boxed{}$$

$$\boxed{} \leq 30$$

$$\frac{-6x}{\boxed{}} \geq \frac{30}{\boxed{}} \qquad \text{Dividing by } -6; \text{ the symbol must be reversed.}$$

$$x \boxed{} -5$$

The solution set is $\left\{ x \mid x \boxed{}\ \boxed{} \right\}$, or $\left[-5, \boxed{} \right)$.

$-10\ -9\ -8\ -7\ -6\ -5\ -4\ -3\ -2\ -1$

YOUR TURN 5

Solve and graph:
$20 - 3x \geq 4x - 8$.

Applications and Problem Solving

ESSENTIALS

Translating "At Least and "At Most"

A quantity x is **at least** some amount q: $x \geq q$.

(If x is at least q, it cannot be less than q.)

A quantity x is **at most** some amount q: $x \leq q$.

(If x is at most q, it cannot be more than q.)

Examples

A few common phrases that are used in application problems involving inequalities are underlined in the examples below.

Sample Sentence	Translation
The temperature of a fully cooked chicken <u>is at least</u> 165°F.	$t \geq 165$
Today's high temperature <u>is at most</u> 75°F.	$t \leq 75$
Your speed on the highway <u>cannot exceed</u> 65 mph.	$s \leq 65$
Your grade <u>must exceed</u> 69.5.	$g > 69.5$
The recommended tire pressure <u>is between</u> 28 psi and 32 psi.	$28 < p < 32$
He can withdraw <u>no more than</u> $1500 from his account.	$w \leq 1500$
The bank requires that you deposit <u>more than</u> $50.	$d > 50$

Textbook   **Instructor**   **Video**

GUIDED LEARNING

EXAMPLE 1	YOUR TURN 1
Greg is playing in a golf tournament. After the first three days of the tournament his scores are 68, 67, and 71. The record low score for this particular golf course is a four-day total of 276. What scores could Greg make on the fourth day in order for his four-day total to be less than the course record?	Sharon manages a small retail clothing store whose 3-day weekend (Friday, Saturday, Sunday) sales record is $5072.53. She has made $1235.67 in sales on Friday and $2790.84 on Saturday. How much money must Sharon make in sales on Sunday so that her total weekend sales will be more than the current weekend record?

1., 2. Familiarize and Translate.

Let f = Greg's score on day #☐ . We want

Greg's total score to be _____
less than / greater than

the record. Thus, total score < course record.

This translates to $68 + 67 + 71 + f$ ☐ 276.

3. **Solve.** We solve the inequality:

$$68 + 67 + 71 + f < 276$$

$$\boxed{} + f < 276$$

$$f < \boxed{}.$$

4. **Check.** As a partial check, substitute a number less than 70 for f. We choose 69. Replacing f with 69, we get $68 + 67 + 71 + 69 = 275$ which is less than the record low of 276.

5. **State.** Greg must score less than 70 on the fourth day of the tournament to have his four day total score less than the course record of 276. The solution set is $\{f \mid f < 70\}$.

EXAMPLE 2	YOUR TURN 2
Samantha is considering two different catering companies to provide the food for her upcoming wedding. Company A charges a flat fee of $500 plus $10 per guest attending the wedding. Company B charges a flat fee of $300 plus $14 per guest. For what number of guests will Company A cost less?	John and Katie are planning a birthday party for their son. Location A charges $8 per guest, whereas location B charges $50 plus $6 per guest. For what number of guests will location B cost less?

1. **Familiarize.** Total cost includes the flat fee plus the cost for guests.

    Company A total:

    $$\$500 + (\$10)(\text{Number of guests})$$

    Company B total:

    $$\boxed{} + (\$14)(\text{Number of guests})$$

2. **Translate.** Let G = the number of guests attending. Thus we want to find all values of G such that

Cost of Company A	is less than	Cost of Company B
↓	↓	↓
$500 + 10G$	$<$	$300 + 14G.$

3. **Solve.** We solve the inequality:

    $$500 + 10G < 300 + 14G$$
    $$\boxed{} + 10G < 14G$$
    $$200 < \boxed{}$$
    $$\boxed{} < G, \text{ or } G > \boxed{}$$

4. **Check.** For $G = 50$, the cost of Company A is $500 + 10 \cdot 50$, or $1000 and the cost of Company B is $300 + 14 \cdot 50$, or $1000. Thus, for 50 guests, the cost of both plans is the same. We can substitute numbers greater than 50 to make a partial check of our result.

5. **State.** The solution set is $\{G \mid G > 50\}$.

    Company A will cost less if more than 50 guests are invited.

Practice Exercises

Readiness Check

For each solution set expressed in set-builder notation, select from the column on the right the equivalent interval notation.

1. $\{x \mid x > 5\}$ a) $(-\infty, 5)$

2. $\{x \mid x < 5\}$ b) $[2, 5)$

3. $\{x \mid 2 < x \le 5\}$ c) $(-\infty, 5]$

4. $\{x \mid x \ge 5\}$ d) $(5, \infty)$

5. $\{x \mid x \le 5\}$ e) $(2, 5]$

6. $\{x \mid 2 \le x < 5\}$ f) $[5, \infty)$

Inequalities

Determine whether the given number is a solution of the inequality.

7. $x + 4 < -9; -10$

8. $2x - 5 \ge 7; 7$

9. $4x + 6 > 1 - x; -1$

Inequalities and Interval Notation

Write interval notation for the given set or graph.

10. $\{x \mid x \ge -2\}$ 11. $\{x \mid -3 \le x \le 5\}$

12.

13.

Solving Inequalities

Solve and graph.

14. $x - 7 > -3$ **15.** $-3x < 18$

16. $2x - 3 \le 5 - 6x$ **17.** $x + 5 \le 5x + 17$

18. $5(x - 2) + 8 < 2(x + 4) - 3x$

Applications and Problem Solving

Solve.

19. This month Terry is expecting to receive a $300 bonus check in addition to his regular pay rate of $15 per hour. How many hours must Terry work this month in order to have a paycheck that is more than $2850?

20. Jennifer must take three tests in her psychology class, each worth 100 points. She scored an 87 on the first test and a 91 on the second test. In order to make an A in the course Jennifer's test grades must total at least 270. What scores on the third test will give Jennifer an A in the course?

21. Hank is planning to repaint his house and has received estimates from two companies. Company A charges $600 plus $10 per hour. Company B charges $25 per hour. How many hours would the job need to take in order for Company B to cost less?

Intersections of Sets and Conjunctions of Inequalities

ESSENTIALS

Compound inequalities consist of two or more inequalities joined by the word *and* or the word *or*.

The **intersection** of two sets A and B is the set of all members common to A and B. We denote the intersection of sets A and B as $A \cap B$. The intersection of two sets can be illustrated as shown below. The region where the sets overlap is the intersection.

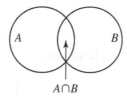

When two or more sentences are joined by the word *and* to make a compound sentence, the new sentence is called a **conjunction** of the sentences.

$a < x$ *and* $x < b$ **can be abbreviated** $a < x < b$;

$b > x$ *and* $x > a$ **can be abbreviated** $b > x > a$.

CAUTION! "$a > x$ *and* $x < b$" cannot be abbreviated as "$a > x < b$".

The word **"and"** corresponds to **"intersection"** and to the symbol "$\cap$". In order for a number to be a solution of a conjunction, it must make each part of the conjunction true.

Sometimes two sets have no elements in common. In such a case, we say that the intersection of the two sets is the **empty set**, denoted $\{\ \}$ or $\varnothing$. Two sets with an empty intersection are said to be **disjoint**. The sets shown below do not overlap, thus they have no elements in common. In this diagram $A \cap B = \varnothing$.

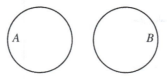

Examples

- The intersection of the sets $\{0, 2, 4, 6, 8\}$ and $\{1, 2, 3, 4, 5\}$ is $\{2, 4\}$.

- The intersection of the sets $\{1, 3, 5, 7, 9\}$ and $\{2, 4, 6, 8, 10\}$ is $\varnothing$.

- The conjunction $3 < x$ *and* $x \leq 7$ can be rewritten as $3 < x \leq 7$.

- Solve $3 < x + 5 \leq 9$.

    $3 < x + 5$ *and* $x + 5 \leq 9$

    $-2 < x$ *and* $x \leq 4$

    The solution set is $\{x \mid -2 < x \leq 4\}$.

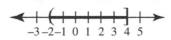

GUIDED LEARNING   🔖 **Textbook**   👤 **Instructor**   ▶ **Video**

EXAMPLE 1	YOUR TURN 1
Find the intersection:	Find the intersection:
$\{2, 4, 6, 8, 10\} \cap \{-4, 0, 4, 8, 12\}$.	$\{-2, 0, 1, 5, 8\} \cap \{0, 1, 2, 8\}$.

4 and 8 are the only numbers common to both sets.

The intersection is $\left\{\boxed{}, \boxed{}\right\}$.

EXAMPLE 2	YOUR TURN 2
Solve and graph: $3 < 4y - 1 \le 23$.	Solve and graph: $-4 \le 8y + 4 < 36$.

Write as the conjunction:

$3 < 4y - 1 \boxed{} 4y - 1 \le 23$.

Solve each inequality separately:

$3 < 4y - 1 \quad and \quad 4y - 1 \le 23$

$\boxed{} < 4y \quad and \quad 4y \le \boxed{} \qquad$ Adding 1

$\boxed{} < y \quad and \quad y \le \boxed{}$. $\quad$ Dividing by 4

The graph is the intersection of these two solution sets.

$\{y | 1 < y\}$

$\{y | y \le 6\}$

$\{y | 1 < y\} \cap \{y | y \le 6\} = \left\{y \middle| \boxed{}\right\}$

The solution set is $\{y | 1 < y \le 6\}$, or in interval notation, $(1, 6]$.

EXAMPLE 3	YOUR TURN 3
Solve and graph: $17 < 2x + 5$ *and* $4x - 3 < 13$.	Solve and graph: $2 < 3x + 5$ *and* $4 - 4x < -12$.

Solve each inequality separately:

$17 < 2x + 5 \quad and \quad 4x - 3 < 13$

$\boxed{} < 2x \quad and \quad 4x < \boxed{}$

$\boxed{} < x \quad and \quad x < \boxed{}.$

The graph is the intersection of these two solution sets.

$\{x \mid 6 < x\}$

$\{x \mid x < 4\}$

$\{x \mid 6 < x\} \cap \{x \mid x < 4\} = \boxed{}$

The solution set is $\varnothing$ because the two sets have no numbers in common.

Unions of Sets and Disjunctions of Inequalities

ESSENTIALS

The **union** of two sets A and B is the collection of all elements belonging to A and/or B. We denote the union of sets A and B by $A \cup B$.

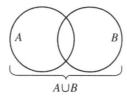

$A \cup B$

When two or more sentences are joined by the word *or* to make a compound sentence, the new sentence is called a **disjunction** of the sentences.

The word **"or"** corresponds to **"union"** and to the symbol **"$\cup$"**. In order for a number to be in the solution set of a disjunction, it must be in *at least one* of the solution sets of the individual sentences.

Examples

- The union of the sets $\{2, 4, 6\}$ and $\{4, 8, 12\}$ is $\{2, 4, 6, 8, 12\}$.

- The disjunction $3 < x \ or \ x \leq -7$ can be rewritten as $(-\infty, -7] \cup (3, \infty)$.

GUIDED LEARNING   📖 **Textbook**   👤 **Instructor**   ▶ **Video**

EXAMPLE 1	YOUR TURN 1
Find the union: $\{1, 2\} \cup \{2, 4, 6\}$.	Find the union: $\{0, 3\} \cup \{1, 3, 9\}$.

Find the union: $\{1, 2\} \cup \{2, 4, 6\}$.

(circle diagram with 1, 2, 4, 6)

The numbers in either or both sets are 1, 2, 4, and 6, so the union is $\{\boxed{}\}$.

EXAMPLE 2	YOUR TURN 2
Solve and graph: $2x - 7 > 3 \ or \ 5 - x \geq 2$.	Solve and graph: $3x + 1 < 7 \ or \ 3 - x < -1$.

Solve each inequality separately:

$$2x - 7 > 3 \quad or \quad 5 - x \geq 2$$
$$2x > \boxed{} \quad or \quad -x \geq \boxed{}$$
$$x > \boxed{} \quad or \quad x \leq \boxed{}.$$

(continued)   (continued)

The graph is the union of these two solution sets.

$\{x \mid x > 5\}$

$\{x \mid x \le 3\}$

$\{x \mid x > 5\} \cup \{x \mid x \le 3\}$

The solution set is written $\{x \mid x \le 3 \text{ or } x > 5\}$, or in interval notation, $(-\infty, 3] \cup (5, \infty)$.

EXAMPLE 3	YOUR TURN 3
Solve and graph: $2x - 15 > -7 \text{ or } 3x + 4 < 25$.	Solve and graph: $x - 4 \le 4 \text{ or } x - 3 \le 2$.

Solve each inequality separately:

$$2x - 15 > -7 \qquad or \qquad 3x + 4 < 25$$
$$2x > \boxed{} \qquad or \qquad 3x < \boxed{}$$
$$x > \boxed{} \qquad or \qquad x < \boxed{}.$$

The graph is the union of these two solution sets.

$\{x \mid x > 4\}$

$\{x \mid x < 7\}$

$\{x \mid x > 4\} \cup \{x \mid x < 7\}$

The solution set is $\{x \mid x \text{ is a real number}\}$, or in interval notation, $(-\infty, \infty)$.

YOUR NOTES   Write your questions and additional notes.

Applications and Problem Solving

EXAMPLE 1	YOUR TURN 1
In order to get a B in her biology class the average of Ruth's midterm exam and final exam must be at least 80 but can be at most 89. If Ruth got a 78 on her midterm what scores could she get on her final exam to get a B in biology?	Rob's cookbook says that chicken is fully cooked when it reaches Celsius temperatures C such that $75° \le C \le 80°$. Use the formula $C = \dfrac{5}{9}(F - 32)$ to find an inequality for the corresponding Fahrenheit temperatures.

1., 2. Familiarize and Translate.

Let f = Ruth's score on the final exam.

Ruth's average could be calculated as follows:

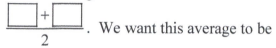

$\dfrac{\boxed{} + \boxed{}}{2}$. We want this average to be greater than or equal to 80 *and* less than or equal to 89. This can be written as the conjunction

$$\boxed{} \le \frac{78 + f}{2} \quad and \quad \frac{78 + f}{2} \le \boxed{}.$$

3. Solve. We solve each inequality separately.

$$80 \le \frac{78 + f}{2} \quad and \quad \frac{78 + f}{2} \le 89$$

$$\boxed{} \le 78 + f \quad and \quad 78 + f \le \boxed{}$$

$$\boxed{} \le f \quad and \quad f \le \boxed{}.$$

4. Check. Suppose Ruth's final exam score was 83. Her average would be $\dfrac{78 + 83}{2} = \dfrac{161}{2} = 80.5$.

Similarly it can be shown that if Ruth scored a 99 on the final exam her average would be 88.5. In either scenario her average is at least 80 and at most 89.

5. State. Ruth's final exam score must be at least 82 and at most 100 which can be written as the solution set $\{f \mid 82 \le f \le 100\}$.

YOUR NOTES   Write your questions and additional notes.

Practice Exercises

Readiness Check

Determine whether each statement is true or false.

1. $\{1, 2, 3, 4, 5\} \cap \{2, 4, 6, 8, 10\} = \varnothing$

2. $\{1, 2, 3, 4, 5\} \cup \{6\} = \{1, 2, 3, 4, 5, 6\}$

3. The compound inequality $3 < x \ or \ x < 7$ can be expressed as $3 < x < 7$.

4. The compound inequality $3 > x \ and \ x < 7$ can be expressed as $3 > x < 7$.

5. The solution set of $x < -2 \ or \ x > 6$ can be written as $(-\infty, -2) \cup (6, \infty)$.

6. The solution set of $x > -2 \ and \ x < 6$ can be written as $(-2, 6)$.

Intersections of Sets and Conjunctions of Inequalities

Find the intersection.

7. $\{1, 2, 3\} \cap \{a, b, c\}$

8. $\{a, b, c, d\} \cap \{c, a, t\}$

Solve and graph.

9. $5 \leq x + 2 \ and \ x + 2 < 9$

10. $-3 < 3x + 6 < 12$

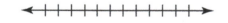

11. $x + 4 \geq -1 \ and \ 5 - x < 8$

12. $-2x + 5 > 9 \ and \ 9 \leq 3x - 6$

Unions of Sets and Disjunctions of Inequalities

Find the union.

13. $\{1, 2, 3\} \cup \{a, b, c\}$

14. $\{1, 2, 3\} \cup \varnothing$

Solve and graph.

15. $7 \le 2x + 1$ *or* $2x + 1 \le 1$

16. $2 - x > 3$ *or* $x + 4 \ge 8$

17. $5x - 3 \ge 7$ *or* $-x + 7 < 2$

18. $x + 8 < 16$ *or* $2x - 3 > 7$

Applications and Problem Solving

Solve.

19. One of the strangest weather days occurred in Browning, Montana, on January 24, 1916. The low temperature for the day was $-49°C$ and the high temperature for the day was $7°C$. Use the inequality $-49° \le C \le 7°$ along with the fact that $C = \dfrac{5}{9}(F - 32)$ to find such an inequality for the corresponding Fahrenheit temperatures.

20. In order to get a grade of A in his psychology class, the average of Rudy's three exams must be greater than or equal to 90. If Rudy got grades of 88 and 96 on his first two exams, what scores could he make on the third exam to get an A in psychology? (Suppose it is possible for him to make higher than 100 on the third exam.)

Properties of Absolute Value

ESSENTIALS

The **absolute value** of a number is the number's distance from zero on the number line.

The **absolute value** of x, denoted $|x|$, is defined as follows:

$$x \geq 0 \rightarrow |x| = x, \qquad x < 0 \rightarrow |x| = -x.$$

Properties of Absolute Value

a) $|ab| = |a| \cdot |b|$, for any real numbers a and b.

(The absolute value of a product is the product of the absolute values.)

b) $\left|\dfrac{a}{b}\right| = \dfrac{|a|}{|b|}$, for any real numbers a and b and $b \neq 0$.

(The absolute value of a quotient is the quotient of the absolute values.)

c) $|-a| = |a|$, for any real number a.

(The absolute value of the opposite of a number is the same as the absolute value of the number.)

Examples

- $|5| = 5$

- $|-5| = 5$

- $|7x| = |7| \cdot |x| = 7|x|$

- $\left|\dfrac{-8}{x}\right| = \dfrac{|-8|}{|x|} = \dfrac{8}{|x|}$

GUIDED LEARNING   📍 **Textbook**   👤 **Instructor**   ▶ **Video**

EXAMPLE 1	YOUR TURN 1
Simplify, leaving as little as possible inside the absolute-value signs. $$|7x^2| = \boxed{} \cdot |x^2| = 7x^2$$ Since x^2 can never be negative for any real number x we can omit the absolute-value signs.	Simplify, leaving as little as possible inside the absolute-value signs. $$|5x^4|$$

EXAMPLE 2	YOUR TURN 2
Simplify, leaving as little as possible inside the absolute-value signs.	Simplify, leaving as little as possible inside the absolute-value signs.
$$\left|\frac{6x^3}{2x^2}\right| = \left|\boxed{} \cdot x\right| = 3\boxed{}$$	$$\left|\frac{16y^6}{4y^5}\right|$$
EXAMPLE 3	YOUR TURN 3
Simplify, leaving as little as possible inside the absolute-value signs.	Simplify, leaving as little as possible inside the absolute-value signs.
$$\left|\frac{5y^7}{-10y^5}\right| = \left|\frac{y^2}{\boxed{}}\right| = \frac{\left|y^2\right|}{\boxed{}} = \frac{\boxed{}}{\boxed{}}$$	$$\left|\frac{6x^9}{-18x^7}\right|$$

YOUR NOTES   Write your questions and additional notes.

Distance on the Number Line

ESSENTIALS

For any real numbers a and b, the **distance** between them is $|a - b|$.

Examples

- The distance between -2 and 5 on the number line is $|-2 - 5| = |-7| = 7$.

- The distance between -13 and -9 on the number line is $|-13 - (-9)| = |-4| = 4$.

- The distance between x and 3 on the number line is $|x - 3|$ or $|3 - x|$.

GUIDED LEARNING	🔲 **Textbook** 👤 **Instructor** ▶ **Video**	

EXAMPLE 1	YOUR TURN 1
Find the distance between 6 and -2 on the number line. $\begin{array}{c}\longleftrightarrow\\ -2\,-1\;0\;1\;2\;3\;4\;5\;6\end{array}$ $\left|\boxed{} - (-2)\right| = \left|\boxed{}\right| = 8$ or $\left|\boxed{} - 6\right| = \left|\boxed{}\right| = 8$ 6 and -2 are 8 units apart on the number line.	Find the distance between 12 and -9 on the number line.
EXAMPLE 2	YOUR TURN 2
Find the distance between -10 and -6 on the number line. $\begin{array}{c}\longleftrightarrow\\ -11\,-10\,-9\,-8\,-7\,-6\,-5\,-4\,-3\end{array}$ $\left|\boxed{} - (-6)\right| = \left|\boxed{}\right| = 4$ or $\left|\boxed{} - (-10)\right| = \left|\boxed{}\right| = 4$ -10 and -6 are 4 units apart on the number line.	Find the distance between -4 and -15 on the number line.

Write your questions and additional notes.

Equations with Absolute Value

ESSENTIALS

The Absolute-Value Principle

For any positive number p and any algebraic expression X:

a) The solution of $|X| = p$ is those numbers that satisfy $X = -p$ or $X = p$.

b) The equation $|X| = 0$ is equivalent to the equation $X = 0$.

c) The equation $|X| = -p$ has no solution.

Examples

- The equation $|x| = 6$ has the solution set $\{-6, 6\}$.
- Solve $|2x - 8| = 0$.

$$2x - 8 = 0 \qquad \text{Using the second part of the absolute-value principle}$$
$$2x = 8 \qquad \text{Adding 8}$$
$$x = 4 \qquad \text{Dividing by 2}$$

The solution set is $\{4\}$.

- Solve $|x - 2| = 7$.

$$|x - 2| = 7$$
$$x - 2 = -7 \quad or \quad x - 2 = 7 \qquad \text{Using the first part of the absolute-value principle}$$
$$x = -5 \quad or \qquad x = 9 \qquad \text{Adding 2}$$

The solution set is $\{-5, 9\}$.

- The equation $|x + 5| = -3$ has no solution. The solution set is the empty set, $\varnothing$.

GUIDED LEARNING	🔖 **Textbook**　　🧑 **Instructor**　　▶ **Video**	
EXAMPLE 1	**YOUR TURN 1**	
Solve: $|4x| = -8$. Since absolute value is always nonnegative, this equation has no solution. The solution set is ☐ .	Solve: $|x - 3| + 5 = 2$.	

EXAMPLE 2	YOUR TURN 2
Solve: $\lvert x+7 \rvert -4=8$.	Solve: $\lvert x \rvert -6=4$.

We first use the addition principle for equations to get $\lvert x+7 \rvert$ by itself.

$$\lvert x+7 \rvert -4=8$$

$$\lvert x+7 \rvert = \boxed{} \quad \text{Adding 4}$$

$$x+7 = \boxed{} \quad or \quad x+7=12 \quad \begin{array}{l}\text{Absolute-value}\\\text{principle}\end{array}$$

$$x = \boxed{} \quad or \quad x=5 \quad \text{Subtracting 7}$$

The solutions are $\boxed{}$ and 5. The solution set is $\left\{ \boxed{}, 5 \right\}$.

EXAMPLE 3	YOUR TURN 3
Solve: $\lvert 5-3y \rvert =7$.	Solve: $\lvert y-1 \rvert =3$.

$$\lvert 5-3y \rvert =7$$

$$5-3y = \boxed{} \quad or \quad 5-3y=7 \quad \begin{array}{l}\text{Absolute-value}\\\text{principle}\end{array}$$

$$-3y = \boxed{} \quad or \quad -3y=2 \quad \text{Subtracting 5}$$

$$y = \boxed{} \quad or \quad y=-\frac{2}{3} \quad \text{Dividing by } -3$$

The solutions are $-\dfrac{2}{3}$ and $\boxed{}$. The solution set is $\left\{ -\dfrac{2}{3}, \boxed{} \right\}$.

YOUR NOTES   Write your questions and additional notes.

Equations with Two Absolute-Value Expressions

ESSENTIALS

If $|a| = |b|$, this means that a and b are the same distance from 0. If a and b are the same distance from 0, then either they are the same number or they are opposites.

Example

- Solve $|3x + 7| = |x - 3|$.

$$3x + 7 = x - 3 \quad or \quad 3x + 7 = -(x - 3)$$
$$3x = x - 10 \quad or \quad 3x + 7 = -x + 3$$
$$2x = -10 \quad or \quad 3x = -x - 4$$
$$x = -5 \quad or \quad 4x = -4$$
$$x = -5 \quad or \quad x = -1$$

The solution set is $\{-5, -1\}$.

	Textbook		Instructor		Video

GUIDED LEARNING

EXAMPLE 1	YOUR TURN 1
Solve: $|x| = |2x - 4|$.	Solve: $|3x + 1| = |x + 3|$.

$$|x| = |2x - 4|$$

$$x = 2x - 4 \quad or \quad x = \boxed{}$$
$$-x = -4 \quad or \quad x = \boxed{}$$
$$x = 4 \quad or \quad 3x = \boxed{}$$
$$x = 4 \quad or \quad x = \frac{\boxed{}}{\boxed{}}$$

The solutions are $\dfrac{\boxed{}}{\boxed{}}$ and 4. The solution set is $\left\{\dfrac{\boxed{}}{\boxed{}}, 4\right\}$.

EXAMPLE 2	YOUR TURN 2
Solve: $\lvert 5x-6 \rvert = \lvert 5x+4 \rvert$.	Solve: $\lvert x+12 \rvert = \lvert x-12 \rvert$.

$$\lvert 5x-6 \rvert = \lvert 5x+4 \rvert$$

$5x-6 = 5x+4$ *or* $5x-6 = \boxed{}$

$-6 = 4$ *or* $5x-6 = \boxed{}$

$-6 = 4$ *or* $10x-6 = \boxed{}$

$-6 = 4$ *or* $10x = \boxed{}$

$-6 = 4$ *or* $x = \dfrac{\boxed{}}{\boxed{}}$

The first equation has no solution. The solution of the second equation is $\dfrac{1}{5}$.

The solution set is $\left\{ \dfrac{1}{5} \right\}$.

YOUR NOTES   Write your questions and additional notes.

Inequalities with Absolute Value

ESSENTIALS

Solutions of Absolute-Value Inequalities:

For any positive number p and any algebraic expression X:

a) The solutions of $|X| < p$ are those numbers that satisfy $-p < X < p$.

b) The solutions of $|X| > p$ are those numbers that satisfy $X < -p$ *or* $X > p$.

Examples

- Solve: $|x| \leq 2$. Then graph.

    $-2 \leq x \leq 2$

    The solution set is $\{x \mid -2 \leq x \leq 2\}$, or $[-2, 2]$;
    $$-4\ -3\ -2\ -1\ \ 0\ \ 1\ \ 2\ \ 3\ \ 4$$

- Solve: $|x| \geq 2$. Then graph.

    $x \leq -2$ *or* $x \geq 2$

    The solution set is $\{x \mid x \leq -2 \text{ or } x \geq 2\}$, or $(-\infty, -2] \cup [2, \infty)$;
    $$-4\ -3\ -2\ -1\ \ 0\ \ 1\ \ 2\ \ 3\ \ 4$$

- Solve: $|x - 1| \leq 3$. Then graph.

    $-3 \leq x - 1 \leq 3$

    $-2 \leq x \leq 4$ Adding 1 to each part of the inequality

    The solution set is $\{x \mid -2 \leq x \leq 4\}$, or $[-2, 4]$;
    $$-3\ -2\ -1\ \ 0\ \ 1\ \ 2\ \ 3\ \ 4\ \ 5$$

GUIDED LEARNING	📖 **Textbook** 👤 **Instructor** ▶ **Video**
EXAMPLE 1	**YOUR TURN 1**
Solve: $|x| \leq 3$. Then graph.	Solve: $|x| < 2$. Then graph.

EXAMPLE 1

Solve: $|x| \leq 3$. Then graph.

$$|X| \leq p$$

$$|x| \leq 3$$

$$\boxed{} \leq x \leq \boxed{}$$

The solution set is $\left\{x \middle| \boxed{} \leq x \leq \boxed{}\right\}$, or

$$\left[\,\boxed{}\,,\,\boxed{}\,\right].$$

The graph is as follows.

$$-4\ -3\ -2\ -1\ \ 0\ \ 1\ \ 2\ \ 3\ \ 4$$

YOUR TURN 1

Solve: $|x| < 2$. Then graph.

EXAMPLE 2	YOUR TURN 2
Solve: $\lvert y \rvert > 3$. Then graph.	Solve: $\lvert y \rvert \geq 1$. Then graph.

$$\lvert X \rvert > p$$
$$\lvert y \rvert > 3$$

$$y < \boxed{} \quad or \quad y > \boxed{}$$

The solution set is $\left\{ y \mid y < \boxed{} \ or \ y > \boxed{} \right\}$, or

$\left(-\infty, \boxed{} \right) \cup \left(\boxed{}, \infty \right)$. The graph is as follows.

EXAMPLE 3	YOUR TURN 3
Solve: $\lvert 2x + 5 \rvert \leq 3$. Then graph.	Solve: $\lvert -3x + 6 \rvert < 9$. Then graph.

$$\lvert X \rvert \leq p$$
$$\lvert 2x + 5 \rvert \leq 3$$

$$\boxed{} \leq 2x + 5 \leq 3$$

$$\boxed{} \leq 2x \leq -2$$

$$\boxed{} \leq x \leq -1$$

The solution set is $\left\{ x \mid \boxed{} \leq x \leq -1 \right\}$, or

$\left[\boxed{}, -1 \right]$. The graph is as follows.

EXAMPLE 4	YOUR TURN 4
Solve: $\left\lvert -x+6 \right\rvert \geq 2$. Then graph.	Solve: $\left\lvert x-4 \right\rvert \geq 3$. Then graph.

$$\left\lvert X \right\rvert \geq p$$

$$\left\lvert -x+6 \right\rvert \geq 2$$

$-x+6 \leq \boxed{}$ or $-x+6 \geq 2$

$-x \leq \boxed{}$ or $-x \geq -4$

$x \geq \boxed{}$ or $x \leq 4$

The solution set is $\left\{ x \,\middle|\, x \leq \boxed{} \ \ or\ \ x \geq \boxed{} \right\}$, or

$\left(-\infty, \boxed{} \right] \cup \left[\boxed{}, \infty \right)$. The graph is as follows.

2 3 4 5 6 7 8 9 10

Practice Exercises

Readiness Check

Determine whether each statement is true or false.

1. Absolute value is never negative.

2. $|4| = -|4|$

3. To find the distance between two numbers on the number line, we simply subtract the two numbers.

4. For any positive number p, the equation $|x| = -p$ has no solution.

5. The inequality $|x| > 4$ has solution set $(-\infty, -4) \cup (4, \infty)$.

Properties of Absolute Value

Simplify, leaving as little as possible inside absolute-value signs.

6. $\left| \dfrac{15x}{5} \right|$

7. $\left| -6x^2 \right|$

Distance on the Number Line

Find the distance between the points on the number line.

8. $-11, 16$

9. $-\dfrac{15}{2}, -\dfrac{5}{2}$

Equations with Absolute Value

Solve.

10. $|x| - 6 = 10$

11. $|-2x + 4| = 16$

12. $|5x+1|+7=8$ **13.** $|7x+2|=0$

Equations with Two Absolute-Value Expressions

Solve.

14. $|2x-1|=|5x+2|$ **15.** $|y+8|=|y|$

16. $\left|\dfrac{3}{4}c\right|=|c+1|$ **17.** $|x+1|=|x+1|$

Inequalities with Absolute Value

Solve and graph.

18. $|x|\le 4$ **19.** $|x+4|<3$

20. $|8x|-4>20$ **21.** $|6x+12|\ge 6$

Plotting Ordered Pairs

ESSENTIALS

On a plane, each point is the graph of a number pair. To form the plane, we use two perpendicular number lines called **axes**, which divide the plane into four **quadrants** which are numbered as shown below. They cross at a point called the **origin**. The arrows show the positive directions.

Example

Consider the **ordered pair** $(5, 2)$, shown in the figure at right. The numbers 5 and 2 are called coordinates. In $(5, 2)$, the **first coordinate** (the **abscissa**) is 5 and the **second coordinate** (the **ordinate**) is 2. To plot $(5, 2)$, we start at the origin and move *horizontally* to the 5. Then we move up *vertically* 2 units and make a "dot." The point $(-4, 3)$ is also plotted at right. We start at the origin and move horizontally to -4, move up 3 units, and then make a "dot." Note that $(-4, 3)$ and $(3, -4)$ represent different points. The order of the numbers in the pair is important.

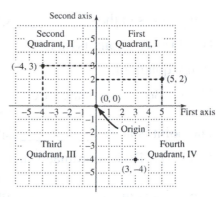

GUIDED LEARNING　　🔲 **Textbook**　　👤 **Instructor**　　▶ **Video**

EXAMPLE 1	YOUR TURN 1
Plot the points $(4, -2)$ and $(-3, 0)$.	Plot the points $(-3, -5)$ and $(0, 2)$.

EXAMPLE 1

Plot the points $(4, -2)$ and $(-3, 0)$.

$(4, -2)$: The first number, 4, is positive.

Starting at the origin, we move ☐ unit(s) to the _____ (left/right). The second number, -2, is negative. We move ☐ unit(s) _____ (up/down).

$(-3, 0)$: Starting at the origin, we move ☐ unit(s) to the _____ (left/right). Then move ☐ unit(s) vertically.

YOUR TURN 1

Plot the points $(-3, -5)$ and $(0, 2)$.

EXAMPLE 2	YOUR TURN 2
In which quadrant, if any, are the points $(-2,6)$, $(0,-1)$, $(5,3)$, and $(-4,-4)$ located? $(-2,6)$: The first coordinate is _____ positive / negative and the second coordinate is _____. positive / negative The point is in the _____ quadrant. $(0,-1)$: The point is on a(n) _____ and is not in any quadrant. $(5,3)$: Both coordinates are _____. positive / negative The point is in the _____ quadrant. $(-4,-4)$: Both coordinates are _____. positive / negative The point is in the _____ quadrant.	In which quadrant, if any, are the points $(-3,-1)$, $(5,-2)$, $(6,0)$, and $(-1,4)$ located?

YOUR NOTES   Write your questions and additional notes.

Solutions of Equations

ESSENTIALS

Unless stated otherwise, to determine whether an ordered pair is a solution of an equation in two variables, we use the first number in the pair to replace the variable that occurs first alphabetically.

In a coordinate plane, the first axis is called the *x*-axis and the second axis, the *y*-axis.

Example

- Determine whether each of the following pairs is a solution of $4x - 3y = 12$: $(0, -4)$ and $(1, 2)$.

For $(0, -4)$:

$$4x - 3y = 12$$
$$\overline{4(0) - 3(-4)} \; ? \; 12$$
$$0 + 12$$
$$12 \qquad \text{TRUE}$$

Thus, $(0, -4)$ *is* a solution.

For $(1, 2)$:

$$4x - 3y = 12$$
$$\overline{4(1) - 3(2)} \; ? \; 12$$
$$4 - 6$$
$$-2 \qquad \text{FALSE}$$

Thus, $(1, 2)$ *is not* a solution

GUIDED LEARNING

 Textbook　　 **Instructor**　　 **Video**

EXAMPLE 1	YOUR TURN 1
Determine whether each of the following pairs is a solution of $3x - 2y = 9$: $(3, 0)$ and $(2, -3)$.	Determine whether each of the following pairs is a solution of $2x - 5y = 10$: $(0, -2)$ and $(5, 1)$.

For $(3, 0)$:

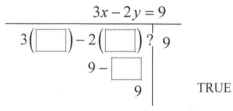

$$3x - 2y = 9$$
$$\overline{3(\boxed{}) - 2(\boxed{})} \; ? \; 9$$
$$9 - \boxed{}$$
$$9 \qquad \text{TRUE}$$

$(3, 0)$ _____ (is/is not) a solution.

For $(2, -3)$:

$$3x - 2y = 9$$
$$\overline{3(\boxed{}) - 2(\boxed{})} \; ? \; 9$$
$$6 + \boxed{}$$
$$12 \qquad \text{FALSE}$$

$(2, -3)$ _____ (is/is not) a solution.

EXAMPLE 2	YOUR TURN 2
Show that the pairs $(1,1)$ and $(0,-2)$ are solutions of $y = 3x - 2$. Then graph the two points and use the graph to determine another pair that is a solution.	Show that the pairs $(2,-1)$ and $(0,-5)$ are solutions of $y = 2x - 5$. Then graph the two points and use the graph to determine another pair that is a solution.

For $(1,1)$:

$$y = 3x - 2$$

$$1 \overset{?}{|} 3(\boxed{}) - 2$$
$$\boxed{} - 2$$
$$1 \qquad\qquad \text{TRUE}$$

For $(0,-2)$:

$$y = 3x - 2$$

$$-2 \overset{?}{|} 3(\boxed{}) - 2$$
$$\boxed{} - 2$$
$$-2 \qquad\qquad \text{TRUE}$$

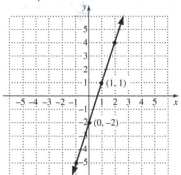

Another solution would be any other point on the line, such as $(2, 4)$ or $\left(-1, \boxed{}\right)$.

These can be checked by substituting the coordinates into $y = 3x - 2$.

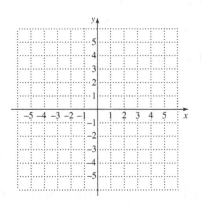

YOUR NOTES   Write your questions and additional notes.

Graphs of Linear Equations

ESSENTIALS

To graph a linear equation:

1. Select a value for one variable and calculate the corresponding value of the other variable. Form an ordered pair using alphabetical order as indicated by the variables.
2. Repeat step (1) to obtain at least two other ordered pairs. Two points are essential to determine a straight line. A third point serves as a check.
3. Plot the ordered pairs and draw a straight line passing through the points.

Example

- Graph: $y = 2x - 2$.

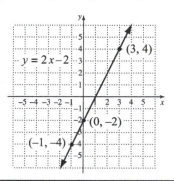

x	$y = 2x - 2$	(x, y)
-1	-4	$(-1, -4)$
0	-2	$(0, -2)$
3	4	$(3, 4)$

GUIDED LEARNING   📖 **Textbook**   👤 **Instructor**   ▶ **Video**

EXAMPLE 1	YOUR TURN 1
Graph: $y = -2x - 2$.	Graph: $y = -2x - 3$.

x	$y = -2x - 2$	(x, y)
-2	☐	$\left(-2, \Box\right)$
0	☐	$\left(0, \Box\right)$
2	☐	$\left(2, \Box\right)$

x	$y = -2x - 3$	(x, y)

Plot the points. We use a ruler or another straightedge to draw a line through the points.

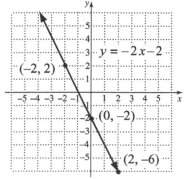

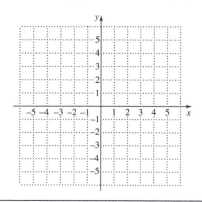

EXAMPLE 2	YOUR TURN 2

EXAMPLE 2

Graph $3x - 6y = 12$.

We first solve for y.

$$3x - 6y = 12$$

$$3x - \boxed{} - 6y = 12 - \boxed{}$$

$$-6y = \boxed{} + 12$$

$$y = \frac{\boxed{}}{\boxed{}} x - 2$$

x	$y = \dfrac{1}{2}x - 2$	(x, y)
0	$\boxed{}$	$\left(0, \boxed{}\right)$
2	$\boxed{}$	$\left(2, \boxed{}\right)$
−2	$\boxed{}$	$\left(-2, \boxed{}\right)$

We then complete and label the graph.

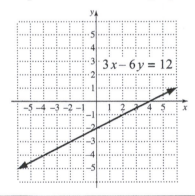

YOUR TURN 2

Graph $2x - 6y = 6$.

x	$y = \boxed{}$	(x, y)

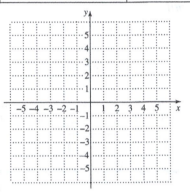

YOUR NOTES   Write your questions and additional notes.

Graphing Nonlinear Equations

ESSENTIALS

The graphs of **nonlinear equations** are not straight lines and they often require us to plot many points to see the general shape of the graph.

Example

- Graph $y = x^2$.

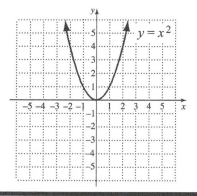

GUIDED LEARNING

 Textbook   **Instructor**   **Video**

EXAMPLE 1	YOUR TURN 1

Graph: $y = -|x|$.

x	$y = -|x|$	(x, y)
-3	-3	$(-3, -3)$
-2	$\square$	$(-2, -2)$
-1	-1	$(\square, \square)$
$\square$	0	$(0, 0)$
1	-1	$(1, -1)$
2	$\square$	$(2, \square)$
3	-3	$(3, -3)$

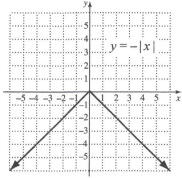

Graph: $y = 2|x|$.

x	$y = 2|x|$	(x, y)
-3		
-2		
-1		
0		
1		
2		
3		

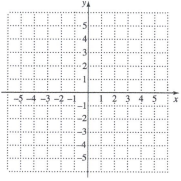

EXAMPLE 2

Graph: $y = x^2 + 1$.

x	$y = x^2 + 1$	(x, y)
-3	10	$(-3, 10)$
-2	☐	$(-2, 5)$
-1	2	$\left(-1, \boxed{}\right)$
☐	1	$(0, 1)$
1	☐	$\left(1, \boxed{}\right)$
2	☐	$\left(2, \boxed{}\right)$
3	☐	$\left(3, \boxed{}\right)$

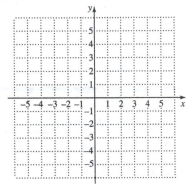

$y = x^2 + 1$

YOUR TURN 2

Graph: $y = -x^2 + 2$.

x	$y = -x^2 + 2$	(x, y)
-2		
-1		
0		
1		
2		

Practice Exercises

Readiness Check

Use the graph of $y = x - 2$ at the right to determine whether each statement is true or false.

1. The point $(-1, -3)$ is in quadrant III.

2. The point $(2, 0)$ is in quadrant I.

3. The ordered pair $(3, 1)$ is a solution of the equation $y = x - 2$.

4. The point $(0, -2)$ is on the y-axis.

5. The graph passes through the origin.

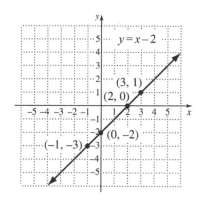

Plotting Ordered Pairs
Plot the following points.

6. $A(2, 5)$, $B(-2, -3)$, $C(-4, -1)$, $D(1, -3)$, $E(0, -4)$, $F(3, 0)$, $G(-2, 1)$, $H(-3, 4)$

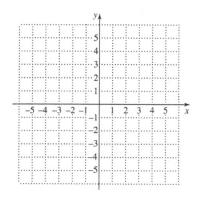

Solutions of Equations

Determine whether the given point is a solution of the equation.

7. $(3, 5)$; $y = 3x - 4$ 8. $(-1, -3)$; $2b = 5a - 1$ 9. $(4, 3)$; $5x - 3y = 10$

10. An equation and two ordered pairs are given. Show that each pair is a solution of the equation. Then graph the equation and use the graph to determine another solution.

$y = 1 - 2x$; $(-2, 5)$, $(1, -1)$

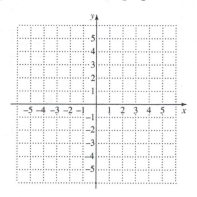

Graphs of Linear Equations

Graph.

11. $y = x - 3$

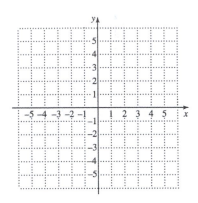

12. $y = \dfrac{1}{2}x + 2$

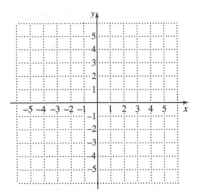

13. $4x + 2y = -2$

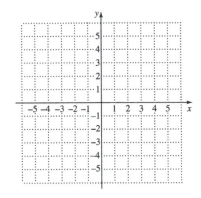

14. $2x - 3y = 6$

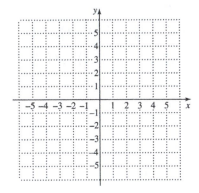

Graphs of Nonlinear Equations

Graph.

15. $y = x^2 - 4$

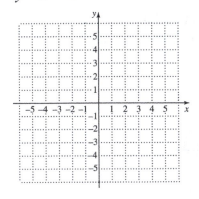

16. $y = |x + 1|$

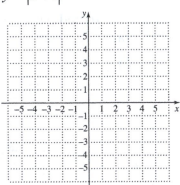

Identifying Functions

ESSENTIALS

A **function** is a correspondence between a first set called the **domain** and a second set called the **range** such that each member of the domain corresponds to *exactly one* member of the range.

Examples

The correspondence *f is* a function because each member of the domain is matched to exactly one member of the range.

The correspondence *g is not* a function because some members of the domain are matched to more than one member of the range.

	Textbook		Instructor		Video

GUIDED LEARNING

EXAMPLE 1	YOUR TURN 1
Determine whether the correspondence is a function.	Determine whether the correspondence is a function.

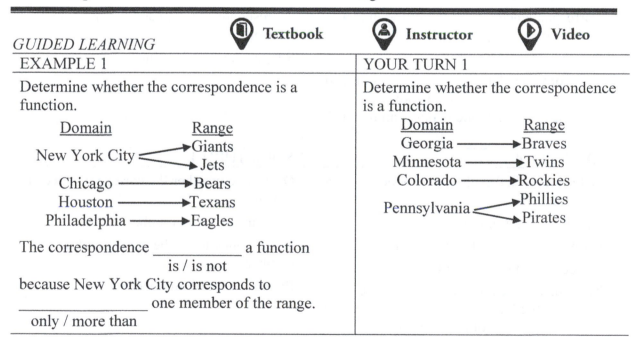

EXAMPLE 1 (continued):

The correspondence _____ a function
               is / is not

because New York City corresponds to

_____ one member of the range.
  only / more than

EXAMPLE 2	YOUR TURN 2
Determine whether the correspondence is a function.	Determine whether the correspondence is a function.

Domain   Range

Domain   Range

A ⟶ a
B ⟶ b
C ⟶ c
D ⟶ d

The correspondence _____ a function
                is / is not

because each member of the domain is
matched to _____ one member
          only / more than

of the range.

EXAMPLE 3	YOUR TURN 3
Determine whether the correspondence is a function.	Determine whether the correspondence is a function.
Domain: A set of occupied residences	Domain: A set of faculty advisors
Correspondence: A person living in that residence	Correspondence: An advisee assigned to that advisor
Range: A set of people	Range: A set of students

This correspondence _____ a function
             is / is not

because more than one person can live in a
residence.

EXAMPLE 4	YOUR TURN 4
Determine whether the correspondence is a function.	Determine whether the correspondence is a function.
Domain: A set of numbers	Domain: A set of numbers
Correspondence: The cube of the number	Correspondence: The opposite of the number
Range: A set of numbers	Range: A set of numbers

This correspondence is a function because
each number has _____ one cube.
             only / more than

YOUR NOTES   Write your questions and additional notes.

Finding Function Values

ESSENTIALS

For the function f, $f(x)$ is read "f of x" or "f at x" or "the value of f at x" and indicates the output that corresponds to the input x.

Example

- For $f(x) = 5x - 2$, $f(-4) = 5 \cdot (-4) - 2 = -20 - 2 = -22$.

GUIDED LEARNING 🔖 **Textbook** 👤 **Instructor** ▶ **Video**	
EXAMPLE 1	YOUR TURN 1
For $g(x) = 3x - 5$, find $g(0)$, $g(-1)$, and $g(a+3)$. $g(0) = 3(\boxed{}) - 5 = 0 - 5 = -5$ $g(-1) = 3(\boxed{}) - 5 = -3 - \boxed{} = \boxed{}$ $g(a+3) = 3(\boxed{}) - 5 = \boxed{} + 9 - 5 = 3a + \boxed{}$	For $g(x) = 3 - 4x$, find $g(0)$, $g(-1)$, and $g(a+2)$.
EXAMPLE 2	YOUR TURN 2
For $f(x) = 13$, find $f(5)$. The function f is a constant function. Every input has the output 13. $f(5) = \boxed{}$	For $f(x) = 7$, find $f(-1)$.
EXAMPLE 3	YOUR TURN 3
For $F(x) = x^2 - x + 1$, find $F(5)$ and $F(-5)$. $F(5) = (\boxed{})^2 - (\boxed{}) + 1$ $\quad = \boxed{} - 5 + 1$ $\quad = \boxed{}$ $F(-5) = (\boxed{})^2 - (\boxed{}) + 1$ $\quad = \boxed{} + 5 + 1$ $\quad = \boxed{}$	For $h(x) = x^2 - 3x$, find $h(1)$ and $h(-1)$.

EXAMPLE 4	YOUR TURN 4
Find the volume of a sphere with radius 5 cm. Use $V(r) = \dfrac{4}{3}\pi r^3$ and 3.14 for π. $$V\left(\boxed{}\right) = \frac{4}{3}\pi \cdot \left(\boxed{}\right)^3$$ $$= \frac{4}{3}\pi \cdot \boxed{} \text{ cm}^3$$ $$= \boxed{} \text{ cm}^3 \approx \boxed{} \text{ cm}^3$$	Find the surface area of a sphere with radius 3 cm. Use $S(r) = 4\pi r^2$ and 3.14 for π.

YOUR NOTES   Write your questions and additional notes.

Graphs of Functions

ESSENTIALS

To graph a function, we find ordered pairs (x, y) or $(x, f(x))$, plot them, and sketch a graph through the points.

Example

- Graph: $f(x) = |x + 2|$.

$f(-3) = |-3 + 2| = |-1| = 1$
$f(-2) = |-2 + 2| = |0| = 0$
$f(-1) = |-1 + 2| = |1| = 1$
$f(0) = |0 + 2| = |2| = 2$
$f(1) = |1 + 2| = |3| = 3$
$f(2) = |2 + 2| = |4| = 4$

x	$f(x)$
-3	1
-2	0
-1	1
0	2
1	3
2	4

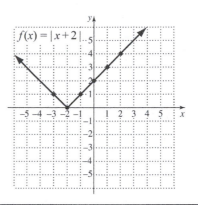

| | Textbook | | Instructor | | Video |

GUIDED LEARNING

EXAMPLE 1	YOUR TURN 1

Graph: $f(x) = 2x - 3$.

First, calculate some function values.

$f(-1) = 2(\boxed{}) - 3 = \boxed{}$

$f(0) = 2(\boxed{}) - 3 = \boxed{}$

$f(1) = 2(\boxed{}) - 3 = \boxed{}$

$f(2) = 2(2) - 3 = 1$

$f(3) = 2(3) - 3 = 3$

x	$f(x)$
-1	$\boxed{}$
0	$\boxed{}$
1	$\boxed{}$
2	1
3	3

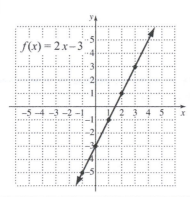

Graph: $f(x) = 5 - x$.

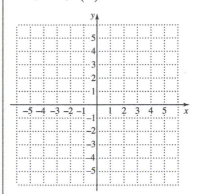

EXAMPLE 2	YOUR TURN 2

EXAMPLE 2

Graph: $g(x) = x^2 - 4x - 5$.

First, calculate some function values.

$g(-2) = (-2)^2 - 4(-2) - 5$

$\qquad = 4 + 8 - 5 = 7$

$g(-1) = (-1)^2 - 4(-1) - 5$

$\qquad = 1 + 4 - 5 = \boxed{}$

$g(0) = (0)^2 - 4(0) - 5 = \boxed{}$

$g(1) = (1)^2 - 4(1) - 5 = -8$

$g(2) = (2)^2 - 4(2) - 5 = \boxed{}$

$g(3) = (3)^2 - 4(3) - 5 = -8$

$g(4) = (4)^2 - 4(4) - 5 = \boxed{}$

$g(5) = (5)^2 - 4(5) - 5 = \boxed{}$

x	$f(x)$
−2	7
−1	$\boxed{}$
0	$\boxed{}$
1	−8
2	$\boxed{}$
3	−8
4	$\boxed{}$
5	$\boxed{}$

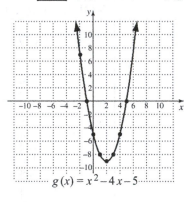

$g(x) = x^2 - 4x - 5$

YOUR TURN 2

Graph: $g(x) = x^2 - 3$.

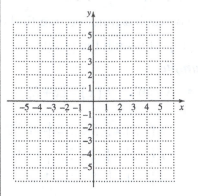

YOUR NOTES   Write your questions and additional notes.

The Vertical-Line Test

ESSENTIALS

The Vertical-Line Test

If it is possible for a vertical line to cross a graph more than once, then the graph is not the graph of a function.

Examples

- A function

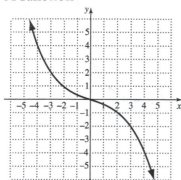

- Not a function

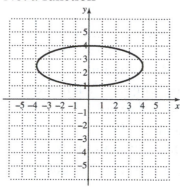

 Textbook   **Instructor**   **Video**

GUIDED LEARNING

EXAMPLE 1	YOUR TURN 1
Determine whether the following graph is that of a function.	Determine whether the following graph is that of a function.

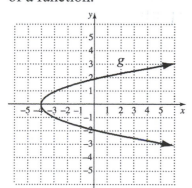

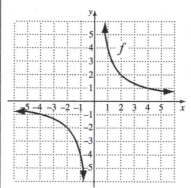

The graph _____ pass the vertical-line
                does / does not

test. It _____ a function.
           is / is not

YOUR NOTES   Write your questions and additional notes.

Applications of Functions and Their Graphs

ESSENTIALS

Example

- The graph shown represents the average number of tweets per day sent by Twitter users.

  The number of tweets is a function *f* of the year *x*.

  To estimate the average number of tweets per day in 2011, locate 2011 on the horizontal axis and move directly up to the graph. Then move across to the vertical axis.

  The average number of tweets per day was about 250 million in 2011. That is, $f(2011) \approx 250,000,000$.

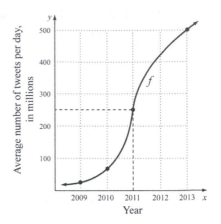

| **Textbook** | **Instructor** | **Video** |

GUIDED LEARNING

EXAMPLE 1	YOUR TURN 1
The following graph represents the number of for-profit hospitals in the United States from 1975 through 2010. The number of hospitals is a function *f* of the year *x*. Use the graph to estimate the number of for-profit hospitals in 2000. That is, find $f(2000)$. 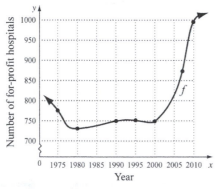 Locate 2000 on the horizontal axis and move up to the graph. Then move across to the vertical axis. In 2000, there were approximately ⬚ for-profit hospitals.	Use the graph in Example 1 to estimate the number of for-profit hospitals in 2010. That is, find $f(2010)$.

YOUR NOTES   Write your questions and additional notes.

Practice Exercises

Readiness Check

Choose the word from the following list that best completes each sentence. Words may be used more than once or not at all.

function   relation   domain   range

horizontal   vertical   input   output

1. A(n) _____ is a correspondence such that each member of the _____ is paired with exactly one member of the _____.

2. The _____-line test is used to determine if a graph represents a(n) _____.

3. For the function given by $f(x) = -8$, every input has the same _____.

Identifying Functions

Determine whether each correspondence is a function.

4.
    Domain   Range
    T-shirt ⟶ $12.99
    Shorts ⟶ $21.99
    Sandals ⟶ $16.99
    Hat ⟶

5. *Domain*   *Correspondence*   *Range*

    A set of artists   The songs the artists have recorded   A set of songs

Finding Function Values

6. Find $f(3)$, for $f(x) = 5x - 1$.

7. Find $g(-3)$, for $g(x) = 2x^2 - x - 7$.

8. Find $g(a-2)$, for $g(x) = 4x - 3$.

9. Find $h(8)$, for $h(x) = 5$.

10. Find the surface area of a cube with a side of 5 cm. Use $A(s) = 6s^2$.

Graphs of Functions

Graph each function.

11. $g(x) = 3x - 1$

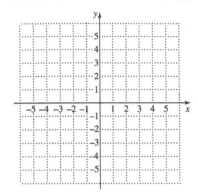

12. $f(x) = x^2 + 1$

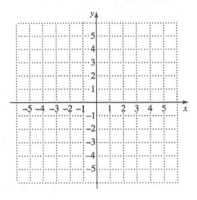

The Vertical-Line Test

Determine whether each of the following is the graph of a function.

13.

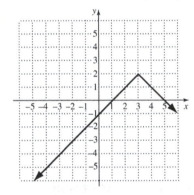

14.

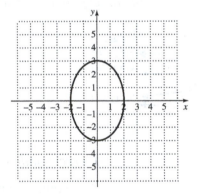

Applications of Functions and Their Graphs

15. The following graph represents the number of hospitals in the United States with 500 beds or more. The number of hospitals is a function g of the year x. Use the graph to estimate the number of hospitals with 500 beds or more in 2000. That is, find $g(2000)$.

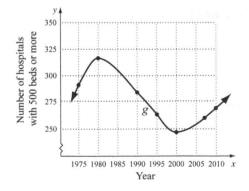

Finding Domain and Range

ESSENTIALS

A **relation** is a set of ordered pairs.

A **function** is a relation in which no two different pairs share a common first coordinate.

The **domain** of a function is the set of all first coordinates.

The **range** of a function is the set of all second coordinates.

Example

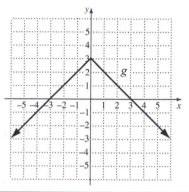

- For the function *g* shown:

    $g(1) = 2$.

    The domain of *g* is the set of all real numbers.

    If $g(x) = 1$, then $x = -2$ *or* $x = 2$.

    The range of *g* is $\{y | y \le 3\}$.

GUIDED LEARNING   **Textbook**   **Instructor**   **Video**

EXAMPLE 1	YOUR TURN 1
Determine the domain and the range of the function *f* whose graph is shown below.	Determine the domain and the range of the function *f* whose graph is shown below.

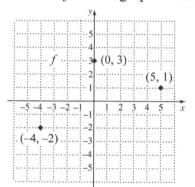

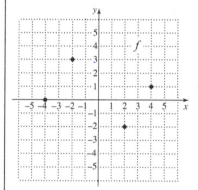

The function *f* can be written

$$f = \{(0, 3), (5, 1), (-4, -2)\}.$$

The domain is the set of all first coordinates:

$\{0, 5, \boxed{}\}$.

The range is the set of all second coordinates:

$\{\boxed{}, 1, -2\}$.

EXAMPLE 2	YOUR TURN 2
Determine the domain and the range of the function.	Determine the domain and the range of the function.

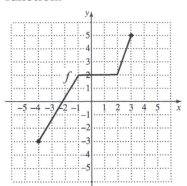

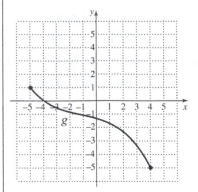

The ⬚ is $\{x \mid -4 \leq x \leq \boxed{}\}$.

The range is $\{y \mid \boxed{} \leq y \leq \boxed{}\}$.

EXAMPLE 3	YOUR TURN 3
For the graph of the function, determine (a) $f(1)$; (b) the domain; (c) any x-values for which $f(x) = 2$; and (d) the range.	For the graph of the function, determine (a) $f(-2)$; (b) the domain; (c) any x-values for which $f(x) = 2$; and (d) the range.

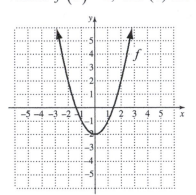

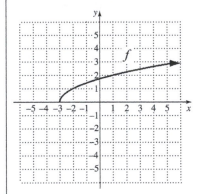

(a) $f(1) = -1$

(b) Domain = $\boxed{}$

(c) $f(\boxed{}) = 2$ and $f(\boxed{}) = 2$

(d) Range = $\{y \mid y \geq \boxed{}\}$

EXAMPLE 4	YOUR TURN 4
Find the domain of $f(x) = \dfrac{3-x}{x+2}$. Find the values for x for which we cannot compute $f(x)$. These are values that make the _____ equal to 0. numerator / denominator $x + 2 = 0$ $x = \boxed{}$ The domain is $\left\{ x \middle\vert x \text{ is a real number } and\ x \neq \boxed{} \right\}.$	Find the domain of $f(x) = \dfrac{5+x}{x-4}$.
EXAMPLE 5	YOUR TURN 5
Find the domain of $f(x) = x^2 - x - 5$. The domain is the set of all possible inputs. Since we can calculate $x^2 - x - 5$ for any real number, the domain of f is _____ _____.	Find the domain of $f(x) = 3x - 1$.

YOUR NOTES   Write your questions and additional notes.

Practice Exercises

Readiness Check

Choose the word from the following list that best completes each sentence.

function   relation   domain   range

1. A _____ is a set of ordered pairs in which no two different pairs share a common first coordinate.

2. Every function is also a _____.

3. The _____ of a function is the set of all first coordinates.

4. The _____ of a function is the set of all second coordinates.

Finding Domain and Range

For the graph of the function shown, determine (a) $f(2)$; (b) the domain; (c) any x-values for which $f(x)=1$; and (d) the range.

5.

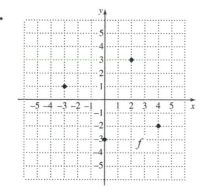

6.

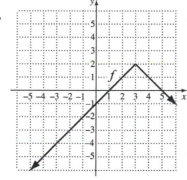

7.

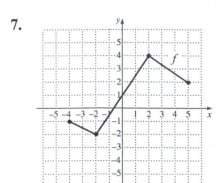

Find the domain of each function.

8. $f(x) = 5 - 7x$

9. $g(x) = \dfrac{4x}{x-5}$

10. $f(x) = \dfrac{8x-1}{2x+3}$

11. $g(x) = |x| + 7$

12. $f(x) = x^2 - x - 6$

The Sum, Difference, Product, or Quotient of Two Functions

ESSENTIALS

The Algebra of Functions

If f and g are functions and x is in the domain of both functions, then:

1. $(f+g)(x) = f(x) + g(x)$;

2. $(f-g)(x) = f(x) - g(x)$;

3. $(f \cdot g)(x) = f(x) \cdot g(x)$;

4. $(f/g)(x) = f(x)/g(x)$, provided $g(x) \neq 0$.

Example

- For $f(x) = 2x^2 - 3$ and $g(x) = x + 1$,

a) $(f+g)(x) = f(x) + g(x)$

$\qquad = 2x^2 - 3 + x + 1$

$\qquad = 2x^2 + x - 2$;

b) $(f-g)(-2) = f(-2) - g(-2)$

$\qquad = 2(-2)^2 - 3 - (-2 + 1)$

$\qquad = 8 - 3 - (-1)$

$\qquad = 8 - 3 + 1 = 6$

c) $(f \cdot g)(1) = f(1) \cdot g(1)$

$\qquad = \left[2(1)^2 - 3 \right] \cdot (1 + 1)$

$\qquad = (2 - 3) \cdot 2$

$\qquad = -1 \cdot 2$

$\qquad = -2$;

d) $(f/g)(x) = f(x)/g(x)$

$\qquad = \dfrac{2x^2 - 3}{x + 1}$, provided $x \neq -1$.

	📖 **Textbook**	👤 **Instructor**	▶ **Video**

GUIDED LEARNING

EXAMPLE 1	YOUR TURN 1
For $f(x) = -x + 4$ and $g(x) = x^2 + 2$, find $(f+g)(x)$. $(f+g)(x) = f(x) + g(x)$ $\qquad = -x + 4 + \boxed{} + 2$ $\qquad = x^2 - x + \boxed{}$	For $f(x) = 3 - x$ and $g(x) = 3x^2 + 1$, find $(f+g)(x)$.

EXAMPLE 2	YOUR TURN 2
For $F(x) = 4 - x^2$ and $G(x) = 2x$, find $(F - G)(-1)$. $F(-1) = 4 - (-1)^2 = 4 - 1 = \boxed{}$ and $G(-1) = 2(-1) = \boxed{}$, so we have $(F - G)(-1) = 3 - (-2) = 3 + 2 = \boxed{}$.	For $F(x) = x^2 - 6$ and $G(x) = -3x$, find $(F - G)(-2)$.
EXAMPLE 3	YOUR TURN 3
For $f(x) = x + 5$ and $g(x) = 2x + 3$, find $(f \cdot g)(a)$. $(f \cdot g)(a) = f(a) \cdot g(a)$ $\qquad = \left(\boxed{} + 5\right)\left(2a + \boxed{}\right)$ $\qquad = 2a^2 + 3a + 10a + \boxed{}$ $\qquad = 2a^2 + \boxed{} + 15$	For $f(x) = 2x - 1$ and $g(x) = x + 4$, find $(f \cdot g)(t)$.
EXAMPLE 4	YOUR TURN 4
For $F(x) = x^2 - 2x$ and $G(x) = x + 3$, find $(F/G)(4)$. $(F/G)(x) = F(x)/G(x) = \dfrac{x^2 - 2x}{x + \boxed{}}$ We assume $x \neq -3$. Then $(F/G)(4) = F(4)/G(4)$ $\qquad = \dfrac{4^2 - 2 \cdot \boxed{}}{4 + \boxed{}}$ $\qquad = \dfrac{16 - \boxed{}}{7} = \dfrac{8}{7}$.	For $F(x) = 2 - x^2$ and $G(x) = x - 1$, find $(F/G)(3)$.

YOUR NOTES   Write your questions and additional notes.

Determining Domain

ESSENTIALS

To find the domain of the sum, the difference, the product, or the quotient of two functions *f* and *g*:

1. Find the domain of *f* and the domain of *g*.
2. The functions $f + g$, $f - g$, and $f \cdot g$ have the same domain. It is the intersection of the domains of *f* and *g*, or, in other words, the set of all values common to the domains of *f* and *g*.
3. Find any values of *x* for which $g(x) = 0$.
4. The domain of f/g is the set found in step (2) (the set of all values common to the domains of *f* and *g*) *excluding* any values of *x* found in step (3).

Example

- Given $f(x) = \dfrac{2}{x-4}$ and $g(x) = x + 7$, find the domains of $f + g$, $f - g$, $f \cdot g$, and f/g.

  The domain of *f* is $\{x \mid x \text{ is a real number } and\ x \neq 4\}$.

  The domain of *g* is $\{x \mid x \text{ is a real number}\}$.

  Then the domains of $f + g$, $f - g$, and $f \cdot g$ are the set of all elements common to the domains of *f* and *g*, or all real numbers except 4.

  The domain of $f + g$ = the domain of $f - g$ = the domain of $f \cdot g$
  $$= \{x \mid x \text{ is a real number } and\ x \neq 4\}.$$

  The domain of f/g must exclude values of *x* for which $g(x) = 0$.

  $$g(x) = 0$$
  $$x + 7 = 0$$
  $$x = -7$$

  The domain of f/g is the domain of $f + g$, $f - g$, and $f \cdot g$, but also excludes -7.

  The domain of $f/g = \{x \mid x \text{ is a real number } and\ x \neq 4\ and\ x \neq -7\}$.

GUIDED LEARNING	📖 **Textbook** 👤 **Instructor** ▶ **Video**

EXAMPLE 1	YOUR TURN 1
For $f(x) = 2x^3$ and $g(x) = \dfrac{4}{x+3}$, determine the domains of $f+g$, $f-g$, and $f \cdot g$.	For $f(x) = x^2 - 1$ and $g(x) = \dfrac{8}{x}$, determine the domains of $f+g$, $f-g$, and $f \cdot g$.
The domain of f is $\{x \mid x \text{ is a real number}\}$.	
The domain of g is $\{x \mid x \text{ is a real number } and \; x \neq \boxed{}\}$.	
The domains of $f+g$, $f-g$, and $f \cdot g$ are the set of all elements common to the domains of f and g, or all real numbers except $\boxed{}$.	
The domain of $f+g =$ the domain of $f-g =$ the domin of $f \cdot g = \{x \mid x \text{ is a real number } and \; x \neq \boxed{}\}$.	

EXAMPLE 2	YOUR TURN 2
For $f(x) = \dfrac{2x}{x-5}$ and $g(x) = 6-x$, determine the domain of f/g.	For $f(x) = \dfrac{4}{x-2}$ and $g(x) = x-7$, determine the domain of f/g.
The domain of f is $\{x \mid x \text{ is a real number } and \; x \neq \boxed{}\}$.	
The domain of g is $\{x \mid x \text{ is a real number}\}$.	
We see that the domain of f/g must exclude $\boxed{}$. Because we cannot divide by 0, the domain of f/g must also exclude x-values for which $g(x) = 0$.	
$$g(x) = 0$$ $$6 - x = 0$$ $$\boxed{} = x$$	
Thus, the domain of $f/g =$ $\{x \mid x \text{ is a real number } and \; x \neq 5 \; and \; x \neq \boxed{}\}$.	

YOUR NOTES   Write your questions and additional notes.

Practice Exercises

Readiness Check

Given that $f(x) = 2x - 1$ *and* $g(x) = x + 4$, *match each expression with an equivalent expression from the column on the right.*

1. $(f + g)(x)$ a) $2x^2 + 7x - 4$

2. $(f - g)(x)$ b) $x^2 + 8x + 16$

3. $(f \cdot g)(x)$ c) $3x + 3$

4. $(g \cdot g)(x)$ d) $x - 5$

The Sum, Difference, Product, or Quotient of Two Functions

Let $f(x) = 3 - x$ *and* $g(x) = x^2 + 1$. *Find each of the following.*

5. $(f + g)(2)$ 6. $(f - g)(-1)$

7. $(f \cdot g)(x)$ 8. $(f/g)(-3)$

9. $(g - f)(4)$ 10. $(f \cdot f)(x)$

11. $(g/f)(a)$ 12. $(f + f)(10)$

Determining Domain

For each pair of functions f and g, determine the domain of the sum, the difference, and the product of the two functions.

13. $f(x) = 2x^2$,

$g(x) = x - 3$

14. $f(x) = 2x + 5$,

$g(x) = 4 - x$

15. $f(x) = \dfrac{1}{x}$,

$g(x) = 4 - x^2$

16. $f(x) = 4x^3$,

$g(x) = \dfrac{3}{x-1}$

For each pair of functions f and g, determine the domain of f/g.

17. $f(x) = -x^3$,

$g(x) = x + 2$

18. $f(x) = 3x + 4$,

$g(x) = 2x - 6$

19. $f(x) = \dfrac{2}{x+4}$,

$g(x) = 8 - x$

20. $f(x) = \dfrac{3}{4x-4}$,

$g(x) = 3x - 9$

The Constant *b*: The *y*-Intercept

ESSENTIALS

The *y*-intercept of the graph of $f(x) = mx + b$ is the point $(0, b)$, or simply *b*.

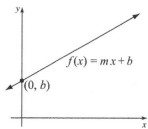

Example

- For $y = 3x + 5$, the *y*-intercept is $(0, 5)$.

GUIDED LEARNING	📖 **Textbook** 👤 **Instructor** ▶ **Video**
EXAMPLE 1	YOUR TURN 1
Find the *y*-intercept: $y = 8x + 1$. $y = mx + b$ $y = 8x + 1$ The *y*-intercept is $(\boxed{}, \boxed{})$.	Find the *y*-intercept: $y = 2x + 13$.
EXAMPLE 2	YOUR TURN 2
Find the *y*-intercept: $f(x) = 2x - \dfrac{1}{3}$. $f(x) = mx + b$ $f(x) = 2x - \dfrac{1}{3}$ $f(x) = 2x + \left(\boxed{}\right)$ The *y*-intercept is $\left(\boxed{}, \boxed{}\right)$.	Find the *y*-intercept: $f(x) = \dfrac{1}{2}x - 9.5$.

EXAMPLE 3	YOUR TURN 3
Find the *y*-intercept: $2x - 3y = 3$. First solve for *y* to write the equation in the form $y = mx + b$. $2x - 3y = 3$ $-3y = 3 - \boxed{}$ $\dfrac{-3y}{-3} = \dfrac{3 - 2x}{-3}$ $y = \boxed{} + \dfrac{2}{3}x$ $y = \dfrac{2}{3}x - \boxed{}$ The *y*-intercept is $\left(\boxed{}, \boxed{} \right)$.	Find the *y*-intercept: $4x + 2y = 6$.

YOUR NOTES   Write your questions and additional notes.

The Constant *m*: Slope

ESSENTIALS

The **slope** of a line containing points (x_1, y_1) and (x_2, y_2) is given by

$$m = \frac{\text{rise}}{\text{run}} = \frac{\text{change in } y}{\text{change in } x} = \frac{y_2 - y_1}{x_2 - x_1} = \frac{y_1 - y_2}{x_1 - x_2}.$$

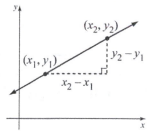

When *m* is negative, the graph slants down from left to right.
When *m* is positive, the graph slants up from left to right.
When *m* is 0, the graph is horizontal.

The **slope** of the line $y = mx + b$ is *m*.

Slope-intercept equation: $y = mx + b$, where *m* is the slope and $(0, b)$ is the *y*-intercept.

Examples

- The slope of the line containing the points $(9, -1)$ and $(8, -7)$ is

$$\frac{\text{rise}}{\text{run}} = \frac{-7 - (-1)}{8 - 9} = \frac{-7 + 1}{8 - 9} = \frac{-6}{-1} = 6.$$

- For $y = 3x + 5$, the slope is 3 and the *y*-intercept is $(0, 5)$.

GUIDED LEARNING	ⓘ **Textbook** 🙎 **Instructor** ▶ **Video**
EXAMPLE 1	**YOUR TURN 1**
Find the slope of the line containing the points $(7, -5)$ and $(3, 2)$.	Find the slope of the line containing the points $(-1, 3)$ and $(4, 5)$.

$$m = \frac{\text{rise}}{\text{run}} = \frac{2 - (\boxed{})}{3 - \boxed{}}$$

$$= \frac{2 + \boxed{}}{-4} = \frac{\boxed{}}{-4} = -\frac{7}{4}$$

EXAMPLE 2	YOUR TURN 2
Determine the slope of the line given by $f(x) = 6x - 3$.	Determine the slope of the line given by $f(x) = 9x - 7$.

$$f(x) = mx + b$$
$$f(x) = 6x - 3$$

The slope is ☐.

EXAMPLE 3	YOUR TURN 3
Determine the slope of the line given by $2x - y = 3$.	Determine the slope of the line given by $x + 2y = 8$.

First solve for y to write the equation in the form $y = mx + b$.

$$2x - y = 3$$
$$-y = -2x + 3$$
$$\frac{-y}{-1} = \frac{-2x + 3}{\boxed{}}$$
$$y = \boxed{} - 3$$

The slope is ☐.

YOUR NOTES   Write your questions and additional notes.

Applications

ESSENTIALS

Slope can be used to represent *rate of change*.

Example

- A plane ascends from 10,000 ft to 15,000 ft in 5 minutes. Its average rate of ascent is $\dfrac{15,000 \text{ ft} - 10,000 \text{ ft}}{5 \text{ min}} = \dfrac{5000 \text{ ft}}{5 \text{ min}}$, or 1000 ft/min.

GUIDED LEARNING   📍 **Textbook**   📍 **Instructor**   📍 **Video**

EXAMPLE 1	YOUR TURN 1
A road rises 3 ft for every horizontal distance of 100 ft. Find the grade of the road. $\text{Grade} = \dfrac{\text{vertical change}}{\text{horizontal change}} = \dfrac{\boxed{} \text{ ft}}{\boxed{} \text{ ft}} = 0.03$ Expressed as a percent, the grade is $\boxed{}$ %.	By 7 p.m., Joe had typed 4 pages of his paper. At 8:30 p.m., he had completed 10 pages. Calculate his typing rate in minutes per page.

EXAMPLE 2	YOUR TURN 2
Find the rate of change for the graph. Remember to use appropriate units.	Find the rate of change for the graph. Remember to use appropriate units.

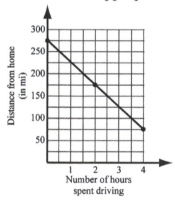

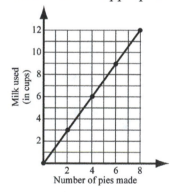

$\text{Rate of change} = \dfrac{175 \text{ mi} - 75 \text{ mi}}{2 \text{ hr} - 4 \text{ hr}} = \dfrac{\boxed{}}{\boxed{}}$

$= -50 \text{ mi per hr, or } -50 \text{ mph}$

YOUR NOTES   Write your questions and additional notes.

Practice Exercises

Readiness Check

Choose the word or phrase from the following list that best completes each sentence.

y-intercept	*x*-intercept	slope	up
slope-intercept form		standard form	down

1. The _____ of the line given by the equation $y = mx + b$ is $(0, b)$.

2. _____ is a ratio that indicates how a change in vertical direction of a graph corresponds to a change in horizontal direction.

3. Lines with negative slope slant _____ from left to right.

4. A linear function in the form $f(x) = mx + b$ is written in _____.

The Constant *b*: The *y*-Intercept

Find the y-intercept of each equation.

5. $f(x) = 2x - \dfrac{1}{2}$

6. $3x - y = 1$

The Constant *m*: Slope

7. Find the slope of the line shown.

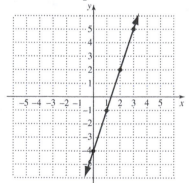

Find the slope of the line containing the given pair of points.

8. $(5, 6)$ and $(1, 2)$

9. $(-3, 2)$ and $(-1, -8)$

The Slope and the *y*-Intercept

Find the slope and the y-intercept of each equation.

10. $f(x) = \dfrac{3}{8}x - 6$

11. $3x - 2y = -6$

12. $x - y = 9$

13. $x + 3y = 2$

Applications

14. At 3:30 p.m., Justin reached highway mile marker 284. At 3:54 p.m., he reached mile marker 312. Assuming a constant rate, find Justin's speed in miles per minute.

15. Find the rate of change.

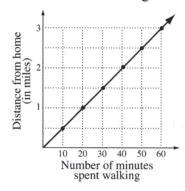

Graphing Using Intercepts

ESSENTIALS

A *y*-intercept of a graph is a point $(0, b)$. To find *b*, let $x = 0$ and solve for *y*.

An *x*-intercept of a graph is a point $(a, 0)$. To find *a*, let $y = 0$ and solve for *x*.

Example

- Find the intercepts of $5x - 4y = 20$.

$$5x - 4 \cdot 0 = 20 \quad \text{Letting } y = 0$$
$$5x = 20$$
$$x = 4$$

The *x*-intercept is $(4, 0)$.

$$5 \cdot 0 - 4y = 20 \quad \text{Letting } x = 0$$
$$-4y = 20$$
$$y = -5$$

The *y*-intercept is $(0, -5)$.

GUIDED LEARNING

 Textbook   **Instructor**   **Video**

EXAMPLE 1	YOUR TURN 1
Find the intercepts of $3x - 4y = -12$ and then graph the line.	Find the intercepts of $-2x + 5y = 10$ and then graph the line.

$$3 \cdot 0 - 4y = -12 \quad \text{Substituting 0 for } x$$

$$\boxed{} = -12$$

$$y = \boxed{}$$

The *y*-intercept is $\left(\boxed{}, \boxed{}\right)$.

$$3x - 4 \cdot 0 = -12 \quad \text{Substituting 0 for } y$$

$$3x = -12$$

$$x = \boxed{}$$

The *x*-intercept is $\left(\boxed{}, \boxed{}\right)$.

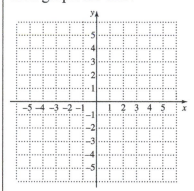

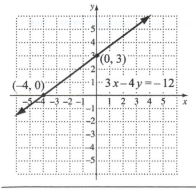

(0, 3)

(-4, 0) $3x - 4y = -12$

EXAMPLE 2	YOUR TURN 2

EXAMPLE 2

Graph $f(x) = 3x - 4$ by using intercepts.

To find the *y*-intercept, let $x = 0$ and find $f(0)$.

$$f(x) = 3x - 4$$

$$f(0) = 3 \cdot 0 - 4$$

$$= \boxed{}$$

The *y*-intercept is $\left(0, \boxed{}\right)$.

To find the *x*-intercept, replace $f(x)$ with 0 and solve for *x*.

$$f(x) = 3x - 4$$

$$0 = 3x - 4$$

$$4 = \boxed{}$$

$$\boxed{} = x$$

The *x*-intercept is $\left(\dfrac{4}{3}, 0\right)$.

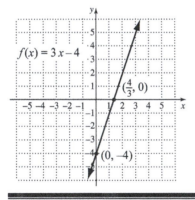

YOUR TURN 2

Graph $f(x) = 2x - 3$ by using intercepts.

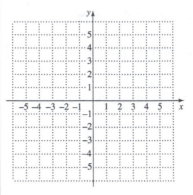

YOUR NOTES   Write your questions and additional notes.

Graphing Using the Slope and the *y*-Intercept

ESSENTIALS

The equation $y = mx + b$ is called the **slope-intercept equation**. The slope is m and the *y*-intercept is $(0, b)$.

Example

- Graph: $f(x) = -\dfrac{3}{4}x - 1$.

  The *y*-intercept is $(0, -1)$.

  The slope is $-\dfrac{3}{4}$, which can be written $\dfrac{-3}{4}$ or $\dfrac{3}{-4}$.

  For $\dfrac{\text{Rise}}{\text{Run}} = \dfrac{-3}{4}$, start at $(0, -1)$ and move down 3 units and to the right 4 units, to the point $(4, -4)$.

  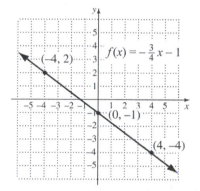

  For $\dfrac{\text{Rise}}{\text{Run}} = \dfrac{3}{-4}$, start at $(0, -1)$ and move up 3 units and to the left 4 units, to the point $(-4, 2)$.

  Draw the graph.

GUIDED LEARNING

📘 **Textbook**   👤 **Instructor**   ▶ **Video**

EXAMPLE 1	YOUR TURN 1
Graph: $y = \dfrac{1}{5}x + 3$.	Graph: $y = \dfrac{1}{2}x - 4$.

The *y*-intercept is $\left(0, \boxed{}\right)$.

The slope is $\boxed{}$, which can also be written $\dfrac{-1}{-5}$.

For $\dfrac{\text{Rise}}{\text{Run}} = \dfrac{1}{5}$, start at $(0, 3)$ and move up $\boxed{}$ unit and to the right $\boxed{}$ units, to the point $\left(5, \boxed{}\right)$.

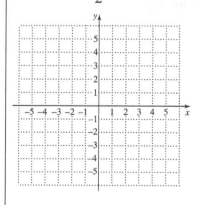

(continued)

For $\dfrac{\text{Rise}}{\text{Run}} = \dfrac{-1}{-5}$, start at $(0, 3)$ and move down ☐ unit

and to the left ☐ units, to the point $\left(-5, \boxed{}\right)$.

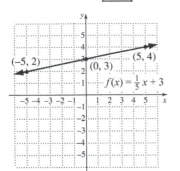

EXAMPLE 2	**YOUR TURN 2**
Graph: $2x - 3y = 6$.	Graph: $3x - 4y = -12$.

First write the equation in the form $y = mx + b$.

$$2x - 3y = 6$$
$$-3y = -2x + 6$$
$$y = \frac{2}{3}x - 2$$

The y-intercept is $\left(0, \boxed{}\right)$.

The slope is $\boxed{}$, which can also be written $\dfrac{-2}{-3}$.

For $\dfrac{\text{Rise}}{\text{Run}} = \dfrac{2}{3}$, start at $(0, -2)$ and move up ☐ units

and to the right ☐ units.

For $\dfrac{\text{Rise}}{\text{Run}} = \dfrac{-2}{-3}$, start at $(0, -2)$ and move down ☐

units and to the left ☐ units.

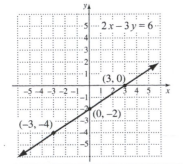

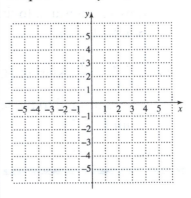

YOUR NOTES   Write your questions and additional notes.

Horizontal Lines and Vertical Lines

ESSENTIALS

The graph of $y = b$ is a **horizontal line** with y-intercept $(0, b)$. Its slope is 0.

The graph of $x = a$ is a **vertical line** with x-intercept $(a, 0)$. Its slope is not defined.

Examples

- Graph $y = 2$. If possible, determine the slope.

  The graph of $y = 2$ is a horizontal line. The slope is 0.

  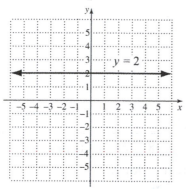

- Graph $x = -3$. If possible, determine the slope.

  The graph of $x = -3$ is a vertical line. The slope is not defined.

  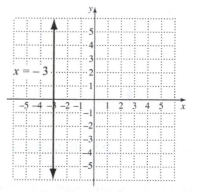

GUIDED LEARNING

 Textbook   **Instructor**   **Video**

EXAMPLE 1	YOUR TURN 1

EXAMPLE 1

Graph: $f(x) = -1$. If possible, determine the slope.

The graph is a _____ line.
            vertical / horizontal

The slope is _____.
            0 / not defined

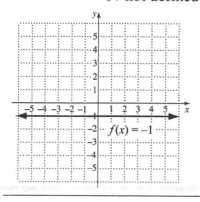

YOUR TURN 1

Graph: $f(x) = 3$. If possible, determine the slope.

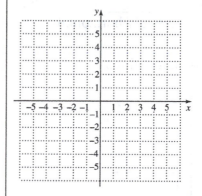

EXAMPLE 2	YOUR TURN 2
Graph: $x = 4$. If possible, determine the slope.	Graph: $x = -2$. If possible, determine the slope.

EXAMPLE 2

Graph: $x = 4$. If possible, determine the slope.

The graph is a _____ line.
                vertical / horizontal

The slope is _____.
              0 / not defined

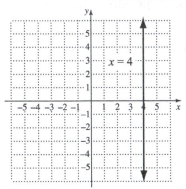

EXAMPLE 3

Graph: $3 + x = 5 - x$. If possible, determine the slope.

The only variable in the equation is x. We write the equation in the form $x = a$.

$$3 + x = 5 - x$$
$$3 + 2x = 5$$
$$2x = 2$$
$$x = 1$$

The graph is a _____ line.
                vertical / horizontal

The slope is _____.
              0 / not defined

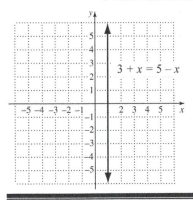

YOUR TURN 2

Graph: $x = -2$. If possible, determine the slope.

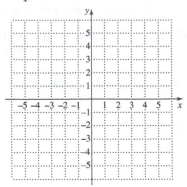

YOUR TURN 3

Graph: $3y - 8 = y + 2$. If possible, determine the slope.

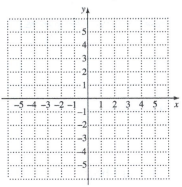

YOUR NOTES   Write your questions and additional notes.

Parallel Lines and Perpendicular Lines

ESSENTIALS

If two lines are vertical, they are parallel.

Two nonvertical lines are parallel if they have the same slope and different y-intercepts.

Two lines are perpendicular if the product of their slopes is -1 or if one line is vertical and the other is horizontal.

Examples

- Determine whether the graphs of $y = -2x - 4$ and $y = -2x + 7$ are parallel.

  The slope of $y = -2x - 4$ is -2. The y-intercept is $(0, -4)$.

  The slope of $y = -2x + 7$ is -2. The y-intercept is $(0, 7)$.

  Since the slopes are the same but the y-intercepts are different, the graphs are parallel.

- Determine whether the graphs of $y = \frac{3}{7}x + 5$ and $y = -\frac{7}{3}x + 2$ are perpendicular.

  The slope of $y = \frac{3}{7}x + 5$ is $\frac{3}{7}$. The slope of $y = -\frac{7}{3}x + 2$ is $-\frac{7}{3}$.

  Since $\frac{3}{7}\left(-\frac{7}{3}\right) = -1$, the graphs are perpendicular.

GUIDED LEARNING 　🔘 **Textbook**　　👤 **Instructor**　　▶ **Video**

EXAMPLE 1	YOUR TURN 1

Determine whether the graphs of $y = 2x - 5$ and $2y - 4x = 3$ are parallel.
 	
The slope of $y = 2x - 5$ is ☐ and the y-intercept is ☐.
 	
We find the slope of $2y - 4x = 3$ by first writing the equation in slope-intercept form.
 	
$2y - 4x = 3$
$2y = 4x + 3$
$y = \boxed{}x + \frac{3}{2}$
 	
The slope is ☐ and the y-intercept is $\left(0, \boxed{}\right)$.
 	
Because the slopes _____ the same and the
    are / are not
y-intercepts are different, the lines are _____ parallel.
    are / are not
	Determine whether the graphs of $y = -4x + 2$ and $4y - x = 5$ are parallel.

EXAMPLE 2	YOUR TURN 2
Determine whether the graphs of $3x - y = 7$ and $y = -\dfrac{1}{3}x + 3$ are perpendicular.	Determine whether the graphs of $5x - 6y = 30$ and $5y + 6x = 0$ are perpendicular.

The slope of $y = -\dfrac{1}{3}x + 3$ is $\boxed{}$.

We find the slope of $3x - y = 7$ by first writing the equation in slope-intercept form.

$$3x - y = 7$$
$$-y = -3x + 7$$
$$y = \boxed{}$$

The slope is $\boxed{}$.

We find the product of the slopes:

$$\left(\boxed{}\right)\left(\boxed{}\right) = \boxed{}$$

Since the product of the slopes _____ -1,
$\quad\quad\quad\quad\quad\quad\quad\quad$ is / is not

the lines _____ perpendicular.
$\quad\quad\quad\quad$ are / are not

YOUR NOTES   Write your questions and additional notes.

Practice Exercises

Readiness Check

Choose the word or number from the following list that best completes each sentence. Not every choice is used.

horizontal	vertical	perpendicular	parallel
0	−1	rise	run

1. The slope of a _____ line is undefined.

2. Two lines are _____ if their slopes are the same and their *y*-intercepts are different.

3. If the product of the slopes of two lines is _____, the lines are perpendicular.

4. A slope of $\dfrac{5}{8}$ corresponds to a _____ of 5 and a _____ of 8.

Graphing Using Intercepts

Find the intercepts and then graph the line.

5. $2x - 3y = -6$

6. $f(x) = x - 4$

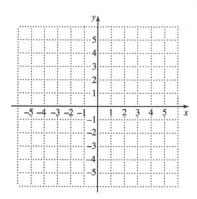
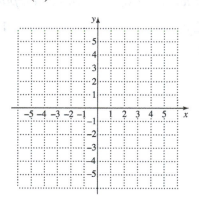

Graphing Using the Slope and the *y*-Intercept

Graph using the slope and the y-intercept.

7. $y = -\dfrac{2}{3}x + 1$

8. $x + 2y = 4$

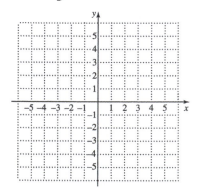
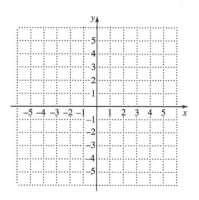

Horizontal Lines and Vertical Lines

Graph and, if possible, determine the slope.

9. $f(x) = -2$

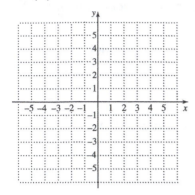

10. $2x = 6$

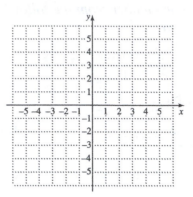

Parallel and Perpendicular Lines

Determine whether the graphs of the given pair of lines are parallel.

11. $10x - 5 = 2y$

$y = 5x + 4$

12. $3x + 4y = -12$

$4y = 3x + 6$

Determine whether the graphs of the given pair of lines are perpendicular.

13. $y = 3x - 1$

$2x + 6y = -5$

14. $x + y = 9$

$x - y = 7$

Finding an Equation of a Line When the Slope and the *y*-Intercept Are Given

ESSENTIALS

Example

- Find an equation for the line with slope $-\dfrac{1}{2}$ and *y*-intercept $(0, 6)$.

  For this line, $m = -\dfrac{1}{2}$ and $b = 6$.

  $y = mx + b$

  $y = -\dfrac{1}{2}x + 6$

GUIDED LEARNING
         📖 **Textbook**   👤 **Instructor**   ▶ **Video**

EXAMPLE 1	YOUR TURN 1
Find an equation for the line with slope 2 and *y*-intercept $(0, -1)$.	Find an equation for the line with slope -7 and *y*-intercept $(0, 5)$.
For this line, $m = \boxed{}$ and $b = \boxed{}$. $y = mx + b$ $y = \boxed{}\, x + \left(\boxed{}\right)$ $y = \boxed{}\, x - \boxed{}$	
EXAMPLE 2	YOUR TURN 2
Find a linear function $f(x) = mx + b$ whose graph has slope $-\dfrac{2}{3}$ and *y*-intercept $(0, 5)$. $f(x) = mx + b$ $f(x) = \boxed{}\, x + \boxed{}$	Find a linear function $f(x) = mx + b$ whose graph has slope 4 and *y*-intercept $(0, -10)$.

Finding an Equation of a Line When the Slope and a Point Are Given

ESSENTIALS

Point-slope equation of a line: $y - y_1 = m(x - x_1)$

The slope is m, and the line passes through (x_1, y_1).

Example

- Find an equation of the line with slope 2 that passes through $(-3, -9)$.

    Using the point-slope equation:

    $$y - y_1 = m(x - x_1)$$
    $$y - (-9) = 2(x - (-3))$$ Substituting 2 for m, -3 for x_1, and -9 for y_1
    $$y + 9 = 2(x + 3)$$
    $$y + 9 = 2x + 6$$
    $$y = 2x - 3$$

    Using the slope-intercept equation:

    $$y = mx + b$$
    $$y = 2x + b$$ Substituting 2 for m
    $$-9 = 2(-3) + b$$ Substituting -3 for x and -9 for y
    $$-9 = -6 + b$$
    $$-3 = b$$ Solving for b
    $$y = 2x - 3$$ Substituting -3 for b

GUIDED LEARNING

 Textbook   **Instructor**   **Video**

EXAMPLE 1	YOUR TURN 1
Use the point-slope equation to find an equation of the line with slope -4 that passes through $(-2, 5)$.	Use the point-slope equation to find an equation of the line with slope $\dfrac{2}{3}$ that passes through $(4, -9)$.

Here, $m = -4$ and $(x_1, y_1) = (-2, 5)$.

$$y - y_1 = m(x - x_1)$$
$$y - \boxed{} = -4\left(x - \left(\boxed{}\right)\right)$$
$$y - 5 = -4\left(x + \boxed{}\right)$$
$$y - 5 = -4x - \boxed{}$$
$$y = -4x - \boxed{}$$

EXAMPLE 2	YOUR TURN 2
Use the slope-intercept equation to find an equation of the line with slope 3 that passes through $(1, -7)$.	Use the slope-intercept equation to find an equation of the line with slope -1 that passes through $(-1, 8)$.

First, substitute 3 for m in the slope-intercept equation.

$$y = mx + b$$

$$y = \boxed{}\, x + b$$

Then, substitute 1 for x and -7 for y to find b.

$$y = 3x + b$$

$$\boxed{} = 3\left(\boxed{}\right) + b$$

$$-7 = \boxed{} + b$$

$$\boxed{} = b$$

Finally, substitute -10 for b.

$$y = 3x + \left(\boxed{}\right), \text{ or } y = \boxed{}$$

YOUR NOTES   Write your questions and additional notes.

Finding an Equation of a Line When Two Points Are Given

ESSENTIALS

Example

- Find an equation of the line containing the points $(3, 8)$ and $(-2, 7)$.

$$m = \frac{y_2 - y_1}{x_2 - x_1} = \frac{7-8}{-2-3} = \frac{-1}{-5} = \frac{1}{5}$$

Using the point-slope equation:

We choose to use $(3, 8)$ for (x_1, y_1).

$$y - y_1 = m(x - x_1)$$

$$y - 8 = \frac{1}{5}(x - 3) \quad \text{Substituting } \frac{1}{5} \text{ for } m, \text{ 3 for } x_1, \text{ and 8 for } y_1$$

$$y - 8 = \frac{1}{5}x - \frac{3}{5}$$

$$y = \frac{1}{5}x + \frac{37}{5}$$

Using the slope-intercept equation:

$$y = mx + b$$

$$y = \frac{1}{5}x + b \quad \text{Substituting } \frac{1}{5} \text{ for } m$$

$$8 = \frac{1}{5}(3) + b \quad \text{Substituting 3 for } x \text{ and 8 for } y$$

$$8 = \frac{3}{5} + b$$

$$\frac{37}{5} = b \qquad\qquad \text{Solving for } b$$

$$y = \frac{1}{5}x + \frac{37}{5} \quad \text{Substituting } \frac{37}{5} \text{ for } b$$

GUIDED LEARNING   **Textbook**   **Instructor**   **Video**

EXAMPLE 1	YOUR TURN 1
Use the point-slope equation to find an equation of the line containing the points $(-2, 1)$ and $(2, 9)$. First, we find the slope of the line. $$m = \frac{y_2 - y_1}{x_2 - x_1} = \frac{9-1}{2-(-2)} = \frac{8}{4} = \boxed{}$$ We choose to use $(2, 9)$ for (x_1, y_1). $$y - y_1 = m(x - x_1)$$ $$y - \boxed{} = 2\left(x - \boxed{}\right)$$ $$y - 9 = 2x - \boxed{}$$ $$y = 2x + \boxed{}$$	Use the point-slope equation to find an equation of the line containing the points $(-5, 3)$ and $(3, -1)$.

EXAMPLE 2	YOUR TURN 2
Use the slope-intercept equation to find an equation of the line containing the points $(4, -1)$ and $(-2, -6)$. First, we find the slope of the line. $$m = \frac{y_2 - y_1}{x_2 - x_1} = \frac{-1-(-6)}{4-(-2)} = \frac{5}{6}$$ $$y = mx + b$$ $$y = \frac{5}{6}x + b \qquad \text{Substituting } \frac{5}{6} \text{ for } m$$ $$-1 = \frac{5}{6}\left(\boxed{}\right) + b \quad \text{Substituting 4 for } x \text{ and } -1 \text{ for } y$$ $$-1 = \frac{10}{3} + b$$ $$-\frac{13}{3} = b \qquad \text{Solving for } b$$ $$y = \boxed{}\,x + \left(\boxed{}\right), \text{ or } y = \frac{5}{6}x - \frac{13}{3}$$	Use the slope-intercept equation to find an equation of the line containing the points $(1, 4)$ and $(-2, 7)$.

YOUR NOTES   Write your questions and additional notes.

Finding an Equation of a Line Parallel or Perpendicular to a Given Line Through a Point Not on the Line

ESSENTIALS

Example

- Find an equation of the line containing the point $(-1, 4)$ and parallel to the line $y = 3x - 5$.

  Since the lines are parallel, both lines have slope 3.

  $$y - y_1 = m(x - x_1)$$
  $$y - 4 = 3[x - (-1)], \text{ or } y = 3x + 7$$

GUIDED LEARNING

📖 **Textbook**   👤 **Instructor**   ▶ **Video**

EXAMPLE 1	YOUR TURN 1
Find an equation of the line containing the point $(8, 11)$ and parallel to the line $3x - 4y = 8$.	Find an equation of the line containing the point $(4, 5)$ and parallel to the line $2x + 5y = -3$.

First, find the equation of the given line in slope-intercept form.

$$3x - 4y = 8$$
$$-4y = -3x + 8$$
$$y = \frac{3}{4}x - 2$$

The slope of the given line is $\boxed{}$. The slope of a line parallel to it is also $\boxed{}$. We have

$$m = \frac{3}{4} \text{ and } (x_1, y_1) = (8, 11).$$

$$y - y_1 = m(x - x_1)$$

$$y - \boxed{} = \boxed{}(x - 8)$$

$$y - 11 = \frac{3}{4}x - \boxed{}$$

$$y = \frac{3}{4}x + \boxed{}$$

EXAMPLE 2	YOUR TURN 2
Find an equation for the line perpendicular to $x = 3y + 5$ and passing through $(4, -2)$.	Find an equation for the line containing the point $(-5, 1)$ and perpendicular to the line $-2x = 3 - 4y$.

EXAMPLE 2

Find an equation for the line perpendicular to $x = 3y + 5$ and passing through $(4, -2)$.

First, find the equation of the given line in slope-intercept form.

$$x = 3y + 5$$

$$x - 5 = 3y$$

$$\frac{1}{3}x - \frac{5}{3} = y$$

The slope of the given line is $\boxed{}$. The slope of a line

perpendicular to it is the opposite of the reciprocal of

$\frac{1}{3}$, or $\boxed{}$. We use the slope-intercept equation.

$$y = mx + b$$

$$y = -3x + b \qquad \text{Substituting } -3 \text{ for } m$$

$$-2 = -3\left(\boxed{}\right) + b \quad \text{Substituting 4 for } x \text{ and } -2 \text{ for } y$$

$$-2 = -12 + b$$

$$\boxed{} = b \qquad \text{Solving for } b$$

Finally, we substitute 10 for b.

$$y = -3x + 10$$

YOUR TURN 2

Find an equation for the line containing the point $(-5, 1)$ and perpendicular to the line $-2x = 3 - 4y$.

YOUR NOTES   Write your questions and additional notes.

Applications of Linear Functions

ESSENTIALS

Given two points, we can model data with a linear function.

GUIDED LEARNING   📘 **Textbook**   👤 **Instructor**   ▶️ **Video**

EXAMPLE 1	YOUR TURN 1
The average monthly revenue of Corp C Enterprise is shown in the table. Use the data from 2010 and 2012 to find a linear function that fits the data. Then use the function to estimate average monthly revenue in 2014.	The average monthly expenses of Corp C Enterprise are shown in the table. Use the data for 2009 and 2011 to find a linear function that fits the data. Let $t =$ the number of years since 2000 and $e =$ the average monthly expenses in thousands of dollars. Then use the function to estimate average monthly expenses in 2014.

Year	Average Monthly Revenue
2010	$20,000
2012	21,000

We let $t =$ the number of years since 2000 and $r =$ the average monthly revenue in thousands of dollars. We find a linear function

containing points $\left(10, \boxed{}\right)$ and $\left(\boxed{}, 21\right)$.

$$m = \frac{21 - \boxed{}}{\boxed{} - 10} = \frac{\boxed{}}{\boxed{}}$$

We use m and $\left(10, \boxed{}\right)$ to find an equation of the line.

$$r - \boxed{} = \boxed{}(t - 10)$$

$$r = \frac{1}{2}t + 15, \text{ or } r(t) = \frac{1}{2}t + 15$$

To estimate the average monthly revenue in 2014, we find $r(14)$.

$$r(14) = \frac{1}{2}\left(\boxed{}\right) + 15 = \boxed{}$$

Assuming constant growth, average monthly revenue in 2014 is estimated to be $\boxed{}$.

Year	Average Monthly Expenses
2009	$20,000
2011	19,000

EXAMPLE 2	YOUR TURN 2
Suppose suppliers are willing to sell 100 handmade headbands when the price is $20 per headband and 60 handmade headbands when the price is $12 per headband. Find a linear function that expresses the number of headbands suppliers are willing to sell as a function of the price per headband. Use the function to predict how many headbands sellers would be willing to sell if the price were $15 per headband.	Suppose buyers are willing to buy 100 handmade headbands when the price is $10 per headband and 70 handmade headbands when the price is $12 per headband. Find a linear function that expresses the number of headbands buyers are willing to buy as a function of the price per headband. Let $p =$ the price and $h =$ the number of headbands. Use the function to predict how many headbands buyers would be willing to buy if the price were $15 per headband.

Let $p =$ the price and $h =$ the number of headbands. We find a linear function containing points $\left(20, \boxed{}\right)$ and $\left(\boxed{}, 60\right)$.

$$m = \frac{\boxed{} - 100}{12 - \boxed{}} = \frac{-40}{\boxed{}} = 5$$

We use m and $(20, 100)$ to find an equation of the line.

$$h - \boxed{} = \boxed{}(p - 20)$$

$$h - 100 = 5p - 100$$

$$h = 5p, \text{ or } h(p) = 5p$$

To estimate the number of headbands sellers would be willing to sell if the price were $15, we find $h(15)$.

$$h(15) = 5 \cdot \boxed{} = \boxed{}$$

The suppliers would be willing to supply $\boxed{}$ headbands.

YOUR NOTES   Write your questions and additional notes.

Practice Exercises

Readiness Check

Pair each item in the first column with the item in the second column that best matches.

1. $y - 4 = -2\left(x - (-6)\right)$

2. $y = 3x + 22$

3. $y = x - 7$ and $y = x + 5$

4. $y = x - 3$ and $y = 3 - x$

a) Parallel lines

b) An equation in slope-intercept form

c) Perpendicular lines

d) An equation in point-slope form

Finding an Equation of a Line When the Slope and the *y*-Intercept Are Given

5. Find an equation of the line with slope -4 and *y*-intercept $(0, 8)$.

6. Find a linear function $f(x) = mx + b$ whose graph has slope $\dfrac{1}{2}$ and *y*-intercept $(0, -1)$.

Finding an Equation of a Line When the Slope and a Point Are Given

7. Find an equation of the line with slope 6 and containing the point $(3, 0)$.

8. Find an equation of the line with slope $-\dfrac{1}{2}$ and containing the point $(-2, -6)$.

Finding an Equation of a Line When Two Points Are Given

9. Find an equation of the line containing the points $(5, 6)$ and $(-4, 3)$.

10. Find an equation of the line containing the points $(0, -5)$ and $(-1, -7)$.

Finding an Equation of a Line Parallel or Perpendicular to a Given Line Through a Point Not on the Line

11. Find an equation for the line containing the point $(5, 3)$ and parallel to the line $x - y = -6.$

12. Find an equation for the line containing the point $(10, -1)$ and perpendicular to the line $5x + 4y = -20.$

Applications of Linear Functions

13. Dan joined a fitness club for $150 and pays a monthly fee of $22.95.

    a) Formulate a linear function that models the total cost $C(t)$ of the club membership for t months.

    b) Use the model to determine the total cost after he has been a member of the club for 16 months.

14. In 2005, the number of students participating in an intramural sport at Castlegate Community College was 150. In 2010, the number had risen to 320.

    a) Find a linear function that fits the data. Let $x =$ the number of years after 2005 and $N(x) =$ the number of students participating in an intramural sport.

    b) Use the function of part (a) to estimate the number of students participating in an intramural sport at Castlegate Community College in 2016.

Solving Systems of Equations Graphically

ESSENTIALS

A **system of equations** in two variables is a set of two or more linear equations that are to be solved simultaneously.

A **solution** of a system of two equations in two variables is an ordered pair that makes both equations true. If we graph a system of equations, the point at which the graphs intersect will be the solution of both equations.

Sometimes the equations in a system have graphs that are parallel lines. In such a case, the graphs will never intersect and as such we say the system has **no solution**.

Sometimes the equations in a system have the same graph. In such a case, the system has **infinitely many solutions**.

Consistent Systems and Inconsistent Systems

If a system of equations has at least one solution, then the system is **consistent**.
If a system of equations has no solution, then the system is **inconsistent**.

Dependent Equations and Independent Equations

If for a system of two equations in two variables:
  the graphs of the equations are the same line, then the equations are **dependent**.
  the graphs of the equations are different lines, then the equations are **independent**.

Examples

- $y = x - 1,$

  $y = -\dfrac{2}{3}x + 4$

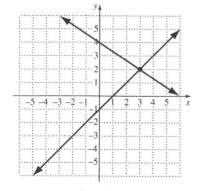

  Solution: $(3, 2)$

  Consistent system
  Independent equations

- $y = \dfrac{1}{2}x - 1,$

  $y = \dfrac{1}{2}x + 2$

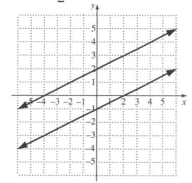

  No solution

  Inconsistent system
  Independent equations

- $y = \dfrac{1}{3}x + 1,$

  $3y = x + 3$

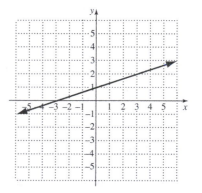

  Infinitely many solutions

  Consistent system
  Dependent equations

GUIDED LEARNING   ⬤ **Textbook**   ⬤ **Instructor**   ⬤ **Video**

EXAMPLE 1	YOUR TURN 1
Solve this system graphically: $2x - y = 3,$ $x + 3y = 5.$	Solve this system graphically: $x - y = -5,$ $x + 2y = 1.$

EXAMPLE 1 (continued)

Draw the graph of each equation and find the coordinates of the point of intersection.

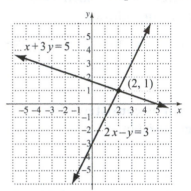

The point of intersection has coordinates that make both equations _____.
true / false

The solution seems to be $\left(\boxed{}, \boxed{}\right)$.

We will check the ordered pair in both equations.
Check:

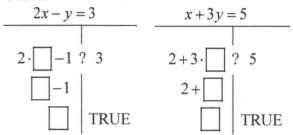

The solution is $\left(\boxed{}, \boxed{}\right)$.

Since this system has at least one solution it is a/an _____ system.
consistent / inconsistent

Since the graphs of the equations are different lines the equations are _____.
dependent / independent

YOUR TURN 1 (continued)

Classify the system as consistent or inconsistent and the equations as dependent or independent.

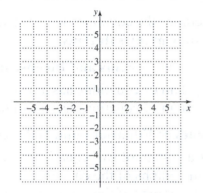

EXAMPLE 2	YOUR TURN 2
Solve this system graphically: $$f(x) = -2x + 4,$$ $$g(x) = -2x + 2.$$	Solve this system graphically: $$f(x) = x + 2,$$ $$g(x) = x - 2.$$

Graph both equations.

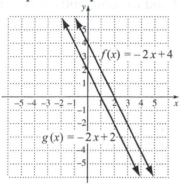

Classify the system as consistent or inconsistent and the equations as dependent or independent.

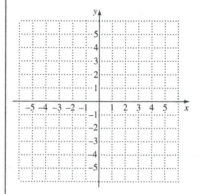

We see that the graphs have the same

_____, but different

slope / *y*-intercept

_____, so the lines are parallel.

slopes / *y*-intercepts

There is no point at which they cross, so the system has

_____.

no solution / infinitely many solutions

The solution set is thus the empty set, denoted ∅, or { }.

Since this system has no solution it is a/an _____ system.

consistent / inconsistent

Since the graphs of the equations are different lines the equations are _____.

dependent / independent

EXAMPLE 3	YOUR TURN 3
Solve this system graphically: $6x - 3y = 9,$ $\quad y = 2x - 3.$	Solve this system graphically: $8x + 2y = 6,$ $\quad y = -4x + 3.$

EXAMPLE 3

Solve this system graphically:

$$6x - 3y = 9,$$
$$y = 2x - 3.$$

Graph both equations.

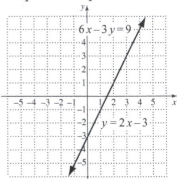

We see that the graphs are _____.

parallel / the same

Thus any solution of one of the equations is a solution of the other. The system has a/an _____ number of solutions.

<u>finite / infinite</u>

Since this system has at least one solution it is a/an _____ system.

consistent / inconsistent

Since the graphs of the equations are the same line the equations are _____.

dependent / independent

YOUR TURN 3

Solve this system graphically:

$$8x + 2y = 6,$$
$$y = -4x + 3.$$

Classify the system as consistent or inconsistent and the equations as dependent or independent.

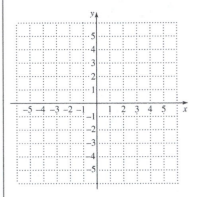

YOUR NOTES   Write your questions and additional notes.

Practice Exercises

Readiness Check

Consider the graphs below when doing Exercises 1-6.

1. Which graph(s) represent(s) a consistent system?

2. Which graph(s) represent(s) an inconsistent system?

3. Which graph(s) represent(s) a system whose equations are dependent?

4. Which graph(s) represent(s) a system whose equations are independent?

5. Which graph(s) represent(s) a system with no solution?

6. Which graph(s) represent(s) a system with a solution at $(2,4)$?

a)

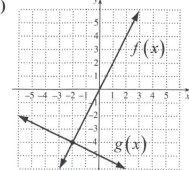

b)

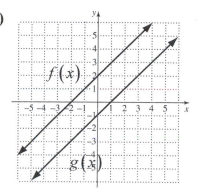

c)

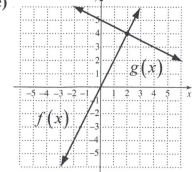

d)

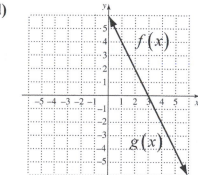

Solving Systems of Equations Graphically

Solve each system of equations graphically. Then classify the system as consistent or inconsistent and the equations as dependent or independent. Complete the check for each.

7. $x - y = 0,$

$x + y = -2$

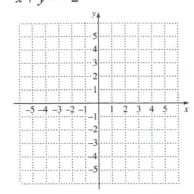

8. $6x - 2y = 4,$

$y = 3x - 2$

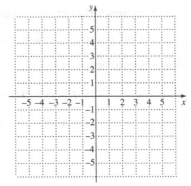

9. $x + 2y = -2,$

$4x - y = 10$

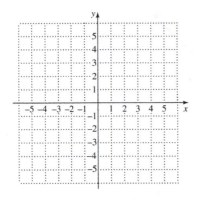

10. $x + y = 5,$

$x + y = 2$

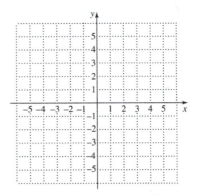

11. $f(x) = -2x + 1,$

$g(x) = 4x - 2$

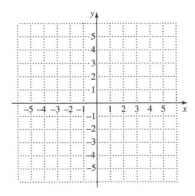

12. $x = 4,$

$x = -\dfrac{3}{2}$

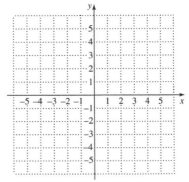

The Substitution Method

ESSENTIALS

To use the **substitution method** to solve systems of two equations:

1. If neither equation has a variable alone on one side, solve one equation for one of the variables.
2. Substitute the expression that is equivalent to the variable into the other equation.
3. You should now have an equation that contains only one variable. Solve this equation.
4. Substitute this number in one of the equations to find the value of the other variable.

No matter the order in which you find your answers, coordinates of ordered pairs must be written in alphabetical order.

Special Cases

When solving a system of two linear equations in two variables:

1. If a false equation is obtained, such as $4 = 5$, then the system has no solution.
2. If a true equation is obtained, then the system has an infinite number of solutions.

Example

- Solve this system by substitution: $\quad 4x + y = 11, \quad (1)$
$$2x - 3y = 23. \quad (2)$$

$$
\begin{aligned}
y &= -4x + 11 && \text{Solving equation (1) for } y \\
2x - 3(-4x + 11) &= 23 && \text{Substituting } -4x + 11 \text{ for } y \text{ in Equation (2)} \\
2x + 12x - 33 &= 23 && \text{Removing parentheses} \\
14x - 33 &= 23 && \text{Collecting like terms} \\
14x &= 56 && \text{Adding 33} \\
x &= 4 && \text{Dividing by 14}
\end{aligned}
$$

Now we substitute 4 for x in one of the equations and solve to find y.

$$y = -4 \cdot (4) + 11 = -16 + 11 = -5$$

The solution is $(4, -5)$.

GUIDED LEARNING

EXAMPLE 1	YOUR TURN 1
Solve this system:	Solve this system:

Solve this system:

$5x + y = 19,$ (1)

$2x + 8y = 0.$ (2)

First, we solve one equation for one of the variables. Since the coefficient of y is 1 in the first equation, it is the easier one to solve for y:

$y = -5x + 19.$ (3)

Next, we substitute $-5x + 19$ for y in the second equation and solve for x:

$2x + 8\left(\boxed{}\right) = 0$

$2x - 40x + \boxed{} = 0$

$\boxed{} + 152 = 0$

$-38x = -152$

$x = \boxed{}$

In order to find y we can return to equation $(1), (2), \text{or} (3)$. It is easiest to use equation (3) since we have already solved for y. We substitute 4 for x and solve for y:

$y = -5 \cdot \boxed{} + 19 = -20 + 19 = \boxed{}$

We obtain the ordered pair $(4, -1)$.

Solve this system:

$4x - y = 10,$

$x + 2y = -2.$

(continued)

Check:

$$5x + y = 19$$

$$5 \cdot (4) + (-1) \ ? \ 19$$
$$20 - 1$$
$$19 \quad | \quad \text{TRUE}$$

$$2x + 8y = 0$$

$$2 \cdot 4 + 8 \cdot (-1) \ ? \ 0$$
$$8 + (-8)$$
$$0 \quad | \quad \text{TRUE}$$

The solution is $(\boxed{}, \boxed{})$.

EXAMPLE 2	YOUR TURN 2
Solve this system:	Solve this system:

EXAMPLE 2

Solve this system:

$$y = -3x - 9,$$
$$9x + 3y = 12.$$

The first equation is already solved for y, so we substitute $-3x - 9$ for y in the second equation.

$$9x + 3\left(\boxed{}\right) = 12$$

$$9x - 9x + \left(\boxed{}\right) = 12$$

$$\boxed{} = 12$$

This is a _____ equation.
   true / false

This means that the system has

_____ .
no solution / an infinite number of solutions

YOUR TURN 2

Solve this system:

$$y = -2x + 4,$$
$$8x + 4y = 6.$$

EXAMPLE 3	YOUR TURN 3
Solve this system:	Solve this system:

Solve this system:

$$x - y = 2,$$
$$3x - 3y = 6.$$

First, we solve one equation for one of the variables. The coefficient of x is 1 in the first equation so it is the easier one to solve for x.

$$x = y + 2.$$

Next, we substitute $y + 2$ for x in the second equation and solve for y:

$$3\left(\boxed{}\right) - 3y = 6$$
$$\boxed{} + 6 - 3y = 6$$
$$\boxed{} = 6$$

This is a _____ equation.
   true / false

This means that the system has

_____ .
no solution / an infinite number of solutions

YOUR TURN 3

Solve this system:

$$x - 3y = 5,$$
$$-2x + 6y = -10.$$

YOUR NOTES   Write your questions and additional notes.

Solving Applied Problems Involving Two Equations

ESSENTIALS

Many applied problems are easier to solve if we first translate to a system of two equations rather than to a single equation.

	📖 **Textbook**	👤 **Instructor**	▶ **Video**

GUIDED LEARNING

EXAMPLE 1	YOUR TURN 1
The perimeter of a football field (including the end zones) is 1040 feet. The length is 40 feet more than twice the width. Find the length and width.	The perimeter of Mr. McGregor's garden is 110 feet. The length is 5 feet less than twice the width. Find the length and width.

1. **Familiarize.** We first make a drawing and label it, using l for length and w for width. Also, we use the formula
$$P = 2l + 2w.$$

w = width ▢

l = length

2. **Translate.** We translate as follows:

The perimeter   is   1040 ft
↓   ↓   ↓

$$2l + 2w \quad = \quad 1040$$

We can also write a second equation:

The length   is   40 feet more than twice the width
↓   ↓   ↓

$$l \quad = \quad 2w + 40$$

We now have a system of equations:
$$2l + 2w = 1040,$$
$$l = 2w + 40.$$

(continued)

3. **Solve.** We substitute $2w+40$ for l in the first equation and solve for w:

$$2\cdot\left(\boxed{}\right)+2w=1040$$

$$4w+\boxed{}+2w=1040$$

$$\boxed{}+80=1040$$

$$6w=960$$

$$w=160$$

Next we substitute 160 for w into either equation and solve for l:

$$l=2\cdot\boxed{}+40$$

$$l=\boxed{}+40$$

$$l=\boxed{}$$

4. **Check.** Consider the dimensions 360 ft and 160 ft. The length is 40 ft more than twice the width: $2\cdot(160)+40=360$. The perimeter is $2\cdot(360)+2\cdot(160)$, or 1040.

5. **State.** The length is 360 ft and the width is 160 ft.

YOUR NOTES   Write your questions and additional notes.

Practice Exercises

Readiness Check

Determine whether each statement is true or false.

1. The substitution method is a graphical method for solving systems of equations.

2. When solving using substitution, if we obtain a true equation, then the system has no solution.

3. When using the substitution method, if neither equation of a system has a variable alone on one side we must solve one equation for one of the variables.

4. When writing the solution of a system of equations, the value that we solved for first is always the first number in the ordered pair.

The Substitution Method

Solve each system of equations by the substitution method.

5. $2x + 5y = 11,$
   $x = 3y$

6. $3x - 2y = -4,$
   $y = 3x - 1$

7. $x - 3y = 4,$
   $-2x + 5y = -5$

8. $2x + y = 5,$
   $4x + 2y = 7$

9. $-3x - 6y = -15,$
    $x = -2y + 5$

10. $3x + y = 2,$
        $6x - 10y = 7$

11. $5x + 7y = 1,$
     $2x + 4y = 4$

12. $3x - 4y = -2,$
        $-6x + 8y = 4$

Solving Applied Problems Involving Two Equations

13. The outer perimeter of a picture frame is 32 inches. The length of the frame is 2 inches longer than the width. Find the length and the width.

14. The sum of the measures of **complementary angles** is 90°. Two complementary angles are such that the measure of one angle is 2° more than seven times the measure of the other. Find the measures of the angles.

The Elimination Method

ESSENTIALS

The **elimination method** for solving systems of equations uses the addition principle for equations. Thus, it is also called the addition method. The object of this method is to add the two equations in such a way that one of the variables is eliminated. To eliminate a variable, we sometimes use the multiplication principle to multiply one or both of the equations by a particular number before adding.

To use the elimination method to solve systems of two equations:

1. Write both equations in the form $Ax + By = C$.
2. Clear any decimals or fractions.
3. Choose a variable to eliminate.
4. Make the chosen variable's terms opposites by multiplying one or both equations by appropriate numbers if necessary.
5. Eliminate a variable by adding the equations and then solve for the remaining variable.
6. Substitute in either of the original equations to find the value of the other variable.

Special Cases

When solving a system of two linear equations in two variables:

1. If a false equation is obtained, such as $0 = 5$, then the system has no solution. The system is inconsistent, and the equations are independent.
2. If a true equation is obtained, such as $0 = 0$, then the system has an infinite number of solutions. The system is consistent, and the equations are dependent.

Example

- Solve this system using the elimination method: $3x + y = 8$, (1)
  $$2x - y = 7. \quad (2)$$

$$
\begin{array}{ll}
3x + y = 8 & (1) \\
\underline{2x - y = 7} & (2) \\
5x + 0y = 15 & \text{Adding the equations} \\
5x + 0 = 15 \\
5x = 15 \\
x = 3 \\
3 \cdot (3) + y = 8 & \text{Substituting 3 for } x \text{ in equation } (1) \\
\left.\begin{array}{l} 9 + y = 8 \\ y = -1 \end{array}\right\} & \text{Solving for } y
\end{array}
$$

The solution is $(3, -1)$.

📖 **Textbook**   👤 **Instructor**   ▶ **Video**

GUIDED LEARNING

EXAMPLE 1	YOUR TURN 1
Solve this system:	Solve this system:

Solve this system:

$2x + y = 2,$ (1)

$x - y = -5.$ (2)

The y in equation (1) is opposite of $-y$ in equation (2). When added, their sum is zero and the variable y is "eliminated."

We add the two equations:

$$2x + y = 2 \quad (1)$$
$$\underline{x - y = -5} \quad (2)$$
$$3x + \boxed{} = -3 \qquad \text{Adding}$$
$$3x + 0 = -3$$
$$\boxed{} = -3$$

We eliminated the variable y and now have an equation with one variable, which we solve for x:

$3x = -3$

$x = \boxed{}$

Next, we substitute -1 for x in either equation and solve for y:

$$2 \cdot \left(\boxed{}\right) + y = 2$$
$$-2 + y = 2$$
$$y = \boxed{}$$

We obtain the ordered pair $\left(\boxed{}, \boxed{}\right)$.

Check:

$$\begin{array}{c|c} 2x + y = 2 & x - y = -5 \\ \hline 2 \cdot (-1) + 4 \;?\; 2 & -1 - 4 \;?\; -5 \\ -2 + 4 & -5 \;\bigl|\; \text{TRUE} \\ 2 \;\bigl|\; \text{TRUE} & \end{array}$$

Since $(-1, 4)$ checks, it is the solution.

YOUR TURN 1

Solve this system:

$4x + 5y = -6,$

$-4x - 3y = 2.$

EXAMPLE 2	YOUR TURN 2
Solve this system: $x + 2y = 4,$ $3x + y = -3.$	Solve this system: $5x + y = 19,$ $x + 4y = 0.$

If we add directly, we will not eliminate a variable. However, note that if the x in the first equation were $-3x$, we could eliminate x.

We multiply the first equation by $\boxed{}$ and then add the equations:

$$\boxed{}\,x - 6y = \boxed{}$$
$$\underline{3x + \ y = -3}$$
$$\boxed{} - 5y = -15 \qquad \text{Adding}$$

$$\left.\begin{array}{l} -5y = -15 \\[4pt] y = \boxed{} \end{array}\right\} \quad \text{Solving for } y$$

Now we substitute 3 for y in one of the equations and solve for x.

$$x + 2 \cdot \boxed{} = 4$$
$$x + \boxed{} = 4$$
$$x = \boxed{}$$

We obtain the ordered pair $\left(\boxed{}, \boxed{} \right)$.

Check:

$$\begin{array}{c|c} \underline{x + 2y = 4} & \underline{3x + y = -3} \\ -2 + 2 \cdot (3) \ \overset{?}{} \ 4 & 3 \cdot (-2) + 3 \ \overset{?}{} \ -3 \\ -2 + 6 & -6 + 3 \\ 4 \mid \text{TRUE} & -3 \mid \text{TRUE} \end{array}$$

Since $(-2, 3)$ checks, it is the solution.

EXAMPLE 3	YOUR TURN 3
Solve this system:	Solve this system:

EXAMPLE 3

Solve this system:

$$5x - 4y = 8, \quad (1)$$

$$2x + 3y = -6. \quad (2)$$

We must first multiply in order to make one pair of terms opposites. We decide to do this with the *y*-terms. We multiply equation (1) by 3 and equation (2) by 4.

The new system is:

$$15x - \boxed{}y = \boxed{}$$

$$\boxed{}x + 12y = \boxed{}$$

$$23x + \boxed{} = \boxed{} \quad \text{Adding}$$

$$\left.\begin{array}{l} 23x = 0 \\ x = 0 \end{array}\right\} \quad \text{Solving for } x$$

Now substitute 0 for *x* in one of the equations and solve for *y*.

$$2 \cdot (0) + 3y = -6$$

$$0 + 3y = -6$$

$$3y = -6$$

$$y = \boxed{}$$

We check the ordered pair $(0, -2)$.

Check:

$$
\begin{array}{c|c}
5x - 4y = 8 & 2x + 3y = -6 \\
\hline
5 \cdot (0) - 4(-2) \; ? \; 8 & 2 \cdot (0) + 3(-2) \; ? \; -6 \\
\quad 0 + 8 & \quad 0 - 6 \\
\quad\quad 8 \;\big|\; \text{TRUE} & \quad\quad -6 \;\big|\; \text{TRUE}
\end{array}
$$

Since $(0, -2)$ checks, it is the solution.

YOUR TURN 3

Solve this system:

$$3x - 2y = 15,$$

$$2x - 3y = 15.$$

EXAMPLE 4	YOUR TURN 4
Solve this system:	Solve this system:

Solve this system:

$$x - 3y = 5,$$
$$-2x + 6y = -10.$$

We must first multiply in order to make one pair of terms opposites. We decide to do this with the x-terms. We multiply equation (1) by 2.

The new system is:

$$2x - \boxed{}\,y = \boxed{}$$
$$\underline{-2x + \quad 6y = -10}$$
$$0x + \boxed{} = \boxed{} \quad \text{Adding}$$
$$\boxed{} = \boxed{}$$

We have eliminated both variables, and what remains is a _____ equation, $0 = 0$.
true / false

If an ordered pair is a solution of one of the original equations, then it will be a solution of the other. The system has _____.
no solution / an infinite number of solutions

The system is _____.
consistent / inconsistent

The equations are _____.
dependent / independent

YOUR TURN 4

Solve this system:

$$3x + y = -9,$$
$$9x + 3y = 12.$$

Solving Applied Problems Using Elimination

ESSENTIALS

Many applied problems are easier to solve if we first translate to a system of two equations rather than to a single equation.

GUIDED LEARNING

 Textbook   **Instructor**   **Video**

EXAMPLE 1	YOUR TURN 1
The Lions basketball team scored 69 points on a combination of two-point shots and three-point shots. If they made a total of 30 shots, how many of each kind of shot was made?	The Tigers basketball team scored 75 points on a combination of two-point shots and three-point shots. If they made a total of 33 shots, how many of each kind of shot was made?

1. Familiarize. We let $x =$ the number of two-point shots and $y =$ the number of three-point shots. The total points scored from two-point shots alone is $2x$. Similarly, the total points scored from three-point shots alone is $\boxed{}$.

2. Translate. We translate to two equations.

$$\underbrace{\text{Total number of points scored}} \quad \text{is} \quad 69$$
$$\downarrow \qquad\qquad \downarrow \quad \downarrow$$
$$2x + 3y \qquad\qquad = \quad 69$$

$$\underbrace{\text{Total number of shots made}} \quad \text{is} \quad 30$$
$$\downarrow \qquad\qquad \downarrow \quad \downarrow$$
$$x + \boxed{} \qquad\qquad = \quad 30$$

We now have a system of equations:
$$2x + 3y = 69, \quad (1)$$
$$x + y = 30. \quad (2)$$

(continued)

3. Solve. First, we multiply by -2 on both sides of equation (2) and add:

$$2x + 3y = 69$$
$$-2x - 2y = \boxed{}$$
$$0x + y = 9 \qquad \text{Adding}$$

$$\left.\begin{array}{l} 0 + y = 9 \\ \quad y = 9 \end{array}\right\} \quad \text{Solving for } y$$

Next we substitute 9 in for y in equation (2) and solve for x:

$$x + \boxed{} = 30$$
$$x = \boxed{}$$

4. Check. If 21 two-point shots and 9 three-point shots were made, there were $2 \cdot 21 + 3 \cdot 9$, or 69, total points scored and $21 + 9$, or 30, shots made. The numbers check in the original problem.

5. State. There were 21 two-point shots made and 9 three-point shots made.

YOUR NOTES   Write your questions and additional notes.

Practice Exercises

Readiness Check

Determine whether each statement is true or false.

1. We call it the elimination method because after the equations are added one variable should have been eliminated.

2. When solving using elimination, if we obtain a false equation, then the equations of the system are dependent.

3. The elimination method is also called the addition method because it uses the addition principle for equations.

4. The first step of the elimination method is to write both equations in the form $Ax + By = C$.

The Elimination Method

Solve each system of equations using the elimination method.

5. $x + 4y = 12,$
      $2x - 4y = 0$

6. $x - y = -4,$
      $-2x + 3y = 9$

7. $2x + 3y = 4,$
      $6x + 9y = 12$

8. $4x - 5y = 7,$
      $5x - 3y = -1$

9. $-3x + 5y = -2,$
    $6x - 10y = 6$

10. $7x + 4y = 9,$
    $14x + 8y = 18$

11. $2x - 3y = 5,$
    $5x + 10y = -12$

12. $x = 2y - 4,$
    $6y = 3x + 5$

Solving Applied Problems Using Elimination

13. The sum of two numbers is 4. The larger number minus the smaller number is 26. Find the numbers.

14. A small concert venue sold a total of 375 tickets for a concert. The number of floor seats was 55 more than three times the number of balcony seats. How many of each type of seat was sold?

Total-Value Problems and Mixture Problems

ESSENTIALS

Examples

- A college basketball team scored 83 points in a game. Of those points, 15 points were from free throws. The remaining 68 points were a result of 30 two-point and three-point baskets. How many baskets of each type were made during the game?

  1. **Familiarize.** Let $x =$ the number of two-pointers made and $y =$ the number of three-pointers made.

  2. **Translate.** A total of 30 baskets were made, so $x + y = 30$.

     We also have a second equation.

     Rewording: $\underbrace{\text{The points scored} \atop \text{from two-pointers}}$ plus $\underbrace{\text{the points scored} \atop \text{from three-pointers}}$ totaled 68.

     Translating: $\qquad\quad x \cdot 2 \qquad\quad + \qquad\quad y \cdot 3 \qquad\quad = \quad 68$

     The system of equations is
     $$x + y = 30, \quad (1)$$
     $$2x + 3y = 68. \quad (2)$$

  3. **Solve.** Multiply equation (1) by -2 and then add the equations.
     $$\begin{array}{ll} -2x - 2y = -60 & (3) \\ \underline{2x + 3y = 68} & (2) \\ y = 8 \end{array}$$

     Substituting 8 for y in equation (1), we get $x + 8 = 30$. Thus $x = 22$.

  4. **Check.** 22 two-pointers and 8 three-pointers = 30 baskets
     $$22 \cdot 2 + 8 \cdot 3 = 44 + 24 = 68 \text{ points}$$
     The answer checks.

  5. **State.** The team made 22 two-pointers and 8 three-pointers.

- A caterer wants to mix almonds that sell for $13.25 per pound and walnuts that sell for $11.25 per pound to make 23 lb of a mixture that sells for $12.75 per pound. How much of each should be used?

  1. **Familiarize.** Let $a =$ the number of pounds of almonds and $w =$ the number of pounds of walnuts in the mixture.

  2. **Translate.** A 23 lb batch is being made, so $a + w = 23$.

     $\underbrace{\text{The value of the almonds}}$ plus $\underbrace{\text{the value of the walnuts}}$ is $\underbrace{\text{the value of the mixture.}}$

     $\qquad a \cdot 13.25 \qquad\qquad + \qquad\quad w \cdot 11.25 \qquad\quad = \qquad\quad 23 \cdot 12.75$

3. **Solve.** When the first equation is solved for a, we have $a = -w + 23$. We then substitute $-w + 23$ for a in the second equation.

$$13.25a + 11.25w = 293.25$$
$$13.25(-w + 23) + 11.25w = 293.25$$
$$-13.25w + 304.75 + 11.25w = 293.25$$
$$-2w + 304.75 = 293.25$$
$$-2w = -11.5$$
$$w = 5.75$$

If $w = 5.75$, then $a = -5.75 + 23 = 17.25$.

4. **Check.** Amount of mixture: $17.25 \text{ lb} + 5.75 \text{ lb} = 23 \text{ lb}$

   Value of mixture: $17.25(\$13.25) + 5.75(\$11.25) = \$293.25$.

   The numbers check.

5. **State.** The 23 lb mixture should be made by combining 17.25 lb of almonds and 5.75 lb of walnuts.

GUIDED LEARNING   🔖 **Textbook**   👤 **Instructor**   ▶ **Video**

EXAMPLE 1	YOUR TURN 1
Matinee ticket prices at Midtown Cinema are $9.50 for adults and $6.50 for children. A total of 90 people purchased tickets for the first showing of a movie. If a total of $681 was collected, how many adults' tickets and how many children's tickets were sold?	A college field hockey team stopped for ice cream on the way home from a game. They ordered 25 cones, some one-scoop at $1.75 and some two-scoop at $2.25. How many of each type of cone was ordered if the total bill was $52.25?

1. **Familiarize.** Let $a =$ the number of adults' tickets and $c =$ the number of children's tickets.

2. **Translate.** 90 people bought tickets, so $a + c = \boxed{}$.

Amount collected for adults **plus** amount collected for children totaled $681.

$$a \cdot \boxed{} \quad + \quad c \cdot \boxed{} \quad = \quad 681$$

We have a system of equations:

$$a + c = 90, \qquad (1)$$
$$9.50a + 6.50c = 681. \qquad (2)$$

(continued)

3. **Solve.** Solve equation (1) for a:

$$a + c = 90$$

$$a = \boxed{} + 90 \quad (3)$$

Next, substitute $\boxed{}$ for a in equation (2):

$$9.50a + 6.50c = 681$$

$$9.50(-c + 90) + 6.50c = 681$$

$$-9.50c + \boxed{} + 6.50c = 681$$

$$\boxed{}c + 855 = 681$$

$$-3c = -174$$

$$c = \boxed{}$$

Substitute 58 for c in equation (1) and solve for a.

$$a + c = 90$$

$$a + 58 = 90$$

$$a = \boxed{}$$

4. **Check.** Number of tickets sold:
32 adults + 58 children = 90 tickets

Cost of tickets: $\$9.50(32) + \$6.50(58) = \$681$

The numbers check.

5. **State.** $\boxed{}$ adults' tickets and $\boxed{}$ children's tickets were sold.

EXAMPLE 2	YOUR TURN 2
Henry wants kelly green paint (18% green pigment). The paint store owner can mix lime green (14.5% pigment) and hunter green (20% pigment) to create kelly green. How much of each color should be mixed to create a 22 gal batch of kelly green?	A chemistry professor has a solution that is 60% base and another solution that is 20% base. How much of each should be used to make 120 L of solution that is 52% base?

1. **Familiarize.** Let l = the number of gallons of lime green paint and h = the number of gallons of hunter green paint.

2. **Translate.** Henry needs 22 gal of paint, so

    $l + h = \boxed{}$ ← Total amount of paint

    To determine the total amount of pigment, we have:

    $0.145l + 0.2h = 0.18 \cdot 22$

    The system of equations is

    $$l + h = 22, \quad (1)$$
    $$0.145l + 0.2h = 3.96. \quad (2)$$

3. **Solve.** When equation (1) is solved for l, we have

    $l = \boxed{} + 22$. We then substitute $-h + 22$ for l in equation (2).

    $$0.145l + 0.2h = 3.96$$
    $$0.145(-h + 22) + 0.2h = 3.96$$
    $$-0.145h + \boxed{} + 0.2h = 3.96$$
    $$\boxed{}h + 3.19 = 3.96$$
    $$0.055h = 0.77$$
    $$h = \boxed{}$$

    If $h = 14$, then $l = -14 + 22 = 8$.

4. **Check.**

    Amount of paint: 8 gal + 14 gal = 22 gal

    Amount of pigment: $0.145 \cdot 8 + 0.2 \cdot 14 = 1.16 + 2.8 = 3.96$ gal

    The answer checks.

5. **State.** Henry needs $\boxed{}$ gal of lime green and $\boxed{}$ gal of hunter green to make 22 gal of kelly green paint.

YOUR NOTES   Write your questions and additional notes.

Motion Problems

ESSENTIALS

Distance, Rate, and Time Equation

If r represents rate, t represents time, and d represents distance, then

$$d = rt.$$

Example

- A small plane flies 2 hr west with a 50-mph tailwind. Returning against the wind takes 3 hr. Find the speed of the plane with no wind.

  1. **Familiarize.** The plane travels the same distance each way. Let d = the distance traveled, in miles. Let r = the speed, in mph, of the plane in still air. Then $r + 50$ = the plane's speed with the wind and $r - 50$ = the plane's speed against the wind.

  2. **Translate.** Construct a table.

	Distance	Rate	Time
With wind	d	$r + 50$	2
Against wind	d	$r - 50$	3

     Using $d = rt$, we have a system of equations.

     $$d = (r + 50)2, \quad (1)$$
     $$d = (r - 50)3 \quad (2)$$

  3. **Solve.** Solve using substitution:

     $$(r + 50)2 = (r - 50)3 \qquad \text{Substituting } (r + 50)2 \text{ for } d \text{ in equation (2)}$$
     $$2r + 100 = 3r - 150$$
     $$-r + 100 = -150$$
     $$-r = -250$$
     $$r = 250.$$

  4. **Check.** When $r = 250$, the speed with the wind is $250 + 50 = 300$ mph and the speed against the wind is $250 - 50 = 200$ mph. The distance with the wind is $300 \cdot 2 = 600$ mi and the distance against the wind is $200 \cdot 3 = 600$ mi. The distances are the same, so the answer checks.

  5. **State.** The speed of the plane with no wind is 250 mph.

GUIDED LEARNING   **Textbook**   **Instructor**   ▶ **Video**

EXAMPLE 1	YOUR TURN 1
A passenger train leaves Hartford, heading to Atlanta, at a speed of 50 mph. Three hours later a freight train leaves Hartford on a parallel track at 70 mph. How far from Hartford will the freight train catch up to the passenger train?	Mary and Susan are training for a triathlon. Mary begins a bike ride at Marker 0 on a bike trail and rides at 6 mph. Two hours later Susan starts at Marker 0 and bikes at 10 mph. How far from Marker 0 will Susan catch up to Mary?

1. **Familiarize.** The distance traveled when the trains meet is the same. Let d = this distance. Let t = the number of hours the passenger train is running before they meet. Then $t-3$ = the number of hours the freight train runs before catching up to the passenger train.

2. **Translate.** Construct a table.

	Distance	Rate	Time
Passenger train	d	50	t
Freight train	d	70	$t-3$

Using $d = rt$, we have a system of equations.

$$d = 50t \qquad (1)$$
$$d = 70(t-3) \quad (2)$$

3. **Solve.** Solve using substitution:

$$50t = 70(t-3)$$
$$50t = 70t - \boxed{}$$
$$\boxed{} = -210$$
$$t = \boxed{}$$

Time for the passenger train: 10.5 hr

Time for the freight train: $10.5 - \boxed{} = 7.5$ hr

4. **Check.** Distance passenger train travels at 50 mph: $50(10.5) = 525$ mi. Distance freight train travels at 70 mph: $70(7.5) = 525$ mi. The numbers check.

5. **State.** The freight train catches up to the passenger train $\boxed{}$ mi from Hartford.

YOUR NOTES   Write your questions and additional notes.

Practice Exercises

Readiness Check

Match each statement with the most appropriate translation.

1. Alice is mixing a 2 lb batch of two kinds of nuts.

2. Tom spent $40 on ice cream cones costing $2 each and milkshakes costing $3 each.

3. A mixture that is 20% acid and another that is 30% acid contains 40 L of acid.

4. Jay spent $20 on candy bars costing $1.25 each and granola bars costing $1.75 each.

a) $2x + 3y = 40$

b) $x + y = 2$

c) $1.25x + 1.75y = 20$

d) $0.2x + 0.3y = 40$

Total-Value Problems and Mixture Problems

Solve. Use the five steps for problem solving.

5. A single day lift ticket at a ski mountain is $75 for adults and $35 for children. A busload of 27 people paid $1305 for lift tickets one day. How many adults and how many children were on the bus?

6. While driving to New York City, Adam stopped and bought gas for $3.89 per gallon and then stopped again and paid $3.49 per gallon. If he purchased a total of 25 gal for $92.85, how many gallons did he buy at each price?

7. A chemist needs to prepare 15 L of a 30% acid solution. How many liters of 20% acid solution and how many liters of 35% acid solution should be mixed to obtain the 30% acid solution?

8. John plans to mix fruit punch costing $1.99 per liter with sparkling water costing $2.19 per liter to make 35 L of punch for a Halloween party. If the punch costs $2.05 per liter, how many liters of each will he use?

Motion Problems

9. Jason drives his boat for 3 hr with a 5-km/h current to reach a campsite. The return trip against the same current took 6 hr. Find the speed of the boat in still water.

10. A small plane leaves Denver flying east at 200 mph. Two hours later, a large jet leaves Denver flying east at 520 mph. How far will the planes be from Denver when the large jet catches up with the small plane?

Solving Systems in Three Variables

ESSENTIALS

Example

- Solve the system of equations:

$$3x - 2y + 6z = 1, \quad (1)$$
$$7x + 3y + z = 4, \quad (2)$$
$$x + y - z = 0. \quad (3)$$

First add equations (2) and (3).

$$
\begin{array}{ll}
7x + 3y + z = 4 & (2) \\
\underline{x + y - z = 0} & (3) \\
8x + 4y \quad\;\; = 4 & (4)
\end{array}
$$

Select a different pair of equations and eliminate z again. Use equations (1) and (3).

$$
\begin{array}{lll}
3x - 2y + 6z = 1 & (1) & \qquad 3x - 2y + 6z = 1 \\
x + y - z = 0 & (3) \quad \text{Multiplying both sides by } 6 \rightarrow & \qquad \underline{6x + 6y - 6z = 0} \\
& & \qquad 9x + 4y \quad\;\; = 1 \quad (5)
\end{array}
$$

Solve the resulting system of equations (4) and (5). Eliminate y first.

$$
\begin{array}{lll}
8x + 4y = 4 & (4) \quad \text{Multiplying both sides by } -1 \rightarrow & \quad -8x - 4y = -4 \\
9x + 4y = 1 & (5) & \quad \underline{9x + 4y = 1} \\
& & \quad x \quad\quad = -3
\end{array}
$$

We will use equation (4) to find y.

$$
\begin{array}{ll}
8x + 4y = 4 & (4) \\
8(-3) + 4y = 4 & \text{Substituting } -3 \text{ for } x \\
-24 + 4y = 4 & \\
4y = 28 & \\
y = 7 &
\end{array}
$$

Use any of the three original equations to find z. We will use equation (3).

$$
\begin{array}{ll}
x + y - z = 0 & (3) \\
-3 + 7 - z = 0 & \text{Substitute } -3 \text{ for } x \text{ and } 7 \text{ for } y \\
4 - z = 0 & \\
4 = z &
\end{array}
$$

The triple $(-3, 7, 4)$ checks in all three equations. It is the solution.

GUIDED LEARNING	🔘 **Textbook** 👤 **Instructor**	▶ **Video**

EXAMPLE 1	YOUR TURN 1

EXAMPLE 1

Solve the system of equations:

$$5a + 3b = 4, \quad (1)$$
$$3b - 4c = 4, \quad (2)$$
$$2a + 2c = 2. \quad (3)$$

Since c is eliminated already from equation 1, eliminate c from equations (2) and (3).

$$3b - 4c = 4 \qquad\qquad\qquad\qquad 3b - 4c = 4$$

$2a + 2c = 2$ Multiplying both sides by $2 \rightarrow 4a + 4c = 4$

$$4a + 3b = \boxed{} \quad (4)$$

Solve the resulting system of equations (1) and (4).

$5a + 3b = 4 \qquad (1)$ Multiplying both sides by $-1 \rightarrow \boxed{} - 3b = -4$

$4a + 3b = \boxed{} \qquad (4) \qquad\qquad\qquad\qquad 4a + 3b = 8$

$$-a = 4$$
$$a = -4$$

Use equation (4) to solve for b.

$$4a + 3b = 8 \qquad (4)$$
$$4(-4) + 3b = 8 \qquad \text{Substituting } -4 \text{ for } a$$
$$\boxed{} + 3b = 8$$
$$3b = 24$$
$$b = \boxed{}$$

Use equation (2) or (3) to solve for c.

$$2a + 2c = 2 \qquad (3)$$
$$2(-4) + 2c = 2 \qquad \text{Substituting } -4 \text{ for } a$$
$$\boxed{} + 2c = 2$$
$$2c = 10$$
$$c = \boxed{}$$

The solution is $\boxed{}$.

YOUR TURN 1

Solve the system of equations:

$$3a + 4c = -2,$$
$$b - 2c = 2,$$
$$a + 2b - c = 5.$$

YOUR NOTES   Write your questions and additional notes.

Practice Exercises

Readiness Check

Each sentence below refers to the following system of equations. Choose the word that best completes each sentence.

$$5x + 3y - z = -1,$$
$$2x + y + 3z = 0,$$
$$x - 2y - 5z = 5$$

1. The system shown is a system of _____ equations in _____ variables.
    two / three   two / three

2. Each equation in the system is a _____ equation in three variables.
    linear / nonlinear

3. The ordered _____ $(1, -2, 0)$ is a solution of the system because it makes
    pair / triple

    _____ true.
    all three equations / at least one equation

Solving Systems In Three Variables

Solve.

4. $x + 2y + 2z = 6,$
    $2x - y + 2z = 11,$
    $4x + 5y + z = 6$

5. $x + 3y - 2z = 0,$
    $-x - y + 3z = -1,$
    $2x + 5y + z = -5$

6.
$$2a - b + c = 0,$$
$$4a + b + c = 7,$$
$$-2a + 2b - c = 3$$

7.
$$2w - x - y = 13,$$
$$w + x + 3y = 5,$$
$$w + 4x + 9y = 11$$

8.
$$x - 6y \quad = 1,$$
$$3x \quad + 2z = 11,$$
$$y - 7z = 4$$

9.
$$p + r = 7,$$
$$p - 2s = 12,$$
$$2r + 7s = -1$$

Using Systems of Three Equations

ESSENTIALS

Example

- Owen invested $15,000 in three mutual funds. His total earnings in interest for the year were $800. He invested $1000 more in Fund B than in Fund A, and he invested the rest in Fund C. The interest rates for Fund A, Fund B, and Fund C are 3.4%, 5%, and 6% respectively.

1. **Familiarize.** Let x = the amount invested in Fund A.

    Let y = the amount invested in Fund B.

    Let z = the amount invested in Fund C.

2. **Translate.** Owen has $15,000 to invest, so

    $x + y + z = 15,000.$

    The interest earned is $800. The interest equals the rate times the amount invested, so

    $0.034x + 0.05y + 0.06z = 800.$

    There is $1000 more in Fund B than in Fund A, so

    $y = x + 1000.$

    We have a system of three equations:

    $$x + y + z = 15,000,$$
    $$0.034x + 0.05y + 0.06z = 800,$$
    $$y = x + 1000.$$

    Rewriting in standard form and clearing decimals, we have:

    $$\begin{aligned} x + \quad y + \quad z &= \ 15,000, \quad (1) \\ 34x + 50y + 60z &= 800,000, \quad (2) \\ -x + \quad y \qquad\quad &= \quad 1000. \quad (3) \end{aligned}$$

3. **Solve.** Equation (3) does not have a z-term. We use equations (1) and (2) to write another equation with no z-term. We multiply equation (1) by -60 and then add the equations.

    $$\begin{aligned} -60x - 60y - 60z &= -900,000 \\ 34x + 50y + 60z &= \ \ 800,000 \\ \hline -26x - 10y \qquad\quad &= -100,000 \quad (4) \end{aligned}$$

Solve the system of equations (3) and (4). Multiply equation (3) by 10 and then add.

$$
\begin{aligned}
-10x + 10y &= 10{,}000 \\
\underline{-26x - 10y} &= \underline{-100{,}000} \\
-36x &= -90{,}000 \\
x &= 2500
\end{aligned}
$$

Use equation (3) to find y.

$$
\begin{aligned}
-x + y &= 1000 \\
-2500 + y &= 1000 \\
y &= 3500
\end{aligned}
$$

Use equation (1) to find z.

$$
\begin{aligned}
x + y + z &= 15{,}000 \\
2500 + 3500 + z &= 15{,}000 \\
6000 + z &= 15{,}000 \\
z &= 9000
\end{aligned}
$$

4. **Check.** Amount of money invested: $\$2500 + \$3500 + \$9000 = \$15{,}000$

The interest earned: $0.034(\$2500) + 0.05(\$3500) + 0.06(\$9000)$
$$= \$85 + \$175 + \$540 = \$800$$

The amount in Fund B, $3500, is $1000 more than the amount in Fund A, $2500.
The answer checks.

5. **State.** Owen invested $2500 in Fund A, $3500 in Fund B, and $9000 in Fund C.

GUIDED LEARNING   **Textbook**   **Instructor**   **Video**

EXAMPLE 1	YOUR TURN 1
The perimeter of a triangle is 63 in. The longest side is 22 in. longer than the shortest side. The longest side is twice as long as the remaining side. Find the lengths of all three sides.	A museum charges $14 admission for adults, $10 for children under 12, and $8 for seniors over 65. During one day, the museum sold 444 tickets and took in $5532. If twice as many adult tickets were sold as the total of children and senior tickets, how many of each type of ticket was sold?

1. **Familiarize.** Let $x =$ the shortest side, $y =$ the middle side, and $z =$ the longest side.

2. **Translate.** The perimeter is 63 in., so we have $x + y + z = 63$.

   The longest side is 22 in. longer than the shortest side, so

   $$z = x + 22.$$

   The longest side is twice as long as the remaining side, so

   $$z = 2y.$$

   We have a system of three equations:

   $$x + y + z = 63,$$
   $$z = x + 22,$$
   $$z = 2y.$$

   Rewriting in standard form, we have:

   $$x + y + z = 63, \quad (1)$$
   $$-x + z = 22, \quad (2)$$
   $$-2y + z = 0. \quad (3)$$

3. **Solve.** Eliminate x by adding equations (1) and (2).

   $$x + y + z = 63$$
   $$\underline{-x + z = 22}$$
   $$y + \boxed{} = 85 \quad (4)$$

   Solve the resulting system of equations (3) and (4).

   $$-2y + z = 0 \qquad\qquad -2y + z = 0$$
   $$y + 2z = 85 \quad \text{Multiplying both sides by 2} \rightarrow \quad \underline{2y + 4z = \boxed{}}$$
   $$5z = 170$$
   $$z = \boxed{}$$

   (continued)

Use equation (4) to find y.

$$y + 2z = 85$$

$$y + 2(34) = 85 \quad \text{Substituting 34 for } z$$

$$y + \boxed{} = 85$$

$$y = 17$$

Use equation (1) to find x.

$$x + y + z = 63$$

$$x + 17 + 34 = 63 \quad \text{Substituting 17 for } y \text{ and 34 for } z$$

$$x + \boxed{} = 63$$

$$x = \boxed{}$$

4. **Check.** Perimeter: $12 \text{ in.} + 17 \text{ in.} + 34 \text{ in.} = 63 \text{ in.}$

The longest side, 34 in., is 22 in. longer than the shortest side, 12 in. The longest side is also twice as long as the remaining side: $2 \cdot 17 \text{ in.} = 34 \text{ in.}$

The answer checks.

5. **State.** The lengths of the sides are $\boxed{}$, $\boxed{}$, and $\boxed{}$.

YOUR NOTES   Write your questions and additional notes.

Practice Exercises

Readiness Check

Match each statement with a translation from the list.

1. The sum of the three numbers is 25.

2. The first number is 25 more than the sum of the other two numbers.

3. The first number minus the second number plus the third is 25.

4. The first number is 25 less than the sum of the other two numbers.

    a) $x - y - z = -25$

    b) $x + y + z = 25$

    c) $x - y + z = 25$

    d) $x - y - z = 25$

Using Systems of Three Equations

Solve.

5. The sum of three numbers is 144. The second number is twice the first number. The third number is three times the first. Find the numbers.

6. A college basketball team scored 84 points in a game. The team scored a combination of 2-point field goals, 3-point field goals, and 1-point foul shots. The number of 2-point field goals was 6 more than the number of foul shots and four times the number of 3-point field goals. How many of each type of shot was made?

7. Mary bought a total of 10 gift cards in denominations of $20, $50, and $100. She spent $370. She bought twice as many $20 gift cards as $50 gift cards. How many of each denomination of gift card did she buy?

8. The sum of three numbers is 40. The second number is four times the first number. The sum of the first number and third number is 16. Find the numbers.

Graphs of Linear Inequalities

ESSENTIALS

A **linear inequality in two variables** is an inequality that can be written in the form

$$Ax + By < C,$$

where A, B, and C are real numbers and A and B are not both zero. The symbol $<$ may be replaced with $\leq$, $>$, or $\geq$.

The **solution set** of a linear inequality is the set of all ordered pairs that make it true. A **graph** of an inequality represents its solutions.

The graph of a **linear inequality in two variables** is a region on one side of a line and may include the boundary line. This region is a called a **half-plane**.

To graph an inequality in two variables:

1. Replace the inequality symbol with an equals sign and graph this related equation. If the inequality symbol is $<$ or $>$, draw the line dashed. If the inequality symbol is $\leq$ or $\geq$, draw the line solid.

2. The graph consists of a half-plane on one side of the line and, if the line is solid, the line as well. To determine which half-plane to shade, test a point not on the line in the original inequality. If that point is a solution, shade the half-plane containing that point. If not, shade the opposite half-plane.

Example

- The solutions of the linear inequality $y > 2x + 3$ are given by all of the ordered pairs in the shaded region of the following graph. Note that any points on the dashed line itself are not solutions.

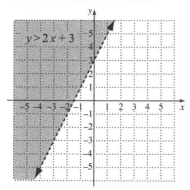

EXAMPLE 1	YOUR TURN 1
Graph $y \geq \dfrac{1}{2}x - 1$.	Graph $y \leq \dfrac{2}{3}x + 1$.

Since the inequality symbol is "greater than or equal to," graph the related equation

$y = \dfrac{1}{2}x - 1$ using a _____ line.

   solid / dashed

Next, we test a point in either region. We choose $(0, 0)$.

$$y \geq \dfrac{1}{2}x - 1$$

$$0 \ ? \ \dfrac{1}{2} \cdot \boxed{} - 1$$

$$\boxed{} - 1$$

$$\boxed{}$$

Since $0 \geq \boxed{}$ is true, the point $(0, 0)$

_____ a solution and we shade the half-

is / is not

plane that _____ contain $(0, 0)$.

           does / does not

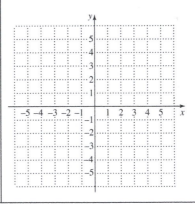

$y \geq \dfrac{1}{2}x - 1$

EXAMPLE 2	YOUR TURN 2
Graph $3x - y > 2$.	Graph $3x - 2y < 6$.

Since the inequality symbol is "greater than," graph the related equation $3x - y = 2$ using a _____ line.
solid / dashed

Next, we test the point $(0, 0)$.

$$3x - y > 2$$

$3 \cdot \boxed{} - 0 \; ? \; 2$

$\boxed{} - 0$

$\boxed{}$

Since $\boxed{} > 2$ is _____, the point
true / false

$(0, 0)$ is not a solution and we shade the

half-plane that _____ contain
does / does not

$(0, 0)$.

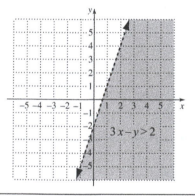

$3x - y > 2$

EXAMPLE 3	YOUR TURN 3
Graph $y < -2$ on a plane.	Graph $x \geq 2$ on a plane.

Since the inequality symbol is "less than," graph the related equation $y = -2$ using a

_____ line.
solid / dashed

Next, we test the point $(0, 0)$.

$$\frac{y < -2}{\Box \,?\, -2}$$

Since $\Box < -2$ is _____, the point
                        true / false

$(0, 0)$ _____ a solution so we shade the
         is / is not

half-plane that _____ contain
                does / does not

$(0, 0)$.

YOUR NOTES   Write your questions and additional notes.

Systems of Linear Inequalities

ESSENTIALS

A **system of linear inequalities** consists of two or more linear inequalities.

A **solution** of a system of linear inequalities is any ordered pair that is a solution of *each* of the individual inequalities.

In order to **graph** the solution set of a system of linear inequalities we graph each inequality and determine the region that is common to all the solution sets. We draw small arrows on the ends of each line indicating which half-plane contains the solution set.

A system of linear inequalities may have a graph that consists of a polygon and its interior. In many applications it is important to find the vertices of the polygon. This requires solving systems of equations.

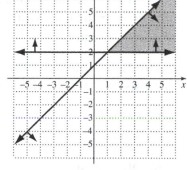

Example

- The solutions of the system

$$y \le x+1,$$

$$y \ge 2$$

are represented by the shaded region to the right. The vertex is the point of intersection of the two lines. Its coordinates are $(1, 2)$.

GUIDED LEARNING   **Textbook**   **Instructor**   **Video**

EXAMPLE 1	YOUR TURN 1
Graph the solution set of the system. Find the coordinates of any vertices formed. $x+3y \le 9,$ $2x-3y \le 0,$ $x \ge -3$ Graph $x+3y=9$ using a solid line. Test point $(0,0)$ is a solution of $x+3y \le 9$, so indicate (with arrows near the ends of the line) the half-plane that _____ does / does not contain $(0, 0)$.	Graph the system. Find the coordinates of any vertices formed. $x-3y \ge 0,$ $x-y \le 2,$ $y \ge -1$ 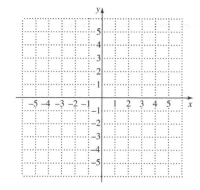

(continued)

Graph $2x - 3y = 0$ using a solid line. Test point $(1, 2)$ _____ a solution of

is / is not

$2x - 3y \leq 0$, so indicate (with arrows near the ends of the line) the half-plane that contains $(1, 2)$.

Graph $x = -3$ using a solid line. Test point $(0, 0)$ _____ a solution of $x \geq -3$, so

is / is not

indicate (with arrows near the ends of the line) the half-plane that contains $(0, 0)$.

The region of solutions is the triangular region, including segments of the boundary lines, where the half-planes indicated by the arrows near the ends of the lines overlap. Shade this region.

To find the vertices, solve the following systems of equations formed by these lines:

$$x + 3y = 9, \qquad x + 3y = 9, \qquad 2x - 3y = 0,$$
$$2x - 3y = 0; \qquad\quad x = -3; \qquad\quad x = -3.$$

The solutions are

$(3, 2)$, $(-3, 4)$, and $(-3, -2)$.

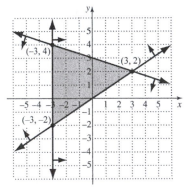

Applications: Linear Programming

ESSENTIALS

Linear programming allows us to find the maximum or minimum value of a linear **objective function** subject to several **constraints** which are expressed as inequalities. The solution set of the system of inequalities composed of the constraints contains all the **feasible solutions**.

It can be shown that the maximum and minimum values of the objective function occur at a vertex of the region of feasible solutions.

Linear Programming Procedure

To find the maximum or minimum value of a linear objective function, subject to constraints:

1. Graph the region of feasible solutions.
2. Determine the coordinates of the vertices of the region.
3. Evaluate the objective function at each vertex. The largest and smallest of those values are the maximum and minimum values of the objective function, respectively.

GUIDED LEARNING	🔖 **Textbook**	👤 **Instructor**	▶ **Video**

EXAMPLE 1	YOUR TURN 1
Aspen Carpentry makes bookcases and desks. Each bookcase requires 5 hr of woodworking and 4 hr of finishing. Each desk requires 10 hr of woodworking and 3 hr of finishing. Each month the shop has 600 hr of labor available for woodworking and 240 hr for finishing. The profit on each bookcase is $75 and on each desk is $140. How many of each product should be made each month in order to maximize profit? What is the maximum profit?	Fashions by Felicity makes tuxedos and prom dresses and can produce at most 200 of these garments per month. It takes 30 hr to make a tuxedo and 10 hr to make a dress. There is at most 3000 hr of labor available each month. The profit on a tuxedo is $80 and is $50 on a dress. How many of each should be made each month in order to maximize the profit? What is the maximum profit?

Let x = the number of bookcases to be produced and sold and let y = the number of desks to be produced and sold.

The profit is given by $P = \boxed{} \cdot x + \boxed{} \cdot y$.

Next we consider the constraints. The total number of hours of woodworking on bookcases and desks must satisfy $5x + 10y \le \boxed{}$.

The total number of hours of finishing on bookcases and desks must satisfy $4x + \boxed{} y \le 240$.

(continued)

Finally, we have $x \geq 0$ and $y \geq 0$.

Maximize

$$P = 75x + \boxed{}y$$

subject to

$$5x + \boxed{}y \leq 600,$$

$$4x + 3y \leq \boxed{},$$

$$x \geq 0,$$

$$y \geq 0.$$

We graph the system of inequalities and find the vertices.

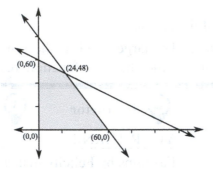

The maximum value will occur at a vertex, so we check the value of P at each one.

Vertex (x, y)	Profit $P = \boxed{}x + \boxed{}y$
$(0, 0)$	$75 \cdot \boxed{} + 140 \cdot \boxed{} = \boxed{}$
$(60, 0)$	$75 \cdot \boxed{} + 140 \cdot 0 = \boxed{}$
$(24, 48)$	$75 \cdot 24 + 140 \cdot \boxed{} = 8520$
$(0, 60)$	$75 \cdot \boxed{} + 140 \cdot \boxed{} = 8400$

We see that the maximum profit of $\boxed{}$ occurs when 24 bookcases and $\boxed{}$ desks are produced and sold.

YOUR NOTES Write your questions and additional notes.

Practice Exercises

Readiness Check

Complete the following sentences.

1. If an ordered pair (x, y) is substituted in an inequality and results in a false statement, then that ordered pair _____ a solution of the inequality.
    is / is not

2. When graphing a linear inequality that contains a > or < symbol, we first graph the related equation using a _____ line.
    solid / dashed

3. When graphing a linear inequality, if the test point (x, y) is a solution, then we shade the half-plane that is on the _____ side of the line as (x, y).
    same / opposite

4. A linear inequality in two variables is graphed on _____.
    a plane / the number line

5. When graphing a system of linear inequalities, we shade the _____ of
    union / intersection
    solution sets of the individual inequalities.

Graphs of Linear Inequalities

Graph each linear inequality on a plane.

6. $y > -\dfrac{1}{3}x + 2$

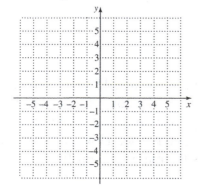

7. $-3x + 2y \le 4$

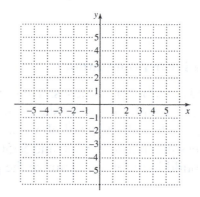

Systems of Linear Inequalities

Graph each system. Find the coordinates of any vertices formed.

8. $y \geq -x - 2,$
$y \leq 1$

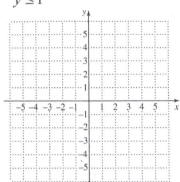

9. $x + y \geq 3,$
$x - y \leq 1$

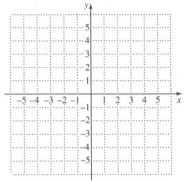

10. $x + y \leq 1,$
$x + 3y \geq -3,$
$-x + y \leq 3$

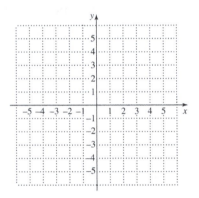

Applications: Linear Programming

11. Corey builds outdoor playhouses. He uses 10 sheets of plywood and 15 two-by-fours for a princess playhouse and 15 sheets of plywood and 45 two-by-fours for each pirate playhouse. He has 60 sheets of plywood and 135 two-by-fours available. The profit on each princess playhouse is $200 and $350 on each pirate playhouse. How many of each should he build in order to maximize the profit? What is the maximum profit?

Polynomial Expressions

ESSENTIALS

A **monomial** is a constant or a constant times some variable or variables raised to powers that are nonnegative integers.

A **polynomial** is a monomial or a combination of sums and/or differences of monomials.
Examples of **polynomials in one variable**: $4a^3$, $2y+1$, x^2+5x-6

Examples of **polynomials in several variables**: $2x^7y^2$, $2x+3y$, $5x+6y^2-4z^3$

The **terms** of a polynomial are separated by + signs.

Names for certain types of polynomials:

Type	Definition: Polynomial of	Examples
Monomial	One term	3, $2x$, $-4y^2$, $7a^4b^8$, $-2x^2y^5z^3$
Binomial	Two terms	$3x-5$, a^2-b^2, $6x^2-7x$, x^2y+4xy^3
Trinomial	Three terms	x^2+6x+9, $2a^2+7ab+3b^2$, $x^2y+3xy-4xy^2$

The **coefficient** is the part of a term that is a constant factor. A **constant term** is a term that contains only a number and no variable.

The **degree of a term** is the sum of the exponents of the variables, if there are variables. The degree of a nonzero constant term is 0. The term 0 has no degree.

The **degree of a polynomial** is the same as the degree of its term of highest degree.

The **leading term** of a polynomial is the term of highest degree. Its coefficient is called the **leading coefficient**.

Descending order: arrangement of terms so that exponents decrease from left to right.

Ascending order: arrangement of terms so that exponents increase from left to right.

Example

- Identify the terms, the degree of each term, and the degree of the polynomial $-6x+4x^3+5-3x^2$. Then identify the leading term, the leading coefficient, and the constant term. Finally, write the polynomial in both descending and ascending order.

Term	$-6x$	$4x^3$	5	$-3x^2$
Degree of term	1	3	0	2
Degree of polynomial		3		
Leading term		$4x^3$		
Leading coefficient		4		
Constant term		5		

Descending order:
$$4x^3-3x^2-6x+5$$

Ascending order:
$$5-6x-3x^2+4x^3$$

GUIDED LEARNING   🔖 **Textbook**   👤 **Instructor**   ▶ **Video**

EXAMPLE 1	YOUR TURN 1
Identify the terms, the degree of each term, and the degree of the polynomial $-6y^5 - 2y - 4y^3 + 12y^9 + 10$. Then identify the leading term, the leading coefficient, and the constant term.	Identify the terms, the degree of each term, and the degree of the polynomial $12y^4 - 7y^7 - 6y^3 - 8$. Then identify the leading term, the leading coefficient, and the constant term.

Term	$-6y^5$	$-2y$	$-4y^3$	$12y^9$	10
Degree of term	☐	☐	3	☐	0
Degree of polynomial			☐		
Leading term			$12y^9$		
Leading coefficient			☐		
Constant term			☐		

Term					
Degree of term					
Degree of polynomial					
Leading term					
Leading coefficient					
Constant term					

EXAMPLE 2	YOUR TURN 2
Consider $-6y^5 + 12y^9 - 4y^3 - 2y + 10$. Arrange in descending order and then in ascending order. Descending order: ☐$-6y^5-$☐$-2y+$☐ Ascending order: $10-$☐$-4y^3-$☐$+$☐	Consider $12y^4 - 7y^7 - 6y^3 - 8$. Arrange in descending order and then in ascending order. Descending order: Ascending order:

YOUR NOTES   Write your questions and additional notes.

Evaluating Polynomial Functions

ESSENTIALS

To **evaluate** a polynomial function, substitute a number for the variable.

Example

- For the polynomial function $P(x) = -2x^4 + 3x - 1$, find $P(4)$.

$$P(x) = -2x^4 + 3x - 1$$

$$P(4) = -2(4)^4 + 3(4) - 1$$

$$= -2(256) + 12 - 1$$

$$= -512 + 12 - 1$$

$$= -501$$

	📖 **Textbook**	👤 **Instructor**	▶ **Video**

GUIDED LEARNING

EXAMPLE 1	YOUR TURN 1
For the polynomial function $f(x) = 3x^2 - 5x + 4$, find $f(-1)$. $$f(x) = 3x^2 - 5x + 4$$ $$f(-1) = 3(-1)^2 - 5(-1) + 4$$ $$= 3(1) - 5(-1) + 4$$ $$= \boxed{} + 5 + 4 = \boxed{}$$	For the polynomial function $f(x) = -6x^3 - 4x^2 + 2x - 10$, find $f(2)$.
EXAMPLE 2	YOUR TURN 2
When an object is launched upward with an initial velocity of 25 m/sec and from the top of a bridge that is 30 m high, its altitude, in meters, after t seconds is given by the polynomial function $A(t) = 30 + 25t - 4.9t^2$. Find the altitude of the object after 3 seconds. We substitute 3 for t. $$A(t) = 30 + 25t - 4.9t^2$$ $$A(3) = 30 + 25(3) - 4.9(3)^2$$ $$= 30 + 25(3) - 4.9(9)$$ $$= 30 + 75 - \boxed{}$$ $$= \boxed{} \text{ m}$$	The polynomial function $$F(x) = -0.00016x^3 + 0.0096x^2 + 0.367x + 10.8$$ can be used to estimate the fuel economy, in miles per gallon (mpg), for a particular vehicle traveling x miles per hour (mph). What is the gas mileage for this vehicle at 40 mph?

YOUR NOTES   Write your questions and additional notes.

Adding Polynomials

ESSENTIALS

Similar, or **like**, **terms** are terms that have the same variable(s) raised to the same power(s).

Polynomials are added by collecting, or combining, like terms.

Example

- Add: $\left(-4x^3 + 5x + 3\right) + \left(6x^3 - 2x^2 - 7\right)$.

$$\left(-4x^3 + 5x + 3\right) + \left(6x^3 - 2x^2 - 7\right) = (-4+6)x^3 - 2x^2 + 5x + (3-7) = 2x^3 - 2x^2 + 5x - 4$$

GUIDED LEARNING   📍 **Textbook**   👤 **Instructor**   ▶ **Video**

EXAMPLE 1	YOUR TURN 1
Collect like terms: $2x^2 + 12x - 11 - 3x^2 - 15x + 6$. $2x^2 + 12x - 11 - 3x^2 - 15x + 6$ $= (2-3)x^2 + (12-15)x + (-11+6)$ $= \boxed{} - 3x - \boxed{}$	Collect like terms: $-4x^2 + 3x - 1 - 8x^2 - 10x + 6$.
EXAMPLE 2	YOUR TURN 2
Add: $\left(4x^3 - 2xy + y^3\right) + \left(5x^3 + 6xy - 5y^3\right)$. $\left(4x^3 - 2xy + y^3\right) + \left(5x^3 + 6xy - 5y^3\right)$ $= (4+5)x^3 + (-2+6)xy + (1-5)y^3$ $= 9x^3 + \boxed{} - \boxed{}$	Add: $\left(12x^2y^3 + 3x^2y - 2x^4y\right) +$ $\qquad\left(5x^2y^3 - 10x^2y - x^4y\right)$.
EXAMPLE 3	YOUR TURN 3
Add using columns: $\left(5y^3 + 6y - 8\right) + \left(-2y^3 + 4y^2 - 4\right)$. In order to use columns to add, we write the polynomials one under the other, listing like terms under one another and leaving spaces for missing terms. $\begin{array}{r} 5y^3 \qquad\quad +6y - 8 \\ -2y^3 + 4y^2 \qquad\; - 4 \\ \hline 3y^3 + \boxed{} + 6y - \boxed{} \end{array}$	Add using columns: $\left(5y^4 + 2y^3 - 5\right) + \left(7y^4 - 3y + 10\right)$.

YOUR NOTES   Write your questions and additional notes.

Subtracting Polynomials

ESSENTIALS

If the sum of two polynomials is 0, the polynomials are **opposites**, or **additive inverses**.

The Opposite of a Polynomial

The *opposite* of a polynomial P can be written as $-P$ or, equivalently, by replacing each term in P with its opposite.

To subtract a polynomial, we add its opposite.

Examples

- Write two equivalent expressions for the opposite of $3x^3 - 2x^2 + 10x - 1$.

  The opposite can be written with parentheses as $-\left(3x^3 - 2x^2 + 10x - 1\right)$.

  The opposite can be written without parentheses by replacing each term by its opposite: $-3x^3 + 2x^2 - 10x + 1$.

- Subtract: $\left(-4x^2 + 3x - 1\right) - \left(10x^2 - 6x - 8\right)$.

  $$\left(-4x^2 + 3x - 1\right) - \left(10x^2 - 6x - 8\right) = \left(-4x^2 + 3x - 1\right) + \left(-10x^2 + 6x + 8\right)$$
  $$= -4x^2 + 3x - 1 - 10x^2 + 6x + 8$$
  $$= -14x^2 + 9x + 7$$

GUIDED LEARNING   📖 **Textbook**   👤 **Instructor**   ▶ **Video**

EXAMPLE 1	YOUR TURN 1
Write two equivalent expressions for the opposite of $3x^2 - 2xy + 4y - 12$.	Write two equivalent expressions for the opposite of $-6xy^2 + 5xy - 16y + 20$.
The opposite can be written with parentheses as $-\left(3x^2 - 2xy + 4y - 12\right)$.	
The opposite can be written without parentheses by replacing each term by its opposite: $-3x^2 + \boxed{} - \boxed{} + 12$.	

EXAMPLE 2	YOUR TURN 2
Subtract: $\left(-7x^3 + 5xy - 1\right) - \left(15xy + 7\right)$.	Subtract: $\left(6x^2 + 5x^3 y^3\right) - \left(8x^2 - 3x^3 y^3 + 11\right)$.
We add the opposite of the polynomial being subtracted.	
$\left(-7x^3 + 5xy - 1\right) - \left(15xy + 7\right)$	
$= \left(-7x^3 + 5xy - 1\right) + \left(-15xy - 7\right)$	
$= -7x^3 + 5xy - 1 - 15xy - 7$	
$= \boxed{} - 10xy - \boxed{}$	

EXAMPLE 3	YOUR TURN 3
Subtract: $\left(\dfrac{1}{4}x^2 - \dfrac{5}{8}x + \dfrac{1}{2}\right) - \left(\dfrac{3}{4}x^2 + \dfrac{3}{8}x - \dfrac{1}{2}\right).$	Subtract: $\left(\dfrac{7}{9}x^2 - \dfrac{2}{3}x + \dfrac{5}{6}\right) - \left(-\dfrac{2}{9}x^2 + \dfrac{1}{3}x - \dfrac{1}{6}\right).$

$$\left(\dfrac{1}{4}x^2 - \dfrac{5}{8}x + \dfrac{1}{2}\right) - \left(\dfrac{3}{4}x^2 + \dfrac{3}{8}x - \dfrac{1}{2}\right)$$

$$= \left(\dfrac{1}{4}x^2 - \dfrac{5}{8}x + \dfrac{1}{2}\right) + \left(-\dfrac{3}{4}x^2 - \dfrac{3}{8}x + \boxed{}\right)$$

$$= \dfrac{1}{4}x^2 - \dfrac{5}{8}x + \dfrac{1}{2} - \boxed{} - \dfrac{3}{8}x + \dfrac{1}{2}$$

$$= -\dfrac{2}{4}x^2 - \dfrac{8}{8}x + 1$$

$$= \boxed{} - x + \boxed{}$$

YOUR NOTES   Write your questions and additional notes.

Practice Exercises

Readiness Check

Choose the word, number, or expression that best completes each statement.

1. The expression $4x^2 + 2$ is a _____ with a leading coefficient of ____.
   binomial / trinomial   2 / 4

   The degree of the term 2 is ____.
                              0 / 2

2. The degree of the polynomial $-3xy^2 + 5xyz^2 - 10xy - 4$ is ____. The leading
                                                              2 / 4

   coefficient is ____. The polynomial has ____ terms.
                 5 / -3   4 / 3

3. The polynomial $-12 + 3x - 15x^2$ is written in _____ order.
                                            ascending / descending

4. To add or subtract polynomials, we add or subtract _____.
                                              like terms / exponents

Polynomial Expressions

Identify the terms, the degree of each term, and the degree of the polynomial. Then identify the leading term, the leading coefficient, and the constant term.

5. $-6x + x^3 + 8 + 3x^4$ 6. $5 - 12x^4 - 6x^7 + 3x^3 - 2x$ 7. $-3xy - 6xyz + 2x^2 - 7$

Arrange in descending powers of x.

8. $15 - 3x^4 + 2x^2 - x$

9. $4x - 3x^2 + 12x^6 - 8x^3 - 10$

Arrange in ascending powers of y.

10. $2y^2 + y^4 - 3y^3 + 9$

11. $3x^2y - 5xy^4 - 2x^3 + 6y^2$

Evaluating Polynomial Functions

Find $f(5)$ and $f(-1)$ for each polynomial function.

12. $f(x) = 4x^3 - 12x^2 + 20$

13. $f(x) = 4 - 3x^2 + 5x^3$

14. A firm determines that, when it sells x tablet computers, its total revenue, in dollars, is $R(x) = 280x - 0.4x^2$. What is the total revenue from the sale of 62 tablet computers?

Adding Polynomials

Collect like terms to write an equivalent expression.

15. $9x + 3 - 4x + 2x^3 - 4x - 7$

16. $-4x^2y^2 + 5x^3 - 9x^2y^2 - 15x^3$

Add.

17. $\left(8xyz - 3x^2y + 5xy\right) + \left(-2xyz + xy\right)$

18. $\left(3x^2 + 2x\right) + \left(-5x^2 - 8\right) + \left(4x^2 + 7\right)$

Subtracting Polynomials

Write two expressions, one with parentheses and one without, for the opposite of each polynomial.

19. $9x^4 - 3x^3 + 2x^2 - 5$

20. $-12xy - 7x^2y^2 + xyz - 6$

Subtract.

21. $\left(4x^2 + 5x - 2\right) - \left(2x^2 - 4x + 9\right)$

22. $\left(7a^3 - 4a^2 + 5\right) - \left(-3a^3 - a^2 + 17\right)$

23. $\left(6x^7yz - x^2y^2z^2 + 2x\right) - \left(3x - 2x^7yz\right)$

24. $\left(-2x^5 - 12\right) - \left(7xy - 15 + 13x^5\right)$

Multiplication of Any Two Polynomials

ESSENTIALS

Multiplying Monomials

To multiply monomials, multiply the coefficients and multiply the variables using the rules for exponents and the commutative and associative laws.

Multiplying Monomials and Binomials

The distributive law is used to multiply polynomials other than monomials.

Multiplying Binomials

To multiply binomials use the distributive law twice, first considering one of the binomials as a single expression and multiplying it by each term of the other binomial.

Multiplying Any Two Polynomials

To find the product of two polynomials P and Q, multiply each term of P by every term of Q and then collect like terms.

Examples

- Multiply: $\left(5x^3 y^4\right)\left(-3x^2 y^2\right)$.

$$\left(5x^3 y^4\right)\left(-3x^2 y^2\right) = 5 \cdot (-3) \cdot x^3 \cdot x^2 \cdot y^4 \cdot y^2$$
$$= -15x^{3+2} \cdot y^{4+2}$$
$$= -15x^5 y^6$$

- Multiply: $4x(5x-2)$.

$$4x(5x-2) = 4x \cdot 5x - 4x \cdot 2 \quad \text{Using the distributive law}$$
$$= 20x^2 - 8x \qquad \text{Multiplying monomials}$$

- Multiply: $\left(x^2 - 1\right)\left(2x^2 + 3\right)$.

$$\left(x^2 - 1\right)\left(2x^2 + 3\right) = \left(x^2 - 1\right)2x^2 + \left(x^2 - 1\right)3 \qquad \text{Using the distributive law}$$
$$= 2x^2\left(x^2 - 1\right) + 3\left(x^2 - 1\right) \qquad \text{Using a commutative law}$$
$$= 2x^2 \cdot x^2 - 2x^2 \cdot 1 + 3 \cdot x^2 - 3 \cdot 1 \quad \text{Using the distributive law twice}$$
$$= 2x^4 - 2x^2 + 3x^2 - 3 \qquad \text{Multiplying monomials}$$
$$= 2x^4 + x^2 - 3 \qquad \text{Collecting like terms}$$

- Multiply: $\left(x^3 - 4x^2 + 5\right)(x+3)$.

$$\left(x^3 - 4x^2 + 5\right)(x+3) = \left(x^3 - 4x^2 + 5\right)(x) + \left(x^3 - 4x^2 + 5\right)(3)$$
$$= x^3(x) - 4x^2(x) + 5(x) + x^3(3) - 4x^2(3) + 5(3)$$
$$= x^4 - 4x^3 + 5x + 3x^3 - 12x^2 + 15$$
$$= x^4 - x^3 - 12x^2 + 5x + 15$$

GUIDED LEARNING	📘 **Textbook** 👤 **Instructor** ▶ **Video**

EXAMPLE 1	YOUR TURN 1
Multiply and simpify: $\left(-7a^2bc^3\right)\left(3abc^5\right)$. $\left(-7a^2bc^3\right)\left(3abc^5\right) = -7\cdot 3\cdot a^2\cdot a\cdot b\cdot b\cdot c^3\cdot c^5$ $= -21\cdot a^{\square+\square}\cdot b^{1+1}\cdot c^{3+5}$ $= \boxed{}$	Multiply and simplify: $\left(12a^3b^5c\right)\left(-4a^4bc\right)$.

EXAMPLE 2	YOUR TURN 2
Multiply: $2a^2\left(4a^3-5a^4\right)$. $2a^2\left(4a^3-5a^4\right)=2a^2\cdot 4a^3-\boxed{}\cdot 5a^4$ $=8a^5-\boxed{}$	Multiply: $9a\left(4a^3-8\right)$.

EXAMPLE 3	YOUR TURN 3
Multiply: $\left(x^2+5\right)\left(x^3-2\right)$. $\left(x^2+5\right)\left(x^3-2\right)=\left(x^2+5\right)x^3+\left(x^2+5\right)(-2)$ $=x^3\left(x^2+5\right)-2\left(x^2+5\right)$ $=x^3\cdot x^2+x^3\cdot\boxed{}-2\cdot\boxed{}-2\cdot 5$ $=\boxed{}$	Multiply: $\left(x^4-1\right)\left(3x^5-6\right)$.

EXAMPLE 4	YOUR TURN 4
Multiply: $\left(3x^3+2x-2\right)\left(-2x^2+4x+5\right)$. $\begin{array}{r} 3x^3 \quad\quad +2x-2 \\ -2x^2+4x+5 \\ \hline 15x^3 \quad\quad +10x-\boxed{} \\ 12x^4 \quad\quad +\boxed{}-8x \\ \boxed{}x^5 \quad -4x^3+4x^2 \\ \hline \boxed{} \end{array}$	Multiply: $\left(2x^3-4x+6\right)\left(-5x^2+x+1\right)$.

YOUR NOTES   Write your questions and additional notes.

Product of Two Binomials Using the FOIL Method

ESSENTIALS

The FOIL Method

To multiply two binomials $(A+B)$ and $(C+D)$:

1. Multiply **F**irst terms: AC
2. Multiply **O**utside terms: AD
3. Multiply **I**nside terms: BC
4. Multiply **L**ast terms: BD

$$\text{FOIL}$$

$$(A+B)(C+D) = AC + AD + BC + BD$$

Example

- Multiply: $(x+6)(x-3)$.

$$(x+6)(x-3) = x^2 - 3x + 6x - 18 \quad \text{FOIL}$$
$$= x^2 + 3x - 18 \qquad \text{Collecting like terms}$$

GUIDED LEARNING　　　📘 **Textbook**　　　👤 **Instructor**　　　▶ **Video**

EXAMPLE 1	YOUR TURN 1
Multiply: $(2x+1)(-4x+5)$.	Multiply: $(3x-6)(6x-2)$.
$(2x+1)(-4x+5) = -8x^2 + \boxed{} - 4x + 5$	
$= \boxed{}$	

EXAMPLE 2	YOUR TURN 2
Multiply: $(m-1)(m+4)(m-7)$.	Multiply: $(m+8)(m+5)(m-4)$.
$(m-1)(m+4)(m-7)$	
$= (m^2 + 4m - m - 4)(m-7)$	
$= (m^2 + \boxed{} - 4)(m-7)$	
$= (m^2 + 3m - 4)\cdot m + (m^2 + 3m - 4)(-7)$	
$= \boxed{} + 3m^2 - 4m - \boxed{} - 21m + 28$	
$= \boxed{}$	

YOUR NOTES　　Write your questions and additional notes.

Squares of Binomials

Squaring a Binomial

$$(A+B)^2 = A^2 + 2AB + B^2$$

$$(A-B)^2 = A^2 - 2AB + B^2$$

Trinomials of the form $A^2 + 2AB + B^2$ or $A^2 - 2AB + B^2$ are called **trinomial squares.**

Example

- Multiply: $(x-4)^2$.

$$(A-B)^2 = A^2 - 2\cdot A\cdot B + B^2$$

$$\downarrow \quad \downarrow \qquad \downarrow \qquad \downarrow \quad \downarrow \quad \downarrow$$

$$(x-4)^2 = x^2 - 2\cdot x\cdot 4 + 4^2$$

$$= x^2 - 8x + 16$$

GUIDED LEARNING	Textbook	Instructor	Video

EXAMPLE 1	YOUR TURN 1
Multiply: $(4y+6)^2$.	Multiply: $(-3y+5)^2$.
$(4y+6)^2 = (4y)^2 + 2\cdot 4y\cdot 6 + 6^2$	
$=\boxed{}$	
EXAMPLE 2	YOUR TURN 2
Multiply: $\left(\dfrac{1}{4}x - 2\right)^2$.	Multiply: $\left(\dfrac{1}{2}x - 5\right)^2$.
$\left(\dfrac{1}{4}x-2\right)^2 = \left(\dfrac{1}{4}x\right)^2 - 2\cdot\dfrac{1}{4}x\cdot 2 + 2^2$	
$=\boxed{}$	

YOUR NOTES   Write your questions and additional notes.

Products of Sums and Differences

ESSENTIALS

The Product of a Sum and a Difference

$$(A+B)(A-B) = A^2 - B^2$$

The product of the sum and the difference of the same two terms is called a **difference of squares**.

Example

- Multiply: $(x+3)(x-3)$.

$$(A+B)(A-B) = A^2 - B^2$$
$$\downarrow \quad \downarrow \quad \downarrow \quad \downarrow \qquad \downarrow \quad \downarrow$$
$$(x+3)(x-3) = x^2 - 3^2$$
$$= x^2 - 9$$

GUIDED LEARNING	📖 Textbook	👤 Instructor	▶ Video

EXAMPLE 1	YOUR TURN 1
Multiply: $(2xy+4x)(2xy-4x)$.	Multiply: $(5x^2y+1)(5x^2y-1)$.
$(2xy+4x)(2xy-4x) = (2xy)^2 - (4x)^2$ $= \boxed{}$	

EXAMPLE 2	YOUR TURN 2
Multiply: $(0.3m+2.1n)(0.3m-2.1n)$.	Multiply: $(0.6x+1.2y)(0.6x-1.2y)$.
$(0.3m+2.1n)(0.3m-2.1n) = (0.3m)^2 - (2.1n)^2$ $= \boxed{}$	

Using Function Notation

ESSENTIALS

Example

- Given $f(x) = x^2 + 6x - 7$, find (a) $f(t) - 10$ and (b) $f(t-10)$.

  a) $f(x) = x^2 + 6x - 7$

  $\quad f(t) - 10 = (t^2 + 6t - 7) - 10$ Evaluating $f(t)$

  $\qquad\qquad = t^2 + 6t - 17$ Simplifying

  b) $f(x) = x^2 + 6x - 7$

  $\quad f(t-10) = (t-10)^2 + 6(t-10) - 7$ Substituting $t - 10$ for x

  $\qquad\qquad = t^2 - 20t + 100 + 6t - 60 - 7$ Multiplying

  $\qquad\qquad = t^2 - 14t + 33$ Simplifying

GUIDED LEARNING   📖 **Textbook**   👤 **Instructor**   ▶ **Video**

EXAMPLE 1	YOUR TURN 1
Given $f(x) = 2x^2 + 3$, find (a) $f(a) + 1$; (b) $f(a+1)$.	Given $f(x) = x^2 + 5x$, find (a) $f(a) - 4$; (b) $f(a-4)$.

a) $f(x) = 2x^2 + 3$

$\quad f(a) + 1 = (2a^2 + 3) + 1$

$\qquad\qquad = 2a^2 + 4$

b) $f(x) = 2x^2 + 3$

$\quad f(a+1) = 2(a+1)^2 + 3$

$\qquad\qquad = 2\left(\boxed{}\right) + 3$

$\qquad\qquad = 2a^2 + 4a + \boxed{} + 3$

$\qquad\qquad = \boxed{}$

EXAMPLE 2	YOUR TURN 2
Given $g(x) = 3x^2y^3 + 2$, find $g(x) \cdot g(x)$. $g(x) \cdot g(x) = (3x^2y^3 + 2)(3x^2y^3 + 2)$ $\qquad = (3x^2y^3 + 2)^2$ $\qquad = \boxed{} + 2 \cdot (3x^2y^3) \cdot 2 + 2^2$ $\qquad = \boxed{}$	Given $g(x) = 4x^2y^6 - 3$, find $g(x) \cdot g(x)$.

YOUR NOTES   Write your questions and additional notes.

Practice Exercises

Readiness Check

Classify each of the following statements as either true or false.

1. FOIL can be used whenever two binomials are multiplied.

2. Given $f(x) = 2x - 1$, $f(y) + 8 = f(y + 8)$.

3. A trinomial square is of the form $A^2 - B^2$.

4. When multiplying polynomials, the distributive law can always be used.

Multiplying Monomials

Multiply.

5. $2x^5 \cdot 4x^3$

6. $\left(-5x^6 y^7\right)\left(-3x^2 y^2\right)$

7. $\left(-9abc^4\right)\left(3a^2 bc\right)$

Multiplying Monomials and Binomials

Multiply.

8. $6(4 - 2x)$

9. $4a\left(-3a^2 - 7a\right)$

10. $5xy\left(-7x^2 y^3 + 6x^4 y\right)$

Multiplying Any Two Polynomials

Multiply.

11. $(2x + 3)(5x - 1)$

12. $(x + 5)\left(x^2 + x - 2\right)$

13. $(x + 4)\left(3x^2 + x - 5\right)$

The Product of Two Binomials: FOIL

Multiply.

14. $\left(x - \dfrac{1}{6}\right)\left(x - \dfrac{1}{2}\right)$

15. $(4x - 2)(3x + 1)$

16. $(1.3x + 3)(2.4x - 5)$

Squares of Binomials

Multiply.

17. $(x-6)^2$

18. $(2x+3y)^2$

19. $(x^3y-2)^2$

Products of Sums and Differences

Multiply.

20. $(a+10)(a-10)$

21. $(3-6x)(3+6x)$

22. $(-4a+b^2)(4a+b^2)$

Function Notation

23. Let $P(x)=2x-5$ and $Q(x)=4x^2-3x+1$. Find $P(x)\cdot Q(x)$.

24. Given $f(x)=x^2+3x-9$, find $f(t-4)$.

25. Given $f(x)=2x^2+3$, find $f(a+h)$.

26. Given $f(x)=6+4x-x^2$, find $f(a)+7$.

Terms with Common Factors

ESSENTIALS

Factoring

To **factor** a polynomial is to find an equivalent expression that is a product of polynomials. An equivalent expression of this type is called a **factorization** of the polynomial.

A polynomial is **completely factored** if it cannot be factored further.
A polynomial that cannot be factored is a **prime polynomial**.

Example

- Write an expression equivalent to $12x^3y^2 - 6x^2y^2 + 4x^4y$ by factoring out the greatest common factor.

  Look for the greatest common factor of the coefficients of $12x^3y^2 - 6x^2y^2 + 4x^4y$.

  $12, -6, 4 \rightarrow$ Greatest common factor $= 2$

  Look for the greatest common factor of the powers of x.

  $x^3, x^2, x^4 \rightarrow$ Greatest common factor $= x^2$

  Look for the greatest common factor of the powers of y.

  $y^2, y^2, y \rightarrow$ Greatest common factor $= y$

  Thus, $2x^2y$ is the greatest common factor of the polynomial.

  $$12x^3y^2 - 6x^2y^2 + 4x^4y = 2x^2y\left(6xy - 3y + 2x^2\right)$$

  Check: $2x^2y\left(6xy - 3y + 2x^2\right) = 12x^3y^2 - 6x^2y^2 + 4x^4y$

GUIDED LEARNING	📖 **Textbook** 👤 **Instructor**	▶ **Video**

EXAMPLE 1	YOUR TURN 1
Write an expression equivalent to $10x^6y^4 + 5x^3y^3 - 5x^4$ by factoring out the greatest common factor. $10x^6y^4 + 5x^3y^3 - 5x^4 = 5x^3\left(2x^3y^4 + \boxed{} - x\right)$ Check: $5x^3\left(2x^3y^4 + y^3 - x\right) = 10x^6y^4 + 5x^3y^3 - 5x^4$	Write an expression equivalent to $8xy^2 + 4x^5y - 2x^3y$ by factoring out the greatest common factor.

EXAMPLE 2	YOUR TURN 2
Write an expression equivalent to $-3x^3 + 6x^2 + 12x$ by factoring out a common factor with a negative coefficient. $-3x^3 + 6x^2 + 12x = -3x\left(\boxed{} - 2x - 4\right)$ Check: $-3x\left(x^2 - 2x - 4\right) = -3x^3 + 6x^2 + 12x$	Write an expression equivalent to $-7x^3 + 14x^2 - 21x$ by factoring out a common factor with a negative coefficient.

EXAMPLE 3	YOUR TURN 3
The number of diagonals of a polygon having n sides is given by the polynomial function $$P(n) = \frac{1}{2}n^2 - \frac{3}{2}n.$$ Find an equivalent expression for $P(n)$ by factoring out $\frac{1}{2}n$. $P(n) = \frac{1}{2}n^2 - \frac{3}{2}n = \boxed{}\left(n - \boxed{}\right)$ Check: $\frac{1}{2}n(n - 3) = \frac{1}{2}n^2 - \frac{3}{2}n$	The surface area of a rectangular prism is represented by the formula $$2lw + 2lh + 2wh.$$ Find an equivalent expression by factoring out 2.

YOUR NOTES   Write your questions and additional notes.

Factoring by Grouping

ESSENTIALS

The largest common factor of a polynomial is sometimes a binomial. In order to identify a common binomial factor in a polynomial with four terms, we must regroup into two groups of two terms each.

Factoring out -1

$$b-a=-1(a-b)=-(a-b)$$

Example

- Factor: $(x+4)(x+2)+(x+4)(x-7)$.

  Here the largest common factor is the binomial $x+4$.

$$(x+4)(x+2)+(x+4)(x-7)=(x+4)\left[(x+2)+(x-7)\right]$$
$$=(x+4)(2x-5)$$

GUIDED LEARNING	🔖 **Textbook**	👤 **Instructor**	▶ **Video**

EXAMPLE 1	YOUR TURN 1
Factor: $(2x+1)(x-1)+(2x+1)(x+6)$. The largest common factor is the binomial $2x+1$. $$(2x+1)(x-1)+(2x+1)(x+6)$$ $$=(2x+1)\left[(x-1)+(\boxed{})\right]$$ $$=(2x+1)\boxed{}$$	Factor: $(3x-2)(x+4)+(3x-2)(x-1)$.
EXAMPLE 2	YOUR TURN 2
Factor: $2x^2+6x+5x+15$. $$2x^2+6x+5x+15$$ $$=(2x^2+6x)+(5x+15)$$ $$=2x(x+3)+5(\boxed{})$$ $$=(x+3)\boxed{}$$	Factor: $2x^3-2x^2+3x-3$.

EXAMPLE 3	YOUR TURN 3
Factor: $6x - 7y + xy - 42$.	Factor: $2a + 4b + ba + 8$.

$6x - 7y + xy - 42$

$= (6x - 42) + (xy - 7y)$

$= 6(x - 7) + y(\boxed{})$

$= \boxed{}$

EXAMPLE 4	YOUR TURN 4
Factor: $3x^2 - 3xy + y^2 - xy$.	Factor: $2x^2 - 2xy + 3y^2 - 3xy$.

$3x^2 - 3xy + y^2 - xy$

$= (3x^2 - 3xy) + (y^2 - xy)$

$= 3x(x - y) + y(\boxed{})$

$= 3x(x - y) + y(-1)(-y + x)$

$= 3x(x - y) - y(x - y)$

$= \boxed{}$

YOUR NOTES   Write your questions and additional notes.

Practice Exercises

Readiness Check

Choose the word or expression that best completes each statement.

1. The largest common factor of $25x^5y^3z^4 + 75x^{10}y^6z^2$ is _____.
   $$25x^{10}y^6z^4 \;/\; 25x^5y^3z^2$$

2. The polynomial $2xy + 3$ is a _____ polynomial.
   prime / factorable

3. The largest common factor of the polynomial $(x+7)(x-2)+(x+8)(x-2)$ is

   _____.
   $x+7 \;/\; x-2$

4. The expression $x - y$ is equivalent to _____.
   $-1(y-x) \;/\; -(x-y)$

Terms with Common Factors

Write an equivalent expression by factoring out the greatest common factor.

5. $12x^2 - 18x + 24$

6. $6a^6 + 9a^5 - 12a^3$

7. $16x^4y^2 - 8x^5y^3 + 4x^6y$

8. $4a^8 - a^5 - a^3$

Write an equivalent expression by factoring out a factor with a negative coefficient.

9. $-9x - 45$

10. $-3x^2 + 9x - 12$

11. $-x^3 - 5x^2 - 6x + 7$

12. $6a^2 - 18ab$

Factoring by Grouping

Write an equivalent expression by factoring.

13. $t(r+2)+s(r+2)$

14. $12a^5 + 8a^3 + 3a^2 + 2$

15. $4x^3 + 3x + 8x^2 + 6$

16. $5x^2 - 5xy + 6y^2 - 6xy$

17. When x hundred DVDs are sold, a company makes a profit of $P(x)$ where $P(x) = x^2 - 4x$, and $P(x)$ is in thousands of dollars. Find an equivalent expression by factoring out a common factor.

18. A company determines that when it sells x units of a product, the total revenue $R(x)$, in dollars, is given by the polynomial function $R(x) = 350x - 0.5x^2$. Find an equivalent expression for $R(x)$ by factoring out a common factor with a negative coefficient.

Factoring Trinomials: $x^2 + bx + c$

ESSENTIALS

To Factor $x^2 + bx + c$

1. If necessary, rewrite the trinomial in descending order.

2. Find a pair of factors that have c as their product and b as their sum.

   - If c is positive, both factors will have the same sign as b.

   - If c is negative, one factor will be positive and the other will be negative. The factor with the larger absolute value will be the factor with the same sign as b.

   - If the sum of the two factors is the opposite of b, changing the signs of both factors will give the desired factors whose sum is b.

3. Check by multiplying.

Examples

- Factor: $x^2 + 5x + 6$.

  There is no common factor. Look for a factorization of 6 in which both factors are positive. Their sum must be 5.

Pairs of factors	Sum of factors
1, 6	7
2, 3	5

  ← The numbers we need are 2 and 3.

  The factorization is $(x+2)(x+3)$.

  Check: $(x+2)(x+3) = x^2 + 3x + 2x + 6 = x^2 + 5x + 6$.

- Factor: $x^2 + 4x - 12$.

  There is no common factor. Look for a factorization of -12 for which the sum of the factors is 4. The positive factor must have the larger absolute value.

Pairs of factors	Sum of factors
−1, 12	11
−2, 6	4
−3, 4	1

  ← The numbers we need are -2 and 6.

  The factorization is $(x-2)(x+6)$.

  Check: $(x-2)(x+6) = x^2 + 6x - 2x - 12 = x^2 + 4x - 12$.

GUIDED LEARNING   📘 **Textbook**   👤 **Instructor**   ▶ **Video**

EXAMPLE 1	YOUR TURN 1

Factor: $x^2 - 3x - 10$.

There is no common factor. Look for a factorization of -10 for which the sum of the factors is -3. The negative factor must have the larger absolute value.

Pairs of factors	Sum of factors
1, −10	☐
2, −5	☐

The numbers we need are 2 and −5. The factorization is $(x+2)(x-5)$.

Check: $(x+2)(x-5) = x^2 - 5x + 2x - 10 = x^2 - 3x - 10$.

Factor: $x^2 - 8x - 9$.

EXAMPLE 2	YOUR TURN 2

Factor: $2x^2 + 22x + 36$.

Factor out the largest common factor, 2.

$$2x^2 + 22x + 36 = 2(x^2 + 11x + 18)$$

Look for a factorization of 18 in which both factors are positive and whose sum is _____.
$$\underline{-11 \,/\, 11}$$

Pairs of factors	Sum of factors
1, 18	19
2, 9	☐
3, 6	☐

Thus, $x^2 + 11x + 18 = (x+2)(x+9)$. The factorization of the original trinomial is $2(x+2)(x+9)$. You can check this by multiplying.

Factor: $3x^2 + 15x + 12$.

YOUR NOTES   Write your questions and additional notes.

Practice Exercises

Readiness Check

Choose from the column on the right the phrase that best completes each statement.

1. To factor $x^2 + 10x + 9$, we look for a factorization of 9 in which _____.

2. To factor $x^2 - 6x - 7$, we look for a factorization of −7 in which _____.

3. To factor $x^2 - 12x + 11$, we look for a factorization of 11 in which _____.

4. To factor $x^2 + 8x - 9$, we look for a factorization of −9 in which _____.

a) both factors are positive.

b) both factors are negative.

c) the positive factor has the greater absolute value.

d) the negative factor has the greater absolute value.

Factoring Trinomials: $x^2 + bx + c$

Factor completely. If a polynomial is not factorable, state this.

5. $x^2 + 7x + 6$

6. $t^2 - 11t - 12$

7. $x^2 - 8x + 15$

8. $x^2 + 4x - 12$

9. $y^2 - 2y + 24$

10. $-48 - 18x + 3x^2$

Factor completely. If a polynomial is not factorable, state this.

11. $y^2 - \dfrac{1}{2}y + \dfrac{1}{16}$

12. $x^7 + 28x^6 + 27x^5$

13. $36 + 5x - x^2$

14. $a^2 + 6ab + 5b^2$

15. $x^4 - 7x^2 - 8$

16. $t^2 + 0.6t + 0.09$

17. $-x^3 + 3x^2 - 2x$

The FOIL Method

ESSENTIALS

To Factor $ax^2 + bx + c$ Using FOIL:

1. Factor out the largest common factor, if one exists. Here we assume that none does.
2. Find two First terms whose product is ax^2.
3. Find two Last terms whose product is c.
4. Repeat steps (2) and (3), if necessary, until a combination is found for which the sum of the Outside and Inside products is bx.
5. Check by multiplying.

Tips:

1. If the largest common factor has been factored out of the original trinomial, then no binomial factor can have a common factor $(\text{other than } 1 \text{ or } -1)$.

2. **a)** If the signs of all the terms are positive, then the signs of all the terms of the binomial factors are positive.

   b) If a and c are positive and b is negative, then the signs of the factors of c are negative.

   c) If a is positive and c is negative, then the factors of c will have opposite signs.

3. Be systematic about your trials. Keep track of those you have tried and those you have not.

4. Changing the signs of the factors of c will change the sign of the middle term.

Example

- Factor using FOIL: $6x^2 + x - 12$.

$$6x^2 + x - 12 = (2x + 3)(3x - 4)$$

GUIDED LEARNING	📖 **Textbook**	👤 **Instructor**	▶ **Video**

EXAMPLE 1	YOUR TURN 1
Factor using FOIL: $3x^2 + 13x - 10$.	Factor using FOIL: $2x^2 + 5x - 12$.
1. There is no common factor other than 1 or −1.	
2. $3x^2$ can be factored as $(3x)(\boxed{})$. The factorization must be of the form $(3x +\ )(x +\ )$.	
3. There are four pairs of factors of −10, and each can be listed in two ways.	
(continued)	

4.

Pairs of factors	Corresponding trial	Product
$-1, 10$	$(3x-1)(x+10) =$	$3x^2\ \boxed{} -10$
$1, -10$	$(3x+1)(x-10) =$	$3x^2\ \boxed{} -10$
$-2, \boxed{}$	$(3x-2)(x+5) =$	$3x^2\ \boxed{} -10$
$2, -5$	$(3x+2)(x-5) =$	$3x^2\ \boxed{} -10$
$10, \boxed{}$	$(3x+10)(x-1) =$	$3x^2\ \boxed{} -10$
$-10, 1$	$(3x-10)(x+1) =$	$3x^2\ \boxed{} -10$
$5, \boxed{}$	$(3x+5)(x-2) =$	$3x^2\ \boxed{} -10$
$-5, 2$	$(3x-5)(x+2) =$	$3x^2\ \boxed{} -10$

The factors -2 and 5 give the correct middle term.
The factorization is $(3x-2)(x+5)$.

EXAMPLE 2

Factor using FOIL: $6 + 7a - 3a^2$.

1. Write in descending order: $-3a^2 + 7a + 6$. Factor out -1 to make the leading coefficient positive.

$$-3a^2 + 7a + 6 = -1\left(\boxed{}\right)$$

2. The only possible factorization of $3a^2$ is $3 \cdot a \cdot a$. Thus, the factorization must be of the form $(3a+\ )(a+\ )$.

3. Pairs of factors of -6 are

$-1, 6 \quad 1, -6 \quad -2, \boxed{} \quad 2, \boxed{}$

$6, -1 \quad -6, \boxed{} \quad 3, \boxed{} \quad -3, \boxed{}$.

4. We conduct trials, stopping when we have the correct factorization.

$$(3a-1)(a+6) = 3a^2\ \boxed{} -6$$

$$(3a-2)(a+3) = 3a^2\ \boxed{} -6$$

$$(3a+2)(a-3) = 3a^2\ \boxed{} -6$$

The factorization of $3a^2 - 7a - 6$ is $(3a+2)(a-3)$.

Thus, the factorization of the original trinomial is $-1(3a+2)(a-3)$. We can also write this as $(-3a-2)(a-3)$ or $(3a+2)(-a+3)$.

YOUR TURN 2

Factor using FOIL: $18 - 15a - 25a^2$.

EXAMPLE 3	YOUR TURN 3
Factor using FOIL: $12x^2 - 22xy + 10y^2$.	Factor using FOIL: $8a^4 + 18a^3b + 9a^2b^2$.

EXAMPLE 3

Factor using FOIL: $12x^2 - 22xy + 10y^2$.

1. $12x^2 - 22xy + 10y^2 = 2(6x^2 - 11xy + 5y^2)$

2. The factorization is of the form
   $(2x + \quad)(3x + \quad)$ or $(6x + \quad)(x + \quad)$.

3. Pairs of factors of $5y^2$ are
   $\begin{matrix} y, 5y \\ -y, -5y \end{matrix}$ and $\begin{matrix} 5y, y \\ -5y, -y. \end{matrix}$

4. Only the factors of $5y^2$ with negative coefficients will give a negative middle term.
   $(2x - y)(3x - 5y) = 6x^2 - \boxed{} + 5y^2$
   $(2x - 5y)(3x - y) = 6x^2 - \boxed{} + 5y^2$
   $(6x - y)(x - 5y) = 6x^2 - \boxed{} + 5y^2$
   $(6x - 5y)(x - y) = 6x^2 - \boxed{} + 5y^2$

Thus, the factorization of $6x^2 - 11xy + 5y^2$ is $(6x - 5y)(x - y)$. The factorization of the original trinomial is $2(6x - 5y)(x - y)$.

The *ac*–Method

ESSENTIALS

To Factor $ax^2 + bx + c$ Using the *ac*–Method:

1. Factor out the largest common factor, if one exists. Here we assume that none does.
2. Multiply the leading coefficient a and the constant c.
3. Find a pair of factors of ac whose sum is b.
4. Rewrite the middle term, bx, as a sum or difference using the factors from step (3).
5. Factor by grouping.
6. Check by multiplying.

This method is also referred to as the **grouping method**.

Example

- Factor: $10x^2 - x - 2$.

$$10x^2 - x - 2 = 10x^2 + 4x - 5x - 2 = 2x(5x + 2) - 1(5x + 2) = (5x + 2)(2x - 1)$$

GUIDED LEARNING	📖 **Textbook**	👤 **Instructor**	▶ **Video**

EXAMPLE 1	YOUR TURN 1
Factor: $6x^2 + 7x - 3$.	Factor: $8x^2 + 2x - 3$.

Factor: $6x^2 + 7x - 3$.

1. There is no common factor (other than 1 or -1).

2. Multiply a and c: $6(-3) = \boxed{}$.

3. Find factors of -18 whose sum is 7.
$$9(-2) = -18 \text{ and } 9 + (-2) = 7.$$

4. Split $7x$ using the results of step (3).
$$7x = 9x - 2x$$

5. Factor by grouping.
$$6x^2 + 7x - 3 = 6x^2 + 9x - 2x - 3$$
$$= 3x(2x + 3) - 1(2x + 3)$$
$$= (2x + 3)(3x - 1)$$

The factorization is $(2x + 3)(3x - 1)$.

YOUR TURN 1

Factor: $8x^2 + 2x - 3$.

EXAMPLE 2	YOUR TURN 2
Factor: $8x^3 + 32x^2 + 30x$.	Factor: $10a^5 - 19a^4 + 6a^3$.

1. Factor out the largest common factor, $2x$.

$$8x^3 + 32x^2 + 30x = 2x\left(4x^2 + \boxed{} + \boxed{}\right)$$

2. To factor $4x^2 + 16x + 15$, first multiply a and c:

$$4(15) = \boxed{}.$$

3. Find factors of 60 whose sum is 16.

Since the middle term is positive, we consider only positive factors.

$$6 \cdot 10 = 60 \text{ and } 6 + 10 = 16.$$

4. Split the middle term, $16x$, using the results of step (3).

$$16x = 6x + 10x$$

5. Factor by grouping.

$$4x^2 + 16x + 15 = 4x^2 + 6x + 10x + 15$$
$$= 2x(2x+3) + 5(2x+3)$$
$$= (2x+3)(2x+5)$$

The factorization of the original trinomial is $2x(2x+3)(2x+5)$.

YOUR NOTES   Write your questions and additional notes.

Practice Exercises

Readiness Check

Choose a word from the list below that will make each statement true. Not all words will be used.

common   negative   first

last   prime   positive

1. When factoring a trinomial, first factor out the largest _____ factor.

2. When factoring a trinomial, $ax^2 + bx + c$, the product of the _____ terms in the binomials is c.

3. When factoring a trinomial $ax^2 + bx + c$, if c is _____, then the signs in both binomial factors are the same.

4. When factoring a trinomial, $ax^2 + bx + c$, the product of the _____ terms in the binomials is ax^2.

The FOIL Method

Factor completely.

5. $10x^2 + x - 3$

6. $6a^2 - 11a - 10$

7. $15a^3 + 13a^2 + 2a$

8. $t - 1 + 2t^2$

The *ac*-Method

Factor completely.

9. $-20x^2 - 5x + 15$

10. $12p^2 - 38pw + 30w^2$

11. $5x^2 + x - 4$

12. $-18z^3 + 24z^2 + 10z$

13. $4r^2 - 35r + 49$

14. $10t^2 - 29t + 10$

Trinomial Squares

ESSENTIALS

An expression of the form $A^2 + 2AB + B^2$ or $A^2 - 2AB + B^2$ is a **trinomial square**, or a **perfect-square trinomial**.

To Recognize a Trinomial Square

Two terms must be squares, such as A^2 and B^2.

The remaining term must be $2AB$ or its opposite.

Factoring a Trinomial Square

$$A^2 + 2AB + B^2 = (A+B)^2$$
$$A^2 - 2AB + B^2 = (A-B)^2$$

Example

- Factor: $x^2 - 20x + 100$.

  We have a trinomial square of the form $A^2 - 2AB + B^2$, with $A = x$ and $B = 10$.

  $$x^2 - 20x + 100 = (x-10)(x-10)$$
  $$= (x-10)^2$$

GUIDED LEARNING 📖 **Textbook**   👤 **Instructor**   ▶ **Video**

EXAMPLE 1	YOUR TURN 1
Factor: $x^2 + 8x + 16$. Two terms, x^2 and 16, are squares. Twice the product of the square roots is $2 \cdot x \cdot 4$, or $8x$, the remaining term of the trinomial. Thus, we have a trinomial square of the form $A^2 + 2AB + B^2$, with $A = x$ and $B = 4$. $x^2 + 8x + 16 = \left(x \,\square\, 4\right)^2$	Factor: $y^2 + 12y + 36$.
EXAMPLE 2	**YOUR TURN 2**
Factor: $9y^2 - 30y + 25$. We have a trinomial square of the form $A^2 - 2AB + B^2$, with $A = 3y$ and $B = 5$. $9y^2 - 30y + 25 = \left(3y \,\square\, 5\right)^2$	Factor: $16y^2 + 56x + 49$.

EXAMPLE 3	YOUR TURN 3
Factor: $-4xy + 4y^2 + x^2$.	Factor: $-6xy + 9y^2 + x^2$.

$$-4xy + 4y^2 + x^2 = x^2 - 4xy + 4y^2$$

We have a trinomial square of the form
$A^2 - 2AB + B^2$, with $A = x$ and $B = 2y$.

$$-4xy + 4y^2 + x^2 = x^2 - 4xy + 4y^2 = \left(x - \boxed{}\right)^2$$

EXAMPLE 4	YOUR TURN 4
Factor: $2y^5 + 48y^4 + 288y^3$.	Factor: $12x^4 + 36x^3 + 27x^2$.

Factor out the largest common factor, $2y^3$.

$$2y^5 + 48y^4 + 288y^3 = 2y^3\left(y^2 + \boxed{} + 144\right)$$

$y^2 + 24y + 144$ is a trinomial square of the form
$A^2 + 2AB + B^2$, with $A = y$ and $B = 12$.

$$2y^5 + 48y^4 + 288y^3$$
$$= 2y^3\left(y^2 + 24y + 144\right)$$
$$= 2y^3\left(\boxed{}\right)^2$$

Check:
$$2y^3\left(y + 12\right)^2 = 2y^3\left(y + 12\right)\left(y + 12\right)$$
$$= 2y^3\left(y^2 + 12y + 12y + 144\right)$$
$$= 2y^3\left(y^2 + 24y + 144\right)$$
$$= 2y^5 + 48y^4 + 288y^3$$

The factorization is $\boxed{}\left(\boxed{}\right)^2$.

YOUR NOTES   Write your questions and additional notes.

Differences of Squares

ESSENTIALS

An expression of the form $A^2 - B^2$ is a **difference of squares**.

Factoring a Difference of Squares

$$A^2 - B^2 = (A+B)(A-B)$$

Example

- Factor: $x^2 - 16$.

    $A = x$ and $B = 4$

    $$x^2 - 16 = x^2 - 4^2 = (x+4)(x-4)$$

GUIDED LEARNING	📖 **Textbook** 👤 **Instructor** ▶ **Video**
EXAMPLE 1	**YOUR TURN 1**
Factor: $y^2 - 64$. $y^2 - 64 = y^2 - 8^2 = (y+8)\left(y - \boxed{}\right)$	Factor: $x^2 - 4$.
EXAMPLE 2	**YOUR TURN 2**
Factor: $25t^2 - 16r^2$. $25t^2 - 16r^2 = (5t)^2 - (4r)^2 = \left(5t + \boxed{}\right)\left(5t - \boxed{}\right)$	Factor: $49t^2 - 9r^2$.
EXAMPLE 3	**YOUR TURN 3**
Factor: $16x^8 - 64x^4y^2$. Factor out the largest common factor, $16x^4$. $16x^8 - 64x^4y^2 = 16x^4\left(x^4 - \boxed{}\right)$ $= 16x^4\left[\left(x^2\right)^2 - \left(2y\right)^2\right]$ $= 16x^4\left(\boxed{} + 2y\right)\left(\boxed{} - 2y\right)$	Factor: $2x^2 - 162y^2$.

YOUR NOTES   Write your questions and additional notes.

More Factoring by Grouping

ESSENTIALS

When factoring by grouping, make sure the result is completely factored. If grouping pairs of terms does not yield a common binomial factor, look to group a trinomial square.

Example

- Factor: $x^3 + 2x^2 - 9x - 18$.

$$x^3 + 2x^2 - 9x - 18 = x^2(x+2) - 9(x+2) \qquad \text{Factoring by grouping}$$
$$= (x+2)(x^2 - 9) \qquad \text{Factoring out } x+2$$
$$= (x+2)(x+3)(x-3) \qquad \text{Factoring } x^2 - 9$$

GUIDED LEARNING   📍 **Textbook**   📍 **Instructor**   📍 **Video**

EXAMPLE 1	YOUR TURN 1
Factor: $a^3 + 2a^2 - 64a - 128$. $a^3 + 2a^2 - 64a - 128 = a^2(a+2) - 64(a+2)$ $= (\boxed{})(a^2 - 64)$ $= (a+2)(\boxed{})(a-8)$	Factor: $a^3 - a^2 - 49a + 49$.
EXAMPLE 2	**YOUR TURN 2**
Factor: $x^2 + 14x + 49 - y^2$. $x^2 + 14x + 49 - y^2 = (x^2 + 14x + 49) - y^2$ $= (x+7)^2 - y^2$ $= ((x+7) + \boxed{})((x+7) - y)$ $= (\boxed{})(x+7-y)$	Factor: $16x^2 + 40x + 25 - y^2$.

YOUR NOTES   Write your questions and additional notes.

Sums or Differences of Cubes

ESSENTIALS

Factoring a Sum or a Difference of Cubes

$$A^3 + B^3 = (A+B)(A^2 - AB + B^2)$$

$$A^3 - B^3 = (A-B)(A^2 + AB + B^2)$$

Example

- Factor completely: $x^3 - 125$.

    This is a difference of cubes: $x^3 - 125 = x^3 - 5^3$.

    $$A^3 \ - \ B^3 \ = \ (A \ - \ B)(A^2 \ + \ AB \ + \ B^2)$$
    $$\downarrow \quad \downarrow \quad\quad \downarrow \quad\quad \downarrow \quad \downarrow \quad\quad \downarrow \quad\quad \downarrow$$
    $$x^3 \ - \ 5^3 \ = \ (x \ - \ 5)(x^2 \ + \ 5x \ + \ 25)$$

    Check: $(x-5)(x^2+5x+25) = x^3 + 5x^2 + 25x - 5x^2 - 25x - 125$

    $$= x^3 - 125.$$

    Thus, $x^3 - 125 = (x-5)(x^2 + 5x + 25)$.

GUIDED LEARNING   **Textbook**   **Instructor**   **Video**

EXAMPLE 1	YOUR TURN 1
Factor completely: $y^9 - 64$.	Factor completely: $t^{15} - 27$.

EXAMPLE 1 (continued)

This is a difference of cubes: $y^9 - 64 = (y^3)^3 - \boxed{}^3$.

$$y^9 - 64 = (y^3 - 4)\left[(y^3)^2 + 4y^3 + 4^2\right]$$

$$= (y^3 - 4)(\boxed{})$$

Check: $(y^3 - 4)(y^6 + 4y^3 + 16)$

$$= y^9 + 4y^6 + 16y^3 - 4y^6 - 16y^3 - 64$$

$$= y^9 - 64.$$

Thus, $y^9 - 64 = (y^3 - 4)(y^6 + 4y^3 + 16)$.

EXAMPLE 2	YOUR TURN 2
Factor completely: $8x^3 + 1$.	Factor completely: $27x^3 + 1$.

EXAMPLE 2 (continued)

This is a sum of cubes: $8x^3 + 1 = (2x)^3 + 1^3$.

$$8x^3 + 1 = (2x+1)(\boxed{})$$

EXAMPLE 3	YOUR TURN 3
Factor completely: $81x^7 + 24xy^6$.	Factor completely: $128x^{12} + 54y^9$.

$$81x^7 + 24xy^6 = 3x\left(27x^6 + \boxed{}\right)$$
$$= 3x\left[\left(3x^2\right)^3 + \left(2y^2\right)^3\right]$$
$$= 3x\left(3x^2 + 2y^2\right)\left(\boxed{}\right)$$

EXAMPLE 4	YOUR TURN 4
Factor: $a^6 - b^6$.	Factor: $t^6 - 1$.

This is a difference of squares and a difference of cubes. In such cases it is generally better to factor as a difference of squares first.

$$a^6 - b^6$$
$$= \left(a^3\right)^2 - \left(b^3\right)^2$$
$$= \left(a^3 + b^3\right)\left(a^3 - \boxed{}\right)$$
$$= \left(a + \boxed{}\right)\left(a^2 - ab + b^2\right)\left(a - \boxed{}\right)\left(a^2 + ab + b^2\right)$$

YOUR NOTES   Write your questions and additional notes.

Practice Exercises

Readiness Check

Match each polynomial with its classification.

1. $9x^2 - 15$
2. $8 + x^3$
3. $x^2 - 16x + 64$
4. $t^{24} - r^{15}$
5. $16x^2 - 81$

a) Difference of squares
b) Polynomial having a common factor
c) Trinomial square
d) Difference of cubes
e) Sum of cubes

Trinomial Squares

Factor completely.

6. $x^2 + 6x + 9$

7. $t^2 - 18t + 81$

8. $-y^3 - 10y^2 - 25y$

9. $9m^2 + 12mn + 4n^2$

10. $4 + 25x^2 - 20x$

11. $3x^2 + 36x + 108$

Differences of Squares

Factor completely.

12. $x^2 - 144$

13. $25a^2 - 64$

14. $100m^2 - 4$

15. $9x^4 - 16x^2$

16. $8x^2y^4 - 8x^2z^4$

17. $\dfrac{1}{25} - x^2$

More Factoring by Grouping

Factor completely.

18. $(c+d)^2 - 36$ **19.** $y^3 + y^2 - 4y - 4$ **20.** $144z^2 - x^2 + 2xy - y^2$

Sums or Differences of Cubes

Factor completely.

21. $x^3 + 27$ **22.** $p^3 - 8y^3$ **23.** $1 - 125a^3$

24. $x^3 y^6 - y^3$ **25.** $a^3 - \dfrac{1}{8}$ **26.** $a^6 - 1$

A General Factoring Strategy

ESSENTIALS

A Strategy for Factoring

A. Always factor out the largest common factor (other than 1 or −1).

B. Once the largest common factor has been factored out, *count the number of terms* in the other factor:

Two terms: Try factoring as a difference of squares first. Next, try factoring as a sum or a difference of cubes. Do *not* try to factor a sum of squares, $A^2 + B^2$.

Three terms: If it is a trinomial square, factor as such. If not, try factoring using FOIL or the *ac*-method.

Four terms: Try factoring by grouping and factoring out a common binomial factor. Next, try grouping into a difference of squares, one of which is a trinomial square.

C. Always *factor completely*. If a factor with more than one term can itself be factored further, do so.

D. Write the complete factorization and check by multiplying. If the original polynomial is not factorable, state this.

Example

• Factor: $27x^5 - x^2y^3$.

A. Factor out the largest common factor: $27x^5 - x^2y^3 = x^2(27x^3 - y^3)$.

B. The factor $27x^3 - y^3$ has two terms and is a difference of cubes.

$$27x^5 - x^2y^3 = x^2(3x - y)(9x^2 + 3xy + y^2)$$

C. No factor with more than one term can be factored further.

D. Check: $x^2(3x - y)(9x^2 + 3xy + y^2) = x^2(27x^3 + 9x^2y + 3xy^2 - 9x^2y - 3xy^2 - y^3)$

$$= x^2(27x^3 - y^3)$$
$$= 27x^5 - x^2y^3.$$

GUIDED LEARNING   **Textbook**    **Instructor**   **Video**

EXAMPLE 1	YOUR TURN 1
Factor: $12x^2 - 48x + 45$.	Factor: $10x^2 - 40x - 120$.

A. Factor out the largest common factor:

$12x^2 - 48x + 45 = \boxed{}\left(4x^2 - 16x + 15\right)$.

B. The trinomial factor is not a perfect square. Try using FOIL or the *ac*-method. We have

$12x^2 - 48x + 45 = 3\left(\boxed{}\right)(2x - 5)$.

C. We cannot factor further.

D. Check: $3(2x - 3)(2x - 5) = 3\left(4x^2 - 16x + 15\right)$

$$= 12x^2 - 48x + 45.$$

EXAMPLE 2	YOUR TURN 2
Factor: $m^2 - 8m + 16 - n^2$.	Factor: $m^2 + 6m + 9 - 25n^2$.

A. There is no common factor (other than 1 or -1).

B. There are four terms. We try grouping to remove a common binomial factor, but find none. We try grouping as a difference of squares, one of which is a trinomial square.

$$m^2 - 8m + 16 - n^2 = \left(m^2 - 8m + 16\right) - n^2$$

$$= \left(\boxed{}\right)^2 - n^2$$

$$= (m - 4 + n)\left(\boxed{}\right)$$

C. No factor with more than one term can be factored further.

D. Check: $(m - 4 + n)(m - 4 - n)$

$$= m(m - 4 - n) - 4(m - 4 - n) + n(m - 4 - n)$$

$$= m^2 - 4m - mn - 4m + 16 + 4n + mn - 4n - n^2$$

$$= m^2 - 8m + 16 - n^2$$

YOUR NOTES   Write your questions and additional notes.

Practice Exercises

Readiness Check

Determine whether each statement is true or false.

1. The largest common factor of the polynomial $36x^3y^2 + 12x^2y - 18x$ is $6x$.

2. The polynomial $25x^2 - 20x + 16$ is a trinomial square.

3. The polynomial $6x^2 + 5x - 6$ is not factorable.

4. The polynomial $x^{15} - 1$ is a difference of cubes.

A General Factoring Strategy

Factor completely.

5. $-7x^3 + 14x^2 + 56x$

6. $18x^2y - 2y$

7. $-2y^2 - 12y - 18$

8. $9m^3 + 18m^2 - 25m - 50$

9. $4x^4 - 32x$

10. $27a^5 - a^2b^3$

Factor completely.

11. $x^2 - 18x + 81 - 36y^2$

12. $8p^5 - 98p^3$

13. $-4t^5 - 12t^2$

14. $24x^3 + 192y^3$

The Principle of Zero Products

ESSENTIALS

A **polynomial equation** is an equation in which two polynomials are set equal to each other.

A **quadratic equation** is a second-degree polynomial equation in one variable.

The Principle of Zero Products

For any real numbers a and b:

If $ab = 0$, then $a = 0$ or $b = 0$ (or both). If $a = 0$ or $b = 0$, then $ab = 0$.

To Use the Principle of Zero Products

1. Write an equivalent equation with 0 on one side, using the addition principle.
2. Factor the nonzero side of the equation.
3. Set each factor that is not a constant equal to 0.
4. Solve the resulting equations.

Example

- Solve: $x^2 + 7x = 18$.

$$x^2 + 7x = 18$$
$$x^2 + 7x - 18 = 0 \qquad \text{Getting 0 on one side}$$
$$(x + 9)(x - 2) = 0 \qquad \text{Factoring}$$
$$x + 9 = 0 \quad or \quad x - 2 = 0 \quad \text{Using the principle of zero products}$$
$$x = -9 \quad or \qquad x = 2$$

The solutions are −9 and 2.

	📖 **Textbook**	👤 **Instructor**	▶ **Video**

GUIDED LEARNING

EXAMPLE 1	YOUR TURN 1
Solve: $5x^2 - 9x - 2 = 0$.	Solve: $3x^2 + 23x + 30 = 0$.

$$5x^2 - 9x - 2 = 0$$
$$(5x + 1)\left(x - \boxed{}\right) = 0$$
$$\boxed{} = 0 \quad or \quad x - 2 = 0$$
$$5x = -1 \quad or \qquad x = \boxed{}$$
$$x = -\frac{1}{5} \quad or \qquad x = 2$$

The solutions are $-\dfrac{1}{5}$ and 2.

EXAMPLE 2	YOUR TURN 2
Let $f(x) = -3x^3 - 147x$ and $g(x) = 42x^2$. Find all x-values for which $f(x) = g(x)$.	Let $f(x) = 2x^2 - 4x$ and $g(x) = -8x + 16$. Find all x-values for which $f(x) = g(x)$.

$$f(x) = g(x)$$
$$-3x^3 - 147x = 42x^2$$
$$-3x^3 - 42x^2 - \boxed{} = 0$$
$$\boxed{}\left(x^2 + 14x + 49\right) = 0$$
$$-3x(x+7)\left(x + \boxed{}\right) = 0$$

$$-3x = 0 \quad or \quad x + 7 = 0 \qquad or \quad x + 7 = 0$$
$$x = 0 \quad or \qquad x = \boxed{} \quad or \qquad x = -7$$

To check, confirm that $f(0) = g(0) = 0$ and $f(-7) = g(-7) = 2058$.

For $x = 0$ or $x = \boxed{}$, we have $f(x) = g(x)$.

EXAMPLE 3	YOUR TURN 3
Find the domain of F if $F(x) = \dfrac{x}{10x^2 - 90}$.	Find the domain of F if $F(x) = \dfrac{10 - 3x}{-3x^3 + 15x^2}$.

To exclude the values for which the denominator equals 0, we solve:

$$10x^2 - 90 = 0$$
$$\boxed{}\left(x^2 - 9\right) = 0$$
$$10(x+3)\left(x - \boxed{}\right) = 0$$
$$x + 3 = 0 \qquad or \quad x - 3 = 0$$
$$x = \boxed{} \quad or \qquad x = 3.$$

The domain of F is $\{x \,|\, x$ is a real number *and* $x \neq -3$ *and* $x \neq \boxed{}\}$.

YOUR NOTES   Write your questions and additional notes.

Applications and Problem Solving

ESSENTIALS

The **Pythagorean Theorem** relates the lengths of the sides of a right triangle.

A **right triangle** has a 90° angle, denoted by the symbol ⌐ or ⌐.

The longest side of a right triangle, called the **hypotenuse**, is opposite the right angle. The other sides are called the **legs**.

The Pythagorean Theorem

In any right triangle, if a and b are the lengths of the legs and c is the length of the hypotenuse, then $a^2 + b^2 = c^2$.

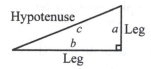

Example

- A wire is stretched from the ground to the top of a pole. The wire is 10 ft long. The height of the pole is 2 ft greater than the distance x from the bottom of the pole to the bottom of the wire. Find the height of the pole and the distance x.

  1. **Familiarize.** The wire, the pole, and the ground form a right triangle. The wire is the hypotenuse. If x is the distance from the bottom of the pole to the bottom of the wire, then $x + 2 = $ the height of the pole.

  2. **Translate.** We use the Pythagorean Theorem.

  $$a^2 + b^2 = c^2$$
  $$x^2 + (x+2)^2 = 10^2$$

  3. **Solve.** We solve the equation.

  $$x^2 + (x+2)^2 = 10^2$$
  $$x^2 + x^2 + 4x + 4 = 100$$
  $$2x^2 + 4x + 4 = 100$$
  $$2x^2 + 4x - 96 = 0$$
  $$2(x^2 + 2x - 48) = 0$$
  $$2(x+8)(x-6) = 0$$
  $$x + 8 = 0 \quad or \quad x - 6 = 0$$
  $$x = -8 \quad or \quad x = 6$$

  4. **Check.** The length cannot be negative, so we check only 6. If $x = 6$, then $x + 2 = 6 + 2 = 8$. Since $6^2 + 8^2 = 100 = 10^2$, the answer checks.

  5. **State.** The height of the pole is 8 ft, and the distance x is 6 ft.

EXAMPLE 1	YOUR TURN 1
A picture frame measures 12 cm by 14 cm, and 48 cm^2 of picture shows. Find the width of the frame.	A picture frame measures 28 in. by 20 in., and 240 in^2 of picture shows. Find the width of the frame.

1. **Familiarize.** Let x represent the width of the frame, in centimeters. Since the frame extends uniformly around the picture, the length of the picture must be $14 - 2x$ and the width must be $12 - 2x$.

2. **Translate.** Translate as follows:

$$\underbrace{\text{Length of picture showing}} \times \underbrace{\text{Width of picture showing}} = \underbrace{\text{Area of picture showing}}$$

$$\downarrow \qquad\qquad \downarrow \qquad\qquad \downarrow$$

$$(14 - 2x) \quad \times \quad (12 - 2x) \quad = \quad \boxed{}$$

3. **Solve.** Solve the equation.

$$(14 - 2x)(12 - 2x) = 48$$
$$168 - 28x - 24x + 4x^2 = 48$$
$$4x^2 - 52x + \boxed{} = 48$$
$$4x^2 - 52x + 120 = 0$$
$$4\left(x^2 - 13x + \boxed{}\right) = 0$$
$$4(x - 3)(x - 10) = 0$$
$$x - 3 = 0 \quad or \quad x - \boxed{} = 0$$
$$x = 3 \quad or \qquad\quad x = \boxed{}$$

4. **Check.** Note that 10 is not a solution because it would yield negative measurements. If the width of the frame is 3 cm, the length of the picture showing is $14 - 2(3) = 14 - 6 = 8$ cm, the width is $12 - 2(3) = 12 - 6 = 6$ cm, and the area of the picture showing is $8 \cdot 6 = 48$ cm^2. The answer checks.

5. **State.** The width of the frame is $\boxed{}$ cm.

YOUR NOTES   Write your questions and additional notes.

Practice Exercises

Readiness Check

Classify each of the following statements as either true or false.

1. The equation $x^2 + 10x + 25 = 0$ has two unique solutions.

2. The longest side of a right triangle is called the hypotenuse.

3. The Pythagorean Theorem relates the lengths of the sides of any triangle.

4. To solve the equation $x(3x+1) = 2$ we would use the principle of zero products first.

The Principle of Zero Products

Solve.

5. $6x(3x-1) = 0$

6. $x^2 + 10x - 24 = 0$

7. $14t^2 - 35t = 0$

8. $(a+3)(a-3) = 7$

9. $15y^2 - 8 = -14y$

10. $a^3 = 2a^2 + 3a$

11. Let $f(x) = x^2 + 14x + 40$. Find a such that $f(a) = -5$.

12. Let $f(y) = y^3 + 7y^2$ and $g(y) = 4y + 28$. Find all y-values for which $f(y) = g(y)$.

Find the domain of the function f given by each of the following.

13. $f(x) = \dfrac{5}{4x^2 - 64}$

14. $f(x) = \dfrac{10 + x}{3x^3 - 15x^2 + 18x}$

Applications and Problem Solving

Solve.

15. An electronics store determines that the revenue R, in thousands of dollars, from the sale of x hundred televisions is given by

$R(x) = \dfrac{4}{25}x^2 + 6x$. If the cost, C, in thousands of dollars, of producing x televisions is given by

$C(x) = \dfrac{1}{5}x^2 + 6x - 1$, how many televisions must be produced and sold to break even (that is, for revenue to equal cost)?

16. Mary's pool is 7 m longer than it is wide and is 17 m diagonally across. How long is Mary's pool?

Rational Expressions and Functions

ESSENTIALS

A **rational expression** consists of the quotient of two polynomials, where the polynomial in the denominator is nonzero.

Example

- Find the domain of f if $f(x) = \dfrac{x-3}{2x+5}$.

  The domain is the set of all replacements for which the rational expression is defined. The rational expression is not defined when the denominator is zero.

  $$2x + 5 = 0 \qquad \text{Setting the denominator equal to 0}$$
  $$2x = -5$$
  $$x = -\frac{5}{2}$$

  We must exclude $-\dfrac{5}{2}$ from the domain.

  The domain of $f = \left\{ x \,\middle|\, x \text{ is a real number } and \ x \neq -\dfrac{5}{2} \right\}$, or $\left(-\infty, -\dfrac{5}{2} \right) \cup \left(-\dfrac{5}{2}, \infty \right)$.

GUIDED LEARNING 📖 **Textbook**   👤 **Instructor**   ▶ **Video**

EXAMPLE 1	YOUR TURN 1
Find all numbers for which the rational expression $\dfrac{3x+5}{x-7}$ is not defined.	Find all numbers for which the rational expression $\dfrac{x}{x+3}$ is not defined.
The rational expression is not defined for replacements that make the denominator 0.	
$x - 7 = \boxed{}$	
$x = \boxed{}$	
The rational expression is not defined for the replacement $\boxed{}$.	

EXAMPLE 2	YOUR TURN 2
Find the domain of f if $f(x) = \dfrac{x^2 - 1}{x^2 + x - 6}$. Find all replacements that make the denominator 0. $\boxed{} = 0$ $(x+3)\left(\boxed{}\right) = 0$ $x + 3 = 0 \quad or \quad \boxed{} = 0$ $x = -3 \quad or \qquad\quad x = \boxed{}$ The expression is not defined for the replacements -3 and $\boxed{}$. The domain of $f = \left\{ x \mid x \text{ is a real number } and \ x \neq \boxed{} \ and \ x \neq \boxed{} \right\}$, or $(-\infty, -3) \cup \left(\boxed{}, 2\right) \cup \left(\boxed{}, \infty\right)$.	Find the domain of g if $g(x) = \dfrac{x^2 - 2x - 10}{x^2 + 3x - 4}$.

YOUR NOTES   Write your questions and additional notes.

Finding Equivalent Rational Expressions

ESSENTIALS

To multiply rational expressions, multiply numerators and multiply denominators:

$$\frac{A}{B} \cdot \frac{C}{D} = \frac{AC}{BD}.$$

Example

- Multiply to obtain an equivalent expression: $\dfrac{x+1}{x+1} \cdot \dfrac{3x}{x+2}$. Do not simplify.

$$\frac{x+1}{x+1} \cdot \frac{3x}{x+2} = \frac{(x+1)(3x)}{(x+1)(x+2)} \qquad \text{Note: } \frac{x+1}{x+1} = 1$$

GUIDED LEARNING	📖 **Textbook** 👤 **Instructor** ▶ **Video**	

EXAMPLE 1	YOUR TURN 1
Multiply to obtain an equivalent expression: $\dfrac{y-2}{y+3} \cdot \dfrac{y+7}{y+7}$. Do not simplify. Multiply numerators and multiply denominators. Use parentheses if the factors contain more than one term. $\dfrac{y-2}{y+3} \cdot \dfrac{y+7}{y+7} = \dfrac{(y-2)(\boxed{})}{(y+3)(\boxed{})}$	Multiply to obtain an equivalent expression: $\dfrac{3t}{3t} \cdot \dfrac{t-1}{2t+5}$.

Simplifying Rational Expressions

ESSENTIALS

To simplify a rational expression, we remove factors that are equal to 1.

Example

- Simplify by removing a factor equal to 1: $\dfrac{2x^2 + 6x}{2x^3}$.

$$\frac{2x^2 + 6x}{2x^3} = \frac{2x(x+3)}{2x \cdot x^2}$$

$$= \frac{2x}{2x} \cdot \frac{x+3}{x^2} \qquad \frac{2x}{2x} = 1$$

$$= \frac{x+3}{x^2}$$

GUIDED LEARNING	📍 **Textbook** 👤 **Instructor** ▶ **Video**
EXAMPLE 1	YOUR TURN 1

EXAMPLE 1

Simplify: $\dfrac{5xy}{30x}$.

$\dfrac{5xy}{30x} = \dfrac{5 \cdot x \cdot y}{5 \cdot x \cdot 6}$ Factoring. The largest common factor is $5x$.

$= \dfrac{5x}{5x} \cdot \boxed{}$

$= 1 \cdot \dfrac{y}{6} \qquad \dfrac{5x}{5x} = 1$

$= \dfrac{y}{6}$ Removing a factor of 1

YOUR TURN 1

Simplify: $\dfrac{12p}{4yp}$.

EXAMPLE 2

Simplify: $\dfrac{4x-20}{10}$.

$\dfrac{4x-20}{10} = \dfrac{2 \cdot 2(x-5)}{2 \cdot 5}$ Factoring

$= \boxed{} \cdot \dfrac{2(x-5)}{5}$

$= \dfrac{2(x-5)}{5}$ Removing a factor of 1

YOUR TURN 2

Simplifying: $\dfrac{x^2 + 4x}{5x}$.

EXAMPLE 3	YOUR TURN 3
Simplify: $\dfrac{x^2-9}{x^2+2x-3}$.	Simplify: $\dfrac{x^2-2x-8}{x^2-9x+20}$.

$\dfrac{x^2-9}{x^2+2x-3}=\dfrac{(x-3)(\boxed{})}{(\boxed{})(x+3)}$ Factoring

$=\dfrac{x-3}{\boxed{}}\cdot\dfrac{\boxed{}}{x+3}$

$=\dfrac{x-3}{\boxed{}}$ Removing a factor of 1

EXAMPLE 4	YOUR TURN 4
Simplify: $\dfrac{3a^2-a}{a}$.	Simplify: $\dfrac{y}{y^2-3y}$.

$\dfrac{3a^2-a}{a}=\dfrac{a\left(\boxed{}\right)}{a\cdot 1}$ $a=a\cdot 1$

$=\dfrac{\cancel{a}(3a-1)}{\cancel{a}\cdot 1}$ Canceling indicates a factor of 1: $\dfrac{a}{a}=1$

$=\dfrac{3a-1}{1}=\boxed{}$

EXAMPLE 5	YOUR TURN 5
Simplify: $\dfrac{p-7}{7-p}$.	Simplify: $\dfrac{y+3}{-y-3}$.

$\dfrac{p-7}{7-p}=\dfrac{p-7}{-(p-7)}$

$=\dfrac{\boxed{}(p-7)}{\boxed{}(p-7)}$

$=\dfrac{\boxed{}}{\boxed{}}\cdot\dfrac{p-7}{p-7}$

$=\dfrac{1}{-1}=\boxed{}$ $\dfrac{p-7}{p-7}=1$

YOUR NOTES   Write your questions and additional notes.

Multiplying and Simplifying

ESSENTIALS

After multiplying rational expressions, we simplify if possible.

Example

- Multiply and simplify: $\dfrac{8}{3x} \cdot \dfrac{5x^2}{2}$.

$$\dfrac{8}{3x} \cdot \dfrac{5x^2}{2} = \dfrac{8 \cdot 5x^2}{3x \cdot 2}$$

$$= \dfrac{\cancel{2} \cdot 2 \cdot 2 \cdot 5 \cdot \cancel{x} \cdot x}{3 \cdot \cancel{x} \cdot \cancel{2}}$$

$$= \dfrac{20x}{3}$$

GUIDED LEARNING   📖 **Textbook**   👤 **Instructor**   ▶ **Video**

EXAMPLE 1	YOUR TURN 1
Multiply and simplify: $\dfrac{5y}{2} \cdot \dfrac{y-3}{20}$.	Multiply and simplify: $\dfrac{8x}{3} \cdot \dfrac{x+3}{4x}$.

$\dfrac{5y}{2} \cdot \dfrac{y-3}{20} = \dfrac{5y(y-3)}{2 \cdot 20}$ Multiplying numerators and multiplying denominators

$= \dfrac{5 \cdot y \cdot (y-3)}{2 \cdot 4 \cdot 5}$ Factoring

$= \dfrac{\cancel{5} \cdot y \cdot (y-3)}{2 \cdot 4 \cdot \cancel{5}}$ Removing a factor of 1: $\dfrac{5}{5} = 1$

$= \dfrac{y(y-3)}{\boxed{}}$

We leave the answer in factored form.

EXAMPLE 2	YOUR TURN 2
Multiply and simplify: $\dfrac{x^3-5x^2}{x^2+6x+5}\cdot\dfrac{x^2-x-2}{x^3-25x}$.	Multiply and simplify: $\dfrac{x^2-3x-40}{x^2-9}\cdot\dfrac{x^2-2x-3}{x^2-25}$.

$\dfrac{x^3-5x^2}{x^2+6x+5}\cdot\dfrac{x^2-x-2}{x^3-25x}=\dfrac{\left(x^3-5x^2\right)\left(x^2-x-2\right)}{\left(x^2+6x+5\right)\left(x^3-25x\right)}$

$\qquad=\dfrac{x^2\left(\boxed{}\right)(x-2)\left(\boxed{}\right)}{(x+5)\left(\boxed{}\right)x\left(x^2-25\right)}$

$\qquad=\dfrac{x\cdot\boxed{}\cdot(x-5)(x-2)(x+1)}{(x+5)(x+1)x(x+5)\left(\boxed{}\right)}$

$\qquad=\dfrac{\cancel{x}\cdot x\cdot\cancel{(x-5)}(x-2)\cancel{(x+1)}}{(x+5)\cancel{(x+1)}\cdot\cancel{x}\cdot(x+5)\cancel{(x-5)}}$

$\qquad=\dfrac{\boxed{}(x-2)}{\left(\boxed{}\right)^2}$

YOUR NOTES   Write your questions and additional notes.

Dividing and Simplifying

ESSENTIALS

To divide by a rational expression, multiply by its reciprocal: $\dfrac{A}{B} \div \dfrac{C}{D} = \dfrac{A}{B} \cdot \dfrac{D}{C} = \dfrac{AD}{BC}$.

Example

- Divide: $\dfrac{3x}{5} \div \dfrac{9}{25y}$.

$\dfrac{3x}{5} \div \dfrac{9}{25y} = \dfrac{3x}{5} \cdot \dfrac{25y}{9}$ Multiplying by the reciprocal of the divisor

$= \dfrac{3x \cdot 25y}{5 \cdot 9}$

$= \dfrac{3 \cdot x \cdot 5 \cdot 5 \cdot y}{5 \cdot 3 \cdot 3}$

$= \dfrac{\cancel{3} \cdot x \cdot \cancel{5} \cdot 5 \cdot y}{\cancel{5} \cdot \cancel{3} \cdot 3}$

$= \dfrac{5xy}{3}$

GUIDED LEARNING   📍 **Textbook**   👤 **Instructor**   ▶ **Video**

EXAMPLE 1	YOUR TURN 1
Divide and simplify: $\dfrac{x^2 + 3x}{x - 1} \div \dfrac{5x}{x + 2}$.	Divide and simplify: $\dfrac{x + 2}{x - 5} \div \dfrac{5x}{2x - 10}$.

$\dfrac{x^2 + 3x}{x - 1} \div \dfrac{5x}{x + 2} = \dfrac{x^2 + 3x}{x - 1} \cdot \dfrac{\boxed{}}{5x}$

$= \dfrac{\left(x^2 + 3x\right)\left(x + 2\right)}{\left(x - 1\right) \cdot 5x}$ Multiplying

$= \dfrac{x\boxed{}\left(x + 2\right)}{\left(x - 1\right) \cdot 5 \cdot x}$ Factoring

$= \dfrac{\cancel{x}\left(x + 3\right)\left(x + 2\right)}{\left(x - 1\right) \cdot 5 \cdot \cancel{x}}$ $\dfrac{x}{x} = 1$

$= \dfrac{\left(x + 3\right)\left(x + 2\right)}{\boxed{}\left(x - 1\right)}$ Simplifying

EXAMPLE 2	YOUR TURN 2
Divide and simplify: $\dfrac{x^2-3x-10}{x+5} \div \dfrac{x^2+2x}{x^2+4x-5}$.	Divide and simplify: $\dfrac{x^2+x-6}{x+1} \div \dfrac{x^2-4}{x^2+4x+3}$.

$\dfrac{x^2-3x-10}{x+5} \div \dfrac{x^2+2x}{x^2+4x-5}$

$= \dfrac{x^2-3x-10}{x+5} \cdot \dfrac{\boxed{}}{\boxed{}}$

$= \dfrac{(x-5)(x+2)(x+5)(x-1)}{(x+5)\,x\,(x+2)}$ Factoring

$= \dfrac{(x-5)\,\cancel{(x+2)}\,\cancel{(x+5)}\,(x-1)}{\cancel{(x+5)}\,x\,\cancel{(x+2)}}$

$= \dfrac{(x-5)(x-1)}{\boxed{}}$ Simplifying

EXAMPLE 3	YOUR TURN 3
Perform the indicated operations and simplify: $\dfrac{a^3-1}{a^2-a-2} \div (a^2-a)\cdot(a+1)^2$. Rewrite the division as multiplication, then multiply.	Perform the indicated operations and simplify: $\dfrac{x^2-y^2}{x^2 y^2} \div (x^3+y^3)\cdot(x^2-xy+y^2)$.

$\dfrac{a^3-1}{a^2-a-2} \div (a^2-a)\cdot(a+1)^2$

$= \dfrac{a^3-1}{a^2-a-2} \cdot \dfrac{1}{\boxed{}} \cdot \dfrac{(a+1)^2}{1}$

$= \dfrac{(a^3-1)(a+1)^2}{(a^2-a-2)(a^2-a)}$

$= \dfrac{(a-1)(\boxed{}+a+1)(a+1)(a+1)}{(a+1)(\boxed{})\,a(\boxed{})}$ Factoring

$= \dfrac{\cancel{(a-1)}(a^2+a+1)(a+1)\cancel{(a+1)}}{\cancel{(a+1)}(a-2)\,a\,\cancel{(a-1)}}$

$= \dfrac{(a^2+a+1)(\boxed{})}{a(\boxed{})}$

YOUR NOTES   Write your questions and additional notes.

Practice Exercises

Readiness Check

Classify each statement as true or false.

1. _____ To divide rational expressions, we multiply by the reciprocal of the divisor.

2. _____ The simplified form of $\dfrac{x+4}{4}$ is x.

3. _____ A rational expression is not defined for numbers that make the numerator zero.

4. _____ To simplify rational expressions, we remove factors equal to 1.

Rational Expressions and Functions

5. Find all the numbers for which the rational expression $\dfrac{t^2-4}{3t+2}$ is not defined.

6. Find the domain of f if $f(x)=\dfrac{x-7}{x^2-10x-11}$.

Finding Equivalent Rational Expressions

7. Multiply to obtain an equivalent expression: $\dfrac{a+3}{a+3}\cdot\dfrac{a-1}{2a+5}$. Do not simplify.

Simplifying Rational Expressions

Simplify.

8. $\dfrac{6a^3}{10a}$

9. $\dfrac{4x-16}{12}$

10. $\dfrac{5x^2 - 5x}{2x - 2}$

11. $\dfrac{t^2 - 3t - 18}{t^2 + 6t + 9}$

Multiplying and Simplifying

Multiply and simplify.

12. $\dfrac{4x + 2}{25y^2} \cdot \dfrac{5y}{2x}$

13. $\dfrac{x^2 - 1}{x^2 + 7x + 12} \cdot \dfrac{x^2 + 3x}{x^2 + 2x + 1}$

Dividing and Simplifying

Divide and simplify.

14. $\dfrac{x - 2}{3} \div \dfrac{4 - 2x}{9}$

15. $\dfrac{y^3 - 8}{y - 5} \div \dfrac{y^2 - 4y + 4}{y^2 - 7y + 10}$

16. $\dfrac{a^2 + 6a}{a^2 + 2a - 3} \div \dfrac{a^2 + 12a + 36}{4a - 4}$

17. Perform the indicated operations and simplify.

$$\dfrac{4x^2 - 25y^2}{2x - y} \div \dfrac{x^2 + 3xy - 10y^2}{5y - 10x} \cdot \dfrac{x^2 - 4xy + 4y^2}{2x + 5y}$$

Finding LCMs by Factoring

ESSENTIALS

To find the LCM (least common multiple) of two or more algebraic expressions, we find their prime factorizations. Then we use each factor the greatest number of times that it occurs in any one prime factorization.

Example

- Find the LCM of $m^2 + 4m + 3$ and $2m^2 + 4m - 6$.

$$m^2 + 4m + 3 = (m+1)(m+3)$$

$$2m^2 + 4m - 6 = 2(m^2 + 2m - 3) = 2(m-1)(m+3)$$

$$LCM = 2(m+1)(m-1)(m+3)$$

GUIDED LEARNING   📖 **Textbook**   👤 **Instructor**   ▶ **Video**

EXAMPLE 1	YOUR TURN 1
Find the LCM of 24 and 50.	Find the LCM of 12 and 30.

$$24 = 2 \cdot 2 \cdot 2 \cdot \boxed{}$$

$$50 = 2 \cdot 5 \cdot \boxed{}$$

$$LCM = 2 \cdot 2 \cdot 2 \cdot \boxed{} \cdot 5 \cdot \boxed{}$$

$$LCM = \boxed{}$$

EXAMPLE 2	YOUR TURN 2
Find the LCM of $12a^2b^4$ and $18a^5b^2$.	Find the LCM of $9x^2y$ and $5xyz$.

Write the prime factorizations.

$$12a^2b^4 = \boxed{} \cdot 2 \cdot 3 \cdot \boxed{} \cdot \boxed{} \cdot b \cdot b \cdot b \cdot b$$

$$18a^5b^2 = 2 \cdot 3 \cdot \boxed{} \cdot a \cdot a \cdot a \cdot a \cdot \boxed{} \cdot b \cdot \boxed{}$$

The factors that appear are 2, 3, a, and b.

We use each factor the greatest number of times that it occurs in any one factorization.

$$LCM = 2 \cdot 2 \cdot 3 \cdot \boxed{} \cdot a \cdot a \cdot a \cdot a \cdot a \cdot b \cdot b \cdot b \cdot \boxed{}$$

$$LCM = \boxed{}$$

EXAMPLE 3	YOUR TURN 3
Find the LCM of $x^2 + 5x + 6$ and $(x+3)^2$. Write the prime factorizations. $x^2 + 5x + 6 = (\boxed{})(x+3)$ $(x+3)^2 = (\boxed{})(x+3)$ $\text{LCM} = (\boxed{})(x+3)(x+3)$, or $(\boxed{})(\boxed{})^2$	Find the LCM of $2x^2 - 9x - 5$ and $x^2 - 25$.
EXAMPLE 4	YOUR TURN 4
Find the LCM of $x+2$ and $2x+6$. Write the prime factorizations. $x + 2 = x + 2$ This expression is prime. $2x + 6 = 2(\boxed{})$ These expressions do not share a common factor other than 1, so the LCM is their product. $\text{LCM} = 2(\boxed{})(\boxed{})$	Find the LCM of $t^2 + t$ and $t - 1$.
EXAMPLE 5	YOUR TURN 5
Find the LCM of $c^2 - 25$, $c^3 + 5c^2$, and $c^2 - 6c + 5$. Write the prime factorizations. $c^2 - 25 = (c+5)(\boxed{})$ $c^3 + 5c^2 = c \cdot c(\boxed{})$ $c^2 - 6c + 5 = (c-1)(\boxed{})$ $\text{LCM} = c^2(c+5)(\boxed{})(\boxed{})$	Find the LCM of $y^6 - y^2$, $2y + 2$, and $y^2 + 2y + 1$.

YOUR NOTES   Write your questions and additional notes.

Adding and Subtracting Rational Expressions

ESSENTIALS

To add or subtract when denominators are the same, add or subtract the numerators and keep the same denominator.

$$\frac{A}{C}+\frac{B}{C}=\frac{A+B}{C} \quad \text{and} \quad \frac{A}{C}-\frac{B}{C}=\frac{A-B}{C}, \text{ where } C \neq 0.$$

To add or subtract rational expressions whose denominators are different,

1. Find the *least common denominator* (LCD) by finding the least common multiple (LCM) of the denominators.
2. Rewrite each of the original rational expressions, as needed, in an equivalent form that has the LCD.
3. Add or subtract the numerators. Write the result over the LCD.
4. Simplify, if possible.

Example

• Add: $\dfrac{5}{4x}+\dfrac{6}{2y}$.

$$\frac{5}{4x}+\frac{6}{2y}=\frac{5}{4x}\cdot\frac{y}{y}+\frac{6}{2y}\cdot\frac{2x}{2x} \qquad \text{The LCD is } 4xy.$$

$$=\frac{5y}{4xy}+\frac{12x}{4xy}$$

$$=\frac{5y+12x}{4xy}$$

GUIDED LEARNING	📑 **Textbook**	🧑 **Instructor**	▶ **Video**

EXAMPLE 1	YOUR TURN 1
Add: $\dfrac{9+3x}{x}+\dfrac{2}{x}$. $\dfrac{9+3x}{x}+\dfrac{2}{x}=\dfrac{9+3x+2}{\boxed{}}$ $=\dfrac{\boxed{}+3x}{\boxed{}}$	Add: $\dfrac{5+2a}{a}+\dfrac{4}{a}$.

EXAMPLE 2	YOUR TURN 2
Add: $\dfrac{5}{27x} + \dfrac{8}{9x^2}$.	Add: $\dfrac{2}{15y} + \dfrac{3}{5y^3}$.

$\dfrac{5}{27x} + \dfrac{8}{9x^2}$

$= \dfrac{5}{27x} \cdot \dfrac{x}{x} + \dfrac{8}{9x^2} \cdot \dfrac{\boxed{}}{\boxed{}}$ Multiplying to obtain a common denominator

$= \dfrac{5x}{27x^2} + \dfrac{24}{27x^2}$

$= \dfrac{5x + 24}{\boxed{}}$

EXAMPLE 3	YOUR TURN 3
Add: $\dfrac{6ac}{(a+c)(a-c)} + \dfrac{a-2c}{a+c}$.	Add: $\dfrac{2xy}{(x+y)(x+2y)} + \dfrac{x-y}{x+2y}$.

$\dfrac{6ac}{(a+c)(a-c)} + \dfrac{a-2c}{a+c}$

$= \dfrac{6ac}{(a+c)(a-c)} + \dfrac{a-2c}{a+c} \cdot \dfrac{\boxed{}}{\boxed{}}$

$= \dfrac{6ac + (a-2c)(a-c)}{(a+c)(a-c)}$

$= \dfrac{6ac + a^2 - ac - 2ac + 2c^2}{(a+c)(a-c)}$ Multiplying

$= \dfrac{\boxed{}}{(a+c)(a-c)}$ Combining like terms

$= \dfrac{(a+2c)(a+c)}{(a+c)(a-c)}$ Factoring

$= \dfrac{(a+2c)\,\cancel{(a+c)}}{\cancel{(a+c)}\,(a-c)}$

$= \boxed{}$

EXAMPLE 4	YOUR TURN 4
Subtract: $\dfrac{5x}{x-7y} - \dfrac{4x-1}{7y-x}$.	Subtract: $\dfrac{3m}{m-5} - \dfrac{2m+1}{5-m}$.

$$\dfrac{5x}{x-7y} - \dfrac{4x-1}{7y-x} = \dfrac{5x}{x-7y} - \dfrac{-1}{-1}\cdot\dfrac{4x-1}{7y-x}$$

$$= \dfrac{5x}{x-7y} - \dfrac{1-4x}{\boxed{}} = \dfrac{5x-(1-4x)}{x-7y}$$

$$= \dfrac{5x-1\;\boxed{}\;4x}{x-7y} = \dfrac{9x-1}{x-7y}$$

EXAMPLE 5	YOUR TURN 5
Subtract: $\dfrac{3x}{x^2+4x-5} - \dfrac{2x}{x^2+6x+5}$.	Subtract: $\dfrac{2x}{x^2+4x-5} - \dfrac{x}{x^2+7x+10}$.

$$\dfrac{3x}{x^2+4x-5} - \dfrac{2x}{x^2+6x+5} = \dfrac{3x}{(x+5)(x-1)} - \dfrac{2x}{(x+5)(x+1)}$$

$$\text{LCD} = (x+5)(x-1)(x+1)$$

$$= \dfrac{3x}{(x+5)(x-1)}\cdot\dfrac{\boxed{}}{\boxed{}} - \dfrac{2x}{(x+5)(x+1)}\cdot\dfrac{x-1}{x-1}$$

$$= \dfrac{3x^2+\boxed{}}{(x+5)(x-1)(x+1)} - \dfrac{2x^2-\boxed{}}{(x+5)(x-1)(x+1)}$$

$$= \dfrac{3x^2+3x-\left(\boxed{}\right)}{(x+5)(x-1)(x+1)} = \dfrac{3x^2+3x-\boxed{}+\boxed{}}{(x+5)(x-1)(x+1)}$$

$$= \dfrac{x^2+\boxed{}}{(x+5)(x-1)(x+1)} = \dfrac{x\left(\boxed{}\right)}{(x+5)(x-1)(x+1)}$$

$$= \dfrac{x\,\cancel{(x+5)}}{\cancel{(x+5)}\,(x-1)(x+1)} = \dfrac{x}{(x-1)(x+1)}$$

Combined Additions and Subtractions

ESSENTIALS

To simplify a combination of additions and subtractions of rational expressions, find the LCD of all the rational expressions, write all the expressions in an equivalent form that has the LCD, and add or subtract as indicated.

GUIDED LEARNING

📖 **Textbook**　　👤 **Instructor**　　▶ **Video**

EXAMPLE 1	YOUR TURN 1
Perform the indicated operations and simplify:	Perform the indicated operations and simplify:

$$\frac{6x}{x^2-1}-\frac{2}{x+1}+\frac{1}{1-x}.$$

YOUR TURN 1:

$$\frac{3x}{x^2-4}+\frac{1}{2-x}+\frac{2}{x-2}.$$

$$\frac{6x}{x^2-1}-\frac{2}{x+1}+\frac{1}{1-x}$$

$$=\frac{6x}{(x+1)(x-1)}-\frac{2}{x+1}+\frac{-1}{-1}\cdot\frac{1}{1-x}$$

$$=\frac{6x}{(x+1)(x-1)}-\frac{2}{x+1}+\frac{-1}{\boxed{}}\qquad LCD=(x+1)(x-1)$$

$$=\frac{6x}{(x+1)(x-1)}-\frac{2}{x+1}\cdot\frac{\boxed{}}{\boxed{}}+\frac{-1}{x-1}\cdot\frac{x+1}{x+1}$$

$$=\frac{6x}{(x+1)(x-1)}-\frac{2x-\boxed{}}{(x+1)(x-1)}+\frac{-x-\boxed{}}{(x-1)(x+1)}$$

$$=\frac{6x-\left(\boxed{}\right)+(-x)-1}{(x+1)(x-1)}$$

$$=\frac{6x-2x+\boxed{}+(-x)-1}{(x+1)(x-1)}$$

$$=\frac{3x+\boxed{}}{(x+1)(x-1)}$$

Practice Exercises

Readiness Check

Classify each statement as true or false.

1. _____ To subtract fractions, they must have a common denominator.

2. _____ When adding rational expressions, we add numerators and add denominators, and then simplify, if possible.

3. _____ The LCM of $8x^3$ and $4x$ is $32x^4$.

4. _____ The LCD of two rational expressions is the LCM of the denominators of the rational expressions.

Finding LCMs by Factoring

Find the LCM.

5. $24, 40$

6. $10x^2y, \ 15x^4y$

7. $y+5, \ y^2-5y$

8. $x^2-1, \ x^2-6x-7$

9. $12a^3-24a^2, \ a^3-4a^2+4a$

10. $6x+18, \ 9x^2-81, \ x^3+6x^2+9x$

Adding and Subtracting Rational Expressions

Add or subtract. Then simplify.

11. $\dfrac{8}{3k}+\dfrac{1}{3k}$

12. $\dfrac{2y-9}{y^2-3y-4}-\dfrac{y+4}{y^2-3y-4}$

13. $\dfrac{x+4}{x+5}+\dfrac{2x-3}{x-6}$

14. $\dfrac{x}{x^2-2x+1}-\dfrac{3}{x^2-6x+5}$

15. $\dfrac{cd}{c^2-d^2}-\dfrac{c+d}{c-d}$

16. $\dfrac{2y}{(y-1)(y-3)}+\dfrac{3}{(y-1)(y+2)}$

Combined Additions and Subtractions

Perform the indicated operations and simplify.

17. $\dfrac{6}{m-2}-\dfrac{m+3}{m^2+m-6}+\dfrac{5}{m+3}$

Divisor a Monomial

ESSENTIALS

To divide a polynomial by a monomial, divide each term of the polynomial by the monomial.

$$\frac{A+B}{C} = \frac{A}{C} + \frac{B}{C}$$

Examples

- $$\frac{4x^2 - 2x + 8}{2x}$$

$$= \frac{4x^2}{2x} - \frac{2x}{2x} + \frac{8}{2x}$$

$$= 2x - 1 + \frac{4}{x}$$

- Divide $8x^3 - 4x^2 + 20x + 40$ by $4x^2$.

$$\left(8x^3 - 4x^2 + 20x + 40\right) \div \left(4x^2\right)$$

$$= \frac{8x^3 - 4x^2 + 20x + 40}{4x^2}$$

$$= \frac{8x^3}{4x^2} - \frac{4x^2}{4x^2} + \frac{20x}{4x^2} + \frac{40}{4x^2}$$

$$= 2x - 1 + \frac{5}{x} + \frac{10}{x^2}$$

GUIDED LEARNING 📖 **Textbook**   👤 **Instructor**   ▶ **Video**

EXAMPLE 1	YOUR TURN 1
Divide $15y^3 - 25y^2 + 5$ by $5y$.	Divide $-48x^4 + 16x^2 - 8$ by $4x$.
Divide each term of the numerator by the monomial, and then perform the three indicated divisions.	

$$\left(15y^3 - 25y^2 + 5\right) \div \left(5y\right)$$

$$= \frac{15y^3 - 25y^2 + 5}{5y}$$

$$= \frac{\boxed{}}{5y} - \frac{\boxed{}}{5y} + \frac{\boxed{}}{5y}$$

$$= 3y^2 - 5y + \frac{1}{y}$$

EXAMPLE 2	YOUR TURN 2
Divide: $\left(20x^5y^4 + 4x^3y^5 - 16x^2y^8\right) \div \left(-2x^2y^3\right)$.	Divide:
We first write a single rational expression, then divide each term of the numerator by the monomial, and finally perform the indicated divisions.	$\left(12x^4y^4 - 18x^5y^2 + 30x^2y^5\right) \div \left(-3x^2y\right)$.

$$\left(20x^5y^4 + 4x^3y^5 - 16x^2y^8\right) \div \left(-2x^2y^3\right)$$

$$= \frac{20x^5y^4 + 4x^3y^5 - 16x^2y^8}{-2x^2y^3}$$

$$= \frac{20x^5y^4}{-2x^2y^3} + \frac{4x^3y^5}{-2x^2y^3} - \frac{16x^2y^8}{-2x^2y^3}$$

$$= \boxed{} - 2xy^2 + \boxed{}$$

YOUR NOTES   Write your questions and additional notes.

Divisor Not a Monomial

ESSENTIALS

In division of polynomials when the divisor has more than one term, we use a procedure very similar to long division in arithmetic.

Tips for Dividing Polynomials

1. Arrange the polynomials in descending order.
2. If there are missing terms in the dividend, write them with 0 coefficients or leave space for them.
3. Perform the long division process until the degree of the remainder is less than the degree of the divisor.

Example

- Divide: $(x^2 - 5x - 24) \div (x - 8)$.

$$
\begin{array}{r}
x + 3 \\
x - 8 \overline{)\, x^2 - 5x - 24} \\
\underline{x^2 - 8x} \\
3x - 24 \\
\underline{3x - 24} \\
0
\end{array}
$$

The quotient is $x + 3$.

GUIDED LEARNING   📖 **Textbook**   👤 **Instructor**   ▶ **Video**

EXAMPLE 1	YOUR TURN 1
Divide $x^2 + 8x + 3$ by $x + 2$.	Divide $x^2 + 4x + 1$ by $x + 5$.

EXAMPLE 1:

$$
\begin{array}{r}
x \\
x + 2 \overline{)\, x^2 + 8x + 3} \\
\underline{x^2 + 2x} \\
\boxed{} + 3
\end{array}
$$

Divide: $x^2/x = x$.

Multiply: $x(x + 2) = x^2 + 2x$.

Subtract: $x^2 + 8x - (x^2 + 2x) = 6x$.

We continue.

(continued)

$$x + \boxed{}$$
$$x + 2 \overline{\smash{\big)}\,x^2 + 8x + 3}$$

$$\underline{x^2 + 2x}$$

$$6x + 3 \qquad \text{Divide: } 6x/x = 6.$$

$$\underline{6x + 12} \qquad \text{Multiply: } 6(x+2) = 6x + 12.$$

$$\boxed{} \qquad \text{Subtract: } 6x + 3 - (6x + 12) = -9.$$

The remainder is −9.

Check: $(x+2)(x+6) + (-9) = x^2 + 8x + 12 + (-9)$

$$= x^2 + 8x + 3$$

We get the dividend, so the answer checks.

The answer is $x + 6$, R −9, or $x + 6 + \dfrac{-9}{x+2}$.

EXAMPLE 2	YOUR TURN 2

EXAMPLE 2

Divide: $\left(n^3 - 6 + 2n^2\right) \div \left(n^2 - 2\right)$.

We rewrite the dividend in descending order and write in the missing term, $0n$.

$$n + 2$$
$$n^2 - 2 \overline{\smash{\big)}\,n^3 + 2n^2 + 0n - 6}$$

$$\underline{n^3 \qquad\quad - 2n}$$

$$2n^2 + 2n - 6$$

$$\underline{2n^2 \qquad - 4}$$

$$2n - \boxed{}$$

The degree of the remainder is less than the degree of the divisor, so we are finished.

The answer is $n + 2 + \dfrac{2n - 2}{\boxed{}}$.

YOUR TURN 2

Divide:

$$\left(3y^2 - 4 + y^3\right) \div \left(y^2 + 1\right).$$

YOUR NOTES   Write your questions and additional notes.

Synthetic Division

ESSENTIALS

To use synthetic division to divide a polynomial by a binomial, the divisor must be in the form $x - a$.

GUIDED LEARNING

 Textbook   **Instructor**   **Video**

EXAMPLE 1	YOUR TURN 1
Use synthetic division to divide: $\left(x^3 + 4x^2 + x - 4\right) \div (x + 1)$.	Use synthetic division to divide: $\left(x^3 - 6x^2 + 2x - 5\right) \div (x + 2)$.

Note: $x + 1 = x - (-1)$. We use the format shown below with -1 in the half-box and the coefficients of the dividend, written in descending order, to the right of it. Next we bring down the first 1, multiply it by -1, write the product below the 4, and then add: $4 + (-1) = 3$. We have

$$
\begin{array}{r|rrrr}
-1 & 1 & 4 & 1 & -4 \\
   &   & -1 &   &   \\
\hline
   & 1 & 3 &   \\
\end{array}
$$

Next we multiply 3 by -1, write the product below the second 1, and add: $1 + (-3) = -2$. We have

$$
\begin{array}{r|rrrr}
-1 & 1 & 4 & 1 & -4 \\
   &   & -1 & -3 &   \\
\hline
   & 1 & 3 & -2 \\
\end{array}
$$

Finally, multiply -2 by -1, write the product below -4 and add: $-4 + 2 = -2$. We have

$$
\begin{array}{r|rrrr}
-1 & 1 & 4 & 1 & -4 \\
   &   & -1 & -3 & \boxed{} \\
\hline
   & 1 & 3 & -2 & -2 \\
\end{array}
$$

The degree of the quotient is one less than the degree of the dividend. The coefficients of the quotient are 1, 3, and -2, and the remainder is -2.

The answer is $x^2 + 3x - 2$, R -2, or

$$x^2 + 3x - 2 + \frac{-2}{\boxed{}}.$$

EXAMPLE 2	YOUR TURN 2
Use synthetic division to divide: $(9x - 5 + 4x^3) \div (x - 3)$.	Use synthetic division to divide: $(2x^2 - 1 + 5x^3) \div (x - 2)$.

Write the dividend in descending order. Include the missing term, $0x^2$:

$$(4x^3 + 0x^2 + 9x - 5) \div (x - 3).$$

We use the procedure outlined in Example 1.

```
3| 4    0    9    -5
        12   36   □
   ─────────────────
   4    □    45 | 130
```

The answer is $4x^2 + \boxed{} + 45 + \dfrac{130}{x - 3}$.

YOUR NOTES   Write your questions and additional notes.

Practice Exercises

Readiness Check

Classify each statement as true or false.

1. _____ When dividing one polynomial by another, we stop dividing when the degree of the remainder is equal to the degree of the divisor.
2. _____ When dividing a polynomial by a monomial, the quotient has the same number of terms as the dividend.
3. _____ To check the result after dividing one polynomial by another, we multiply the divisor by the remainder and add the quotient.
4. _____ When dividing one polynomial by another, if the remainder is zero then the divisor is a factor of the dividend.

Divisor a Monomial

Divide and check.

5. $\dfrac{42x^5 + 14x^3 - 35x^2}{7x^2}$

6. $\dfrac{24m^4n^7 - 15m^3n^3 + 18m^4n^2}{3m^3n^2}$

7. $\left(12y^8 - 3y^2 + 6y\right) \div \left(-3y\right)$

8. $\left(a^4b^4 + 2a^2b^2 - 10ab\right) \div \left(2ab^2\right)$

Divisor Not a Monomial

Divide and check.

9. $\left(x^2 + 2x - 15\right) \div \left(x - 3\right)$

10. $\left(z^2 - 5z - 20\right) \div \left(z + 6\right)$

11. $\left(4x^3 + 8x^2 - 3x + 5\right) \div \left(2x - 1\right)$ **12.** $\left(y^3 - 5 + y\right) \div \left(y + 2\right)$

Synthetic Division

Use synthetic division to divide.

13. $\left(x^3 - 5x^2 + 2x - 1\right) \div \left(x - 3\right)$ **14.** $\left(4x^3 + 3x^2 - 5x - 3\right) \div \left(x + 4\right)$

15. $\left(y^3 - 2y + 8\right) \div \left(y + 1\right)$ **16.** $\left(n^5 - 32\right) \div \left(n + 2\right)$

17. $\left(9x^4 + 3x^3 - 23x^2 + 13x - 2\right) \div \left(x - \dfrac{1}{3}\right)$

Simplifying Complex Rational Expressions

ESSENTIALS

A **complex rational expression** is a rational expression that contains rational expressions within its numerator and/or its denominator.

To simplify a complex rational expression by multiplying by 1,

1. Find the LCM of all the denominators of all the rational expressions within the complex rational expression.

2. Multiply by 1 using $\dfrac{\text{LCM}}{\text{LCM}}$.

3. If possible, simplify.

To simplify a complex rational expression by adding or subtracting in the numerator and the denominator,

1. Add or subtract, as necessary, to get one rational expression in the numerator.

2. Add or subtract, as necessary, to get one rational expression in the denominator.

3. Divide the numerator by the denominator.

4. If possible, simplify.

Example

$$\frac{4-\dfrac{1}{2x}}{\dfrac{4}{x}+\dfrac{1}{x^2}}=\frac{4-\dfrac{1}{2x}}{\dfrac{4}{x}+\dfrac{1}{x^2}}\cdot\frac{2x^2}{2x^2}$$

The LCM of all the denominators is $2x^2$.

Multiplying by 1 using $\dfrac{2x^2}{2x^2}$

$$=\frac{4\cdot 2x^2-\dfrac{1}{2x}\cdot 2x^2}{\dfrac{4}{x}\cdot 2x^2+\dfrac{1}{x^2}\cdot 2x^2}$$

Using the distributive law

$$=\frac{8x^2-x}{8x+2}$$

Simplifying in the numerator and denominator

$$=\frac{x(8x-1)}{2(4x+1)}$$

We can't simplify further.

GUIDED LEARNING   📍 **Textbook**   👤 **Instructor**   ▶ **Video**

EXAMPLE 1	YOUR TURN 1
Simplify: $\dfrac{\dfrac{3}{2x^3}+\dfrac{4}{x^2}}{\dfrac{5}{6x}-\dfrac{1}{x^2}}$.	Simplify: $\dfrac{\dfrac{6}{am^4}+\dfrac{m}{2a^2}}{\dfrac{6}{am^2}+\dfrac{2m}{a}}$.

We use the method of multiplying by 1.

The LCM of all the denominators is $6x^3$. We multiply the rational expression by 1 using the form $\dfrac{6x^3}{6x^3}$.

$$\frac{\dfrac{3}{2x^3}+\dfrac{4}{x^2}}{\dfrac{5}{6x}-\dfrac{1}{x^2}}=\frac{\dfrac{3}{2x^3}+\dfrac{4}{x^2}}{\dfrac{5}{6x}-\dfrac{1}{x^2}}\cdot\frac{\boxed{}}{\boxed{}}$$

$$=\frac{\left(\dfrac{3}{2x^3}+\dfrac{4}{x^2}\right)6x^3}{\left(\dfrac{5}{6x}-\dfrac{1}{x^2}\right)6x^3}$$

$$=\frac{\dfrac{3}{2x^3}\cdot 6x^3+\dfrac{4}{x^2}\cdot 6x^3}{\dfrac{5}{6x}\cdot 6x^3-\dfrac{1}{x^2}\cdot 6x^3}\quad\text{Distributing }6x^3$$

$$=\frac{\boxed{}+\boxed{}}{\boxed{}-\boxed{}}\quad\text{Simplifying}$$

$$=\frac{3(3+8x)}{x(5x-6)}\quad\text{Factoring}$$

We cannot simplify further.

EXAMPLE 2	YOUR TURN 2
Simplify: $\dfrac{\dfrac{3}{2x^3}+\dfrac{4}{x^2}}{\dfrac{5}{6x}-\dfrac{1}{x^2}}$.	Simplify: $\dfrac{\dfrac{2}{4k}-\dfrac{k}{2}}{\dfrac{6}{k^2}-\dfrac{1}{k}}$.

We use the method of adding or subtracting in the numerator and the denominator.

We multiply to obtain a common denominator in the numerator and in the denominator.

$$\frac{\dfrac{3}{2x^3}+\dfrac{4}{x^2}}{\dfrac{5}{6x}-\dfrac{1}{x^2}}=\frac{\dfrac{3}{2x^3}+\dfrac{4}{x^2}\cdot\dfrac{2x}{2x}}{\dfrac{5}{6x}\cdot\dfrac{x}{x}-\dfrac{1}{x^2}\cdot\dfrac{6}{6}}$$

$$=\frac{\dfrac{3}{2x^3}+\dfrac{8x}{2x^3}}{\dfrac{5x}{6x^2}-\dfrac{6}{6x^2}}$$

$$=\frac{\dfrac{\boxed{}}{2x^3}}{\dfrac{\boxed{}}{6x^2}} \qquad \text{Adding in the numerator; subtracting in the denominator}$$

$$=\frac{3+8x}{2x^3}\cdot\frac{\boxed{}}{\boxed{}} \qquad \text{Multiplying by the reciprocal of the divisor}$$

$$=\frac{(3+8x)\cdot3\cdot\cancel{2}\cdot\cancel{x^2}}{\cancel{2}\cdot\cancel{x^2}\cdot x(5x-6)}$$

$$=\frac{3(3+8x)}{x(5x-6)} \qquad \frac{2x^2}{2x^2}=1$$

EXAMPLE 3	YOUR TURN 3
Simplify: $\dfrac{\dfrac{1}{x}-\dfrac{1}{y}}{\dfrac{1}{x^2}-\dfrac{1}{y^2}}$.	Simplify: $\dfrac{\dfrac{1}{m^3}-\dfrac{1}{n^3}}{\dfrac{1}{m^2}-\dfrac{1}{n^2}}$.

$$\frac{\dfrac{1}{x}-\dfrac{1}{y}}{\dfrac{1}{x^2}-\dfrac{1}{y^2}} = \frac{\dfrac{1}{x}\cdot\dfrac{y}{y}-\dfrac{1}{y}\cdot\dfrac{\boxed{}}{\boxed{}}}{\dfrac{1}{x^2}\cdot\dfrac{y^2}{y^2}-\dfrac{1}{y^2}\cdot\dfrac{\boxed{}}{\boxed{}}}$$

$$= \frac{\dfrac{y-x}{xy}}{\dfrac{y^2-x^2}{x^2y^2}} \qquad \text{Writing a single rational expression in the numerator and in the denominator}$$

$$= \frac{y-x}{xy}\cdot\frac{x^2y^2}{y^2-x^2}$$

$$= \frac{\cancel{y-x}}{\cancel{x}\cdot\cancel{y}}\cdot\frac{\cancel{x}\cdot x\cdot\cancel{y}\cdot y}{\cancel{(y-x)}(y+x)} \qquad \text{Factoring}$$

$$= \boxed{}$$

YOUR NOTES   Write your questions and additional notes.

Difference Quotients

ESSENTIALS

The expression

$$\frac{f(x+h)-f(x)}{h}$$

is the **difference quotient**, or the **average rate of change**, of a function *f*.

Examples

- For the function *f* given by $f(x)=3x-2$, find and simplify the difference quotient.

$$\frac{f(x+h)-f(x)}{h} = \frac{3(x+h)-2-(3x-2)}{h}$$

$$= \frac{3x+3h-2-3x+2}{h}$$

$$= \frac{3h}{h}=3$$

- For the function *f* given by $f(x)=\dfrac{3}{x}$, find and simplify the difference quotient.

$$\frac{f(x+h)-f(x)}{h} = \frac{\dfrac{3}{x+h}-\dfrac{3}{x}}{h}$$

$$= \frac{\dfrac{3}{x+h}\cdot\dfrac{x}{x}-\dfrac{3}{x}\cdot\dfrac{x+h}{x+h}}{h}$$

$$= \frac{\dfrac{3x}{x(x+h)}-\dfrac{3x+3h}{x(x+h)}}{h} = \frac{\dfrac{3x-(3x+3h)}{x(x+h)}}{h}$$

$$= \frac{\dfrac{3x-3x-3h}{x(x+h)}}{h} = \frac{\dfrac{-3h}{x(x+h)}}{h} = \frac{-3\cdot h}{x(x+h)}\cdot\frac{1}{h}$$

$$= \frac{-3}{x(x+h)}, \text{ or } -\frac{3}{x(x+h)}$$

GUIDED LEARNING

🅣 **Textbook**   🅘 **Instructor**   ▶ **Video**

EXAMPLE 1	YOUR TURN 1

For the function f given by $f(x) = 2x + 5$, find and simplify the difference quotient.

$$\frac{f(x+h) - f(x)}{h} = \frac{2(x+h) + \boxed{} - (2x+5)}{h}$$

$$= \frac{2x + 2h + 5 - 2x - \boxed{}}{h}$$

$$= \frac{\boxed{}}{h} = \boxed{}$$

For the function f given by $f(x) = 4x - 3$, find and simplify the difference quotient.

EXAMPLE 2	YOUR TURN 2

For the function f given by $f(x) = x^2 - 4$, find and simplify the difference quotient.

$$\frac{f(x+h) - f(x)}{h} = \frac{(x+h)^2 - 4 - \left(\boxed{} - 4\right)}{h}$$

$$= \frac{x^2 + 2xh + h^2 - 4 - x^2 + \boxed{}}{h}$$

$$= \frac{2xh + h^2}{\boxed{}}$$

$$= \frac{h\left(2x + \boxed{}\right)}{h}$$

$$= \boxed{} + h$$

For the function f given by $f(x) = 3 - x^2$, find and simplify the difference quotient.

YOUR NOTES   Write your questions and additional notes.

Practice Exercises

Readiness Check

Choose the word or expression that best completes the sentence.

numerator   denominator   LCM   complex rational expression

1. A _____ contains one or more rational expressions within its numerator and/or its denominator.

2. To simplify a complex rational expression, we may begin by multiplying both the numerator and the denominator by the _____ of all the denominators of all the rational expressions within the complex rational expression.

3. To simplify a complex rational expression, we may begin by obtaining a single rational expression in the _____ and in the _____.

4. If we have a single rational expression in the numerator and in the denominator of a complex rational expression, we multiply the rational expression in the _____ by the reciprocal of the rational expression in the _____.

Simplifying Complex Rational Expressions

Simplify.

5. $\dfrac{\dfrac{3}{4}-\dfrac{4}{5}}{\dfrac{5}{8}-\dfrac{1}{2}}$

6. $\dfrac{\dfrac{3}{x}+1}{\dfrac{9}{x^2}-1}$

7. $\dfrac{\dfrac{2x+4}{x+5}}{\dfrac{3x+6}{x-6}}$

8. $\dfrac{\dfrac{x}{x^2-2x+1}-\dfrac{1}{x-1}}{\dfrac{3}{x^2-6x+5}+\dfrac{x}{x-5}}$

9. $\dfrac{\dfrac{1}{c^2}-\dfrac{1}{d^2}}{\dfrac{1}{c^3}+\dfrac{1}{d^3}}$

10. $\dfrac{\dfrac{2}{x^2-3x-4}-\dfrac{1}{x^2+3x+2}}{\dfrac{2}{x^2-2x-8}-\dfrac{1}{x^2-4}}$

Difference Quotients

For each function f, find and simplify the difference quotient.

11. $f(x)=4x+5$

12. $f(x)=3x^2+1$

13. $f(x)=\dfrac{1}{6x}$

Rational Equations

ESSENTIALS

A **rational equation** is an equation that contains one or more rational expressions.

To solve a rational equation,

1. Clear fractions by multiplying both sides of the equation by the LCM of all the denominators.
2. Solve the resulting equation.
3. Check all possible solutions in the original equation.

Example

* Solve: $\dfrac{1}{x} + \dfrac{3}{4x} = 1$.

Multiply both sides of the equation by the LCM of the denominators, $4x$.

$$4x\left(\frac{1}{x} + \frac{3}{4x}\right) = 4x \cdot 1$$

$$4x \cdot \frac{1}{x} + 4x \cdot \frac{3}{4x} = 4x$$

$$4 + 3 = 4x$$

$$7 = 4x$$

$$\frac{7}{4} = x$$

Check:

$$\dfrac{\dfrac{1}{x} + \dfrac{3}{4x} = 1}{\begin{array}{c|c} \dfrac{1}{\dfrac{7}{4}} + \dfrac{3}{4 \cdot \dfrac{7}{4}} & \overset{?}{{}} 1 \\[4ex] 1 \cdot \dfrac{4}{7} + \dfrac{3}{7} & \\[3ex] \dfrac{4}{7} + \dfrac{3}{7} & \\[3ex] 1 & \text{TRUE} \end{array}}$$

The solution is $\dfrac{7}{4}$.

GUIDED LEARNING

Textbook   **Instructor**   **Video**

EXAMPLE 1	YOUR TURN 1

Solve: $\dfrac{1}{2x}+\dfrac{3}{x^2}=1$.

Solve: $\dfrac{2}{x}-\dfrac{1}{x^2}=-3$.

$$\dfrac{1}{2x}+\dfrac{3}{x^2}=1$$

$$2x^2\left(\dfrac{1}{2x}+\dfrac{3}{x^2}\right)=2x^2\,(1) \qquad \text{Multiplying by the LCM}$$

$$2x^2\cdot\dfrac{1}{2x}+2x^2\cdot\dfrac{3}{x^2}=2x^2$$

$$\boxed{}+\boxed{}=2x^2$$

$$0=2x^2-x-6$$

$$0=(2x+3)(x-2)$$

$$x=\boxed{} \quad or \quad x=\boxed{}$$

Check:

For $x=-\dfrac{3}{2}$:

$$\dfrac{1}{2x}+\dfrac{3}{x^2}=1$$

$$\dfrac{1}{2\left(-\dfrac{3}{2}\right)}+\dfrac{3}{\left(-\dfrac{3}{2}\right)^2}\ ?\ 1$$

$$\dfrac{1}{-3}+\dfrac{3}{\dfrac{9}{4}}$$

$$-\dfrac{3}{9}+\dfrac{12}{9}$$

$$1 \qquad \text{TRUE}$$

For $x=2$:

$$\dfrac{1}{2x}+\dfrac{3}{x^2}=1$$

$$\dfrac{1}{2\cdot2}+\dfrac{3}{2^2}\ ?\ 1$$

$$\dfrac{1}{4}+\dfrac{3}{4}$$

$$1 \qquad \text{TRUE}$$

The solutions are $-\dfrac{3}{2}$ and 2.

EXAMPLE 2	YOUR TURN 2
Solve: $6 + \dfrac{2x}{x-3} = \dfrac{6}{x-3}$.	Solve: $1 + \dfrac{2x}{x+5} = \dfrac{-10}{x+5}$.

We multiply both sides of the equation by the LCM of the denominators, $x - 3$, and solve the resulting equation.

$$\boxed{}\left(6 + \frac{2x}{x-3}\right) = \boxed{} \cdot \frac{6}{x-3}$$

$$(x-3)\cdot 6 + (x-3)\cdot \frac{2x}{x-3} = (x-3)\cdot \frac{6}{x-3}$$

$$6x - 18 + 2x = 6$$

$$\boxed{} - 18 = 6$$

$$8x = \boxed{}$$

$$x = 3$$

Check:

$$
\begin{array}{c|c}
6 + \dfrac{2x}{x-3} & = \dfrac{6}{x-3} \\
\hline
6 + \dfrac{2\cdot 3}{3-3} \ \ ? & \dfrac{6}{3-3} \\
6 + \dfrac{6}{0} & \dfrac{6}{0} \qquad \text{NOT DEFINED}
\end{array}
$$

There is no solution.

EXAMPLE 3	YOUR TURN 3
Given that $f(x) = \dfrac{x+2}{x-1}$, find all values of x for which $f(x) = \dfrac{1}{2}$.	Given that $f(x) = \dfrac{x-2}{x+4}$ find all values of x for which $f(x) = -\dfrac{1}{2}$.

$$f(x) = \frac{1}{2}$$

$$\frac{x+2}{\boxed{}} = \frac{1}{2}$$

$$2(x-1)\left(\frac{x+2}{x-1}\right) = 2(x-1)\frac{1}{2}$$

$$2(x+2) = \boxed{}$$

$$2x + \boxed{} = x-1$$

$$x = \boxed{}$$

Check: $f(-5) = \dfrac{-5+2}{-5-1} = \dfrac{-3}{-6} = \dfrac{1}{2}$.

The solution is $\boxed{}$.

YOUR NOTES   Write your questions and additional notes.

Practice Exercises

Readiness Check

Classify each statement as true or false.

1. If an equation contains a rational expression with a variable in the denominator, then zero cannot be a solution of the equation.

2. Multiplying both sides of a rational equation by the LCM of the denominators might produce an equation that is not equivalent to the original equation.

3. $\dfrac{x+3}{x-2}$ is an example of a rational equation.

4. Numbers that cause a denominator in a rational equation to be zero cannot be solutions of the equation.

Rational Equations

Solve. Don't forget to check!

5. $\dfrac{m}{2} + \dfrac{m}{6} = 5$

6. $\dfrac{n+4}{n-3} = \dfrac{4}{5}$

7. $\dfrac{1}{y} - \dfrac{1}{6y} = \dfrac{8}{9}$

8. $\dfrac{5}{6x+2} + \dfrac{3}{4} = \dfrac{6}{x+3}$

9. $\dfrac{7}{8} - \dfrac{5}{4x} = \dfrac{14}{16}$

10. $\dfrac{k+4}{k-2} = \dfrac{k-3}{k+2}$

11. $\dfrac{x+2}{x-3} = \dfrac{5}{x-3}$

12. $\dfrac{4}{x-5} = \dfrac{x}{x-2}$

13. $\dfrac{a}{a-4} - \dfrac{2}{a} = \dfrac{16}{a^2 - 4a}$

14. $\dfrac{n}{n-2} = \dfrac{5}{n+2} - \dfrac{n-1}{2-n}$

15. For $f(x) = x - \dfrac{18}{x}$, find all values of a for which $f(a) = 3$.

Work Problems

ESSENTIALS

The Work Principle

If a = the time it takes A to do a job,

        b = the time it takes B to do the same job, and

        t = the time it takes A and B to do the job working together,

then $\dfrac{t}{a} + \dfrac{t}{b} = 1$.

GUIDED LEARNING

 Textbook   **Instructor**   **Video**

EXAMPLE 1	YOUR TURN 1

EXAMPLE 1

The Harrells' pool can be filled in 20 hr using their spring alone and in 15 hr using their well alone. How long will it take to fill the pool if both the spring and the well are used at the same time?

1. **Familiarize.** In the work equation, $\dfrac{t}{a} + \dfrac{t}{b} = 1$, $a = 20$, the number of hours required to fill the pool using the spring alone, and $b = 15$, the number of hours required to fill the pool using the well alone.

2. **Translate.**

$$\frac{t}{20} + \frac{t}{\boxed{}} = \boxed{}$$

3. **Solve.**

$$\frac{t}{20} + \frac{t}{15} = 1$$

$$\boxed{}\left(\frac{t}{20} + \frac{t}{15}\right) = \boxed{} \cdot 1 \qquad \text{Multiplying by the LCM of the denominators}$$

$$60 \cdot \frac{t}{20} + 60 \cdot \frac{t}{15} = 60$$

$$\boxed{} + \boxed{} = 60 \qquad \text{Simplifying}$$

$$7t = 60$$

$$t = \frac{60}{7}, \text{ or } 8\frac{4}{7} \text{ hr}$$

4. **Check.** A check confirms the result.

5. **State.** It would take $8\dfrac{4}{7}$ hr to fill the pool using both the spring and the well at the same time.

YOUR TURN 1

It takes Becky 3 hr to clean out a barn. It takes Jon 4 hr to do the same job. How long would it take if they worked together to clean out the barn?

EXAMPLE 2	YOUR TURN 2
It takes Kori 20 hr longer to paint a townhouse than it takes Chris. Working together, they can do the job in 24 hr. How long would it take each, working alone, to paint a townhouse?	Working together, Gavin and Jeff can groom a shelter's dogs in 6 hr. Working alone, it would take Gavin 5 hr longer than it would take Jeff. How long would it take each, working alone, to groom the shelter's dogs?

1. **Familiarize.** We let $c =$ the number of hr it takes Chris, working alone, to paint the townhouse. Then $c + 20 =$ the number of hr it would take Kori, working alone, to paint the townhouse.

2. **Translate.** Using the work equation, $\dfrac{t}{a} + \dfrac{t}{b} = 1$, we have

$$\frac{24}{c+20} + \frac{24}{\boxed{}} = 1.$$

3. **Solve.** We multiply both sides by the LCM of the denominators to clear fractions.

$$c(c+20)\left(\frac{24}{c+20} + \frac{24}{c}\right) = c(c+20) \cdot 1$$

$$c \cdot 24 + (c+20) \cdot 24 = c(c+20)$$

$$24c + \boxed{} + \boxed{} = c^2 + 20c \qquad \text{Simplifying}$$

$$0 = c^2 - 28c - 480 \qquad \begin{array}{l}\text{Getting 0 on}\\ \text{one side}\end{array}$$

$$0 = \boxed{}(c-40)$$

$$c = \boxed{} \ \text{ or } \ c = \boxed{} \qquad \begin{array}{l}\text{Using the}\\ \text{principle of}\\ \text{zero products}\end{array}$$

4. **Check.** Since time cannot be negative in this situation, we check only 40. If it takes Chris 40 hr alone, it takes Kori $40 + 20$, or 60 hr alone. In 24 hr, they would have finished $\dfrac{24}{60} + \dfrac{24}{40} = \dfrac{2}{5} + \dfrac{3}{5} = 1$ complete townhouse. The answer checks.

5. **State.** It would take Kori 60 hr to paint the townhouse working alone, and it would take Chris 40 hr.

YOUR NOTES   Write your questions and additional notes.

Applications Involving Proportions

ESSENTIALS

Any rational expression $\frac{a}{b}$ represents a **ratio**.

The ratio of two different kinds of measure is called a **rate**.

A **proportion** is an equality of ratios, $\frac{A}{B} = \frac{C}{D}$, read "*A* is to *B* as *C* is to *D*."

The numbers named in a true proportion are **proportional** to each other.

If $\frac{A}{B} = \frac{C}{D}$, then the **cross products** *AD* and *BC* are equal.

Example

- The dimensions on an architect's drawing are proportional to the dimensions of the building represented. A wall that is 4 cm long in the drawing represents a 15-ft wall in the building. How long is a building wall that is represented by 6 cm in the drawing?

Let $w =$ the building wall length.

Drawing length → $\frac{4}{15} = \frac{6}{w}$ ← Drawing length
Building length → $\frac{4}{15} = \frac{6}{w}$ ← Building length

$$4w = 15 \cdot 6 \qquad \text{Equating cross products}$$

$$w = \frac{15 \cdot 6}{4} = 22.5$$

The building wall is 22.5 ft long.

GUIDED LEARNING   📖 **Textbook**   👤 **Instructor**   ▶ **Video**

EXAMPLE 1	YOUR TURN 1
In city driving, a Subaru Outback can travel 171 mi on 6 gal of gas. How much gas will be required for 228 mi of city driving? 1. **Familiarize.** Let $x =$ the number of gallons of gas required to drive 228 mi. 2. **Translate.** We write a proportion. Miles → $\frac{171}{6} = \frac{228}{x}$ ← Miles Gas → $\quad$ ← Gas <div align="center">(continued)</div>	On the highway, a Ford F150 can travel 111 mi on 6 gal of gas. How much gas will be required for 185 mi of highway driving?

3. **Solve.**

$$\frac{171}{6} = \frac{228}{x}$$

$$171x = 6 \cdot 228$$

$$\frac{171x}{171} = \frac{1368}{171}$$

$$x = \boxed{}$$

4. **Check.** The answer checks.

5. **State.** The Outback will require $\boxed{}$ gal of gas for 228 mi of city driving.

EXAMPLE 2	YOUR TURN 2
To estimate the number of trout in Jones' Pond, a ranger catches and tags 20 trout and releases them. Later, 30 trout are caught, and 4 of them are tagged. Estimate how many trout are in the pond.	To estimate the number of whales in a pod, researchers identify the tail markings of 6 whales. Days later, they identify 10 whales, 2 of which had been identified earlier. Estimate how many whales are in the pod.

1. **Familiarize.** Let t = the number of trout in the pond.

2. **Translate.**

Trout tagged originally $\rightarrow$ $\boxed{}$ $=$ $\boxed{}$ $\leftarrow$ Tagged trout caught

Trout in pond $\rightarrow$ t $\quad$ 30 $\leftarrow$ Trout caught

3. **Solve.**

$$\frac{20}{t} = \frac{4}{30}$$

$$20 \cdot 30 = 4 \cdot \boxed{}$$

$$\boxed{} = 4t$$

$$\boxed{} = t$$

4. **Check.** Since $20 \cdot 30 = 600 = 4 \cdot 150$, the answer checks.

5. **State.** We estimate there are $\boxed{}$ trout in the pond.

YOUR NOTES   Write your questions and additional notes.

Motion Problems

ESSENTIALS

Distance = Rate (or speed) × Time

$$d = rt; \qquad r = \frac{d}{t}; \qquad t = \frac{d}{r}$$

GUIDED LEARNING

 Textbook   **Instructor**   **Video**

EXAMPLE 1	YOUR TURN 1
The speed of the current in Little River is 2 mph. Brent can paddle 12 mi upstream in the same time it takes him to paddle 20 mi downstream. How fast does he paddle in still water?	A moving sidewalk moves at a rate of 1.5 ft/sec. Walking on the moving sidewalk, Karen can travel 240 ft forward in the same time it takes to travel 120 ft in the opposite direction. What is Karen's rate on a nonmoving sidewalk?

1. **Familiarize.** Let b = Brent's speed in still water.

	Distance	Speed	Time
Upstream	12	$b-2$	$12/(b-2)$
Downstream	20	$b+2$	$20/(b+2)$

2. **Translate.** The times are the same, so we equate the two expressions in the last column of the table.

3. **Solve.**

$$\frac{12}{b-2} = \frac{20}{b+2}$$

$$(b+2)(b-2)\left(\frac{12}{b-2}\right) = (b+2)(b-2)\left(\frac{20}{b+2}\right)$$

$$\boxed{} \cdot 12 = \boxed{} \cdot 20$$

$$12b + 24 = 20b - 40$$

$$\boxed{} = 8b$$

$$8 = b$$

4. **Check.** The speed upstream is $8-2=6$ mph, so Brent travels 12 mi in $\frac{12}{6} = 2$ hr. The speed downstream speed is $8+2=10$ mph, so he travels 20 mi in $\frac{20}{10} = 2$ hr. The times are the same, so the answer checks.

5. **State.** Brent paddles 8 mph in still water.

EXAMPLE 2	YOUR TURN 2
Auree drove 100 mi to the ocean. On the way home, she had to travel 10 mph slower due to traffic. Her total driving time was 4.5 hr. How fast did she drive on the way to the ocean?	Seb traveled a total of 320 mi in 5 hr. The first 280 mi of the trip he was on the highway and was able to travel 30 mph faster than the last 40 mi which was on country roads. How fast did he travel on each segment?

1. Familiarize. Let r = Auree's speed on the way to the ocean.

	Distance	Speed	Time
To ocean	100	r	$100/r$
Home	100	$r-10$	$100/(r-10)$

2. Translate. The total time was 4.5 hr, or $\dfrac{9}{2}$ hr, so the sum of the two expressions in the last column of the table is $\dfrac{9}{2}$.

$$\frac{100}{r}+\frac{100}{r-10}=\frac{9}{2}$$

3. Solve.

$$\frac{100}{r}+\frac{100}{r-10}=\frac{9}{2}$$

$$2r(r-10)\left(\frac{100}{r}+\frac{100}{r-10}\right)=2r(r-10)\cdot\frac{9}{2} \qquad \text{Clearing fractions}$$

$$2(r-10)\cdot100+2r\cdot100=9r(r-10)$$

$$200r-\boxed{}+200r=9r^2-\boxed{}$$

$$0=9r^2-490r+2000 \qquad \text{Getting 0 on one side}$$

$$0=(9r-40)(r-50)$$

$$r=\boxed{} \quad or \quad r=\boxed{} \qquad \text{Using the principle of zero products}$$

4. Check. If $r=\dfrac{40}{9}$, then $r-10$, the speed home, is negative, so we reject $\dfrac{40}{9}$. A check confirms that $r=50$ checks.

5. State. Auree's speed to the ocean was 50 mph.

YOUR NOTES   Write your questions and additional notes.

Practice Exercises

Readiness Check

Complete each statement using the following words or expressions. Choices may be used more than once or not at all.

rate	distance	work	time	0	1	$\dfrac{A+B}{2}$

1. Distance = _____ × _____ .

2. Rate = _____ ÷ _____ .

3. Time = _____ ÷ _____ .

4. $\dfrac{\text{Time required for A and B to complete the job together}}{\text{Time required for A to complete the job alone}} + \dfrac{\text{Time required for A and B to complete the job together}}{\text{Time required for B to complete the job alone}} = $ _____ .

Work Problems

5. It takes Clementine 12 hr to cut and piece a full size quilt. It takes Brianna 10 hr to do the same job. How long would it take them working together to cut and piece a full size quilt?

6. Robert can chop and stack the winter firewood in 3 fewer days than it would take Georgia. If they work together, it takes $6\dfrac{2}{3}$ days to cut and stack the wood they need for the winter. How long would it take each to chop and stack the firewood if working alone?

7. Penny can type twice as fast as her younger sister. Working together, it takes them 12 hr to type the annual report for the organization at which they volunteer. How long would it take Penny to type the report working alone?

Applications Involving Proportions

8. A baseball player gets 4 hits in 12 games. At this rate, how many hits will he get in 162 games?

9. To estimate the number of deer in a game preserve, a conservationist catches and tags 18 deer and releases them. Later, 30 deer are caught; 4 of them are tagged. How many deer are in the preserve?

Motion Problems

10. Jackson's running speed is 7 ft/sec slower than Lucia's. Lucia can run 74 ft in the time it takes Jackson to run 60 ft. How fast does Jackson run?

11. A boat moves 15 km/h in still water. It travels 40 km upriver and 40 km downriver in a total time of 6 hours. What is the speed of the current?

Formulas

ESSENTIALS

To solve a rational formula for a given letter, identify the letter, and:

1. Multiply on both sides to clear fractions, if necessary.
2. Multiply to remove parentheses, if necessary.
3. Get all terms with the specified letter alone on one side.
4. Factor out the specified letter if it is in more than one term.
5. Solve for the letter in question, using the multiplication principle.

Example

- Solve $r = \dfrac{d}{t}$ for t.

$$r = \frac{d}{t}$$

$$t \cdot r = t \cdot \frac{d}{t} \quad \text{Clearing fractions}$$

$$tr = d \quad \text{Removing a factor of 1: } \frac{t}{t} = 1$$

$$\frac{tr}{r} = \frac{d}{r} \quad \text{Dividing to isolate } t$$

$$t = \frac{d}{r} \quad \text{Removing a factor of 1: } \frac{r}{r} = 1$$

GUIDED LEARNING   🔖 **Textbook**   👤 **Instructor**   ▶ **Video**

EXAMPLE 1	YOUR TURN 1
Solve $\dfrac{1}{m} + \dfrac{1}{x} = \dfrac{1}{z}$ for m.	Solve $\dfrac{2}{k} + \dfrac{1}{y} = 3$ for y.

We multiply by the LCM of the denominators to clear fractions.

$$\boxed{} \cdot \left(\frac{1}{m} + \frac{1}{x} \right) = \boxed{} \cdot \frac{1}{z}$$

$$mxz \cdot \frac{1}{m} + mxz \cdot \frac{1}{x} = mxz \cdot \frac{1}{z} \quad \begin{array}{l}\text{Multiplying to}\\ \text{remove parentheses}\end{array}$$

$$xz + \boxed{} = \boxed{}$$

We need all terms involving m alone on one side.

(continued)

$xz = mx - mz$ Subtracting mz from both sides

$xz = m(x - z)$ Factoring out m

$\boxed{} = m$ Dividing both sides by $x - z$

Since m is isolated, we have solved for m.

EXAMPLE 2	YOUR TURN 2
Solve $S = \dfrac{H}{m(t_1 - t_2)}$ for t_1.	Solve $S = \dfrac{H}{m(t_1 - t_2)}$ for t_2.

We multiply by $m(t_1 - t_2)$ to clear the fraction.

$$m(t_1 - t_2) \cdot S = m(t_1 - t_2) \cdot \frac{H}{m(t_1 - t_2)}$$

$$mS(t_1 - t_2) = \boxed{}$$

Since t_1 is inside parentheses, we multiply to remove the parentheses as the next step in isolating t_1.

$mSt_1 - mSt_2 = H$

$mSt_1 = H + mSt_2$ Adding mSt_2 to both sides

$\dfrac{mSt_1}{mS} = \dfrac{H + mSt_2}{\boxed{}}$ Dividing by mS

$\boxed{} = \dfrac{H + mSt_2}{mS}$ Removing a factor of 1 on the left

YOUR NOTES   Write your questions and additional notes.

Practice Exercises

Readiness Check

For each of Exercises 1–3, indicate which equation is not equivalent to the others.

1. $y = kx$

$k = \dfrac{y}{x}$

$\dfrac{y}{k} = x$

$yx = k$

2. $ab = ac + c$

$b = ac + c - b$

$c = \dfrac{ab}{a + 1}$

$a(b - c) = c$

3. $\dfrac{1}{w} + \dfrac{1}{p} = \dfrac{1}{t}$

$pt + wt = pw$

$w + p = t$

$p(w - t) = wt$

Formulas

Solve.

4. $\dfrac{a}{b} = \dfrac{x}{y}$, for b

5. $\dfrac{1}{w} + \dfrac{1}{a} = \dfrac{1}{c}$, for a

6. $x = \dfrac{ab}{b + ac}$, for c

7. $A = \dfrac{(x_1 + x_2)w}{2}$, for x_2

8. $\dfrac{G}{g} = \dfrac{a + A}{A}$, for g

9. $\dfrac{1}{t} = y_1 + y_2$, for t

10. $r = \dfrac{A - P}{Pt}$, for P

Equations of Direct Variation

ESSENTIALS

Direct Variation

If a situation gives rise to a linear function $f(x) = kx$, or $y = kx$, where k is a positive constant, we say that we have **direct variation**, or that *y* **varies directly as** *x*, or that *y* **is directly proportional to** *x*. The number *k* is called the **variation constant**, or **constant of proportionality**.

Example

- Find the variation constant and an equation of variation in which *y* varies directly as *x*, and $y = 3$ when $x = 8$.

We know that $(8, 3)$ is a solution of $y = kx$.

$$y = kx$$

$$3 = k \cdot 8$$

$$\frac{3}{8} = k$$

The variation constant is $\frac{3}{8}$. The equation of variation is $y = \frac{3}{8}x$.

GUIDED LEARNING   📖 **Textbook**   👤 **Instructor**   ▶ **Video**

EXAMPLE 1	YOUR TURN 1
Find the variation constant and an equation of variation in which *y* varies directly as *x* and $y = 20$ when $x = 4$. We know that $(4, 20)$ is a solution of $y = kx$. $y = kx$ *y* varies directly as *x*. $\boxed{} = k \cdot \boxed{}$ Substituting 20 for *y* and 4 for *x* $\boxed{} = k$ Solving for *x* The variation constant is $\boxed{}$. The equation of variation is $y = \boxed{}$.	Find the variation constant and an equation of variation in which *y* varies directly as *x* and $y = 300$ when $x = 50$.

YOUR NOTES   Write your questions and additional notes.

Applications of Direct Variation

ESSENTIALS

Example

- Mike's pay P varies directly as the number of hours worked H. If Mike works 25 hr, his pay is $305. Find the pay for 40 hr of work.

$$P = kH$$

$$305 = k \cdot 25$$

$$\frac{305}{25} = k, \text{ or } k = 12.2$$

The equation of variation is $P = 12.2H$. When $H = 40$,

$$P = 12.2H$$

$$P = 12.2(40) \qquad \text{Substituting 40 for } H$$

$$P = 488.$$

If Mike works 40 hr, he will be paid $488.

GUIDED LEARNING	📖 **Textbook** 🧑 **Instructor** ▶ **Video**	

EXAMPLE 1	YOUR TURN 1

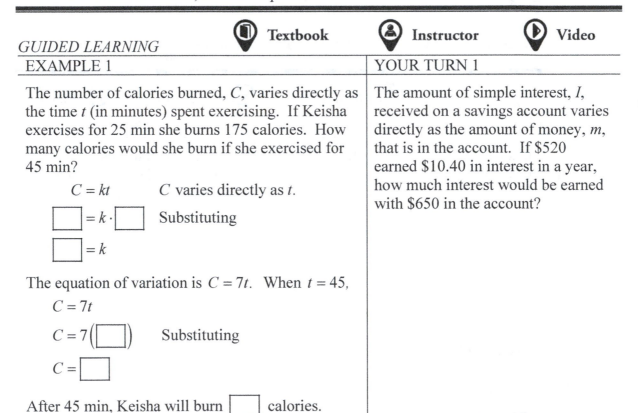

The number of calories burned, C, varies directly as the time t (in minutes) spent exercising. If Keisha exercises for 25 min she burns 175 calories. How many calories would she burn if she exercised for 45 min?

$$C = kt \qquad C \text{ varies directly as } t.$$

$$\boxed{} = k \cdot \boxed{} \qquad \text{Substituting}$$

$$\boxed{} = k$$

The equation of variation is $C = 7t$. When $t = 45$,

$$C = 7t$$

$$C = 7\left(\boxed{}\right) \qquad \text{Substituting}$$

$$C = \boxed{}$$

After 45 min, Keisha will burn $\boxed{}$ calories.

The amount of simple interest, I, received on a savings account varies directly as the amount of money, m, that is in the account. If $520 earned $10.40 in interest in a year, how much interest would be earned with $650 in the account?

YOUR NOTES   Write your questions and additional notes.

Equations of Inverse Variation

ESSENTIALS

Inverse Variation

If a situation gives rise to a function $f(x) = k/x$, or $y = k/x$, where k is a positive constant, we say that we have **inverse variation**, or that *y* **varies inversely as** *x*, or that *y* **is inversely proportional to** *x*. The number k is called the **variation constant**, or **constant of proportionality**.

Example

- Find the variation constant and an equation of variation in which *y* varies inversely as *x*, and $y = 130$ when $x = 0.6$.

We know that $(0.6, 130)$ is a solution of $y = k/x$.

$$y = \frac{k}{x}$$

$$130 = \frac{k}{0.6}$$

$$(0.6)130 = k, \text{ or } k = 78$$

The variation constant is 78. The equation of variation is $y = \dfrac{78}{x}$.

GUIDED LEARNING	📘 **Textbook** 👤 **Instructor** ▶ **Video**	

EXAMPLE 1	YOUR TURN 1
Find the variation constant and an equation of variation in which *y* varies inversely as *x* and $y = 0.4$ when $x = 8$. We know that $(8, 0.4)$ is a solution of $y = k/x$. $y = \dfrac{k}{x}$ *y* varies inversely as *x*. $\boxed{} = \dfrac{k}{\boxed{}}$ Substituting 0.4 for *y* and 8 for *x* $8(0.4) = k$ Solving for *k* $\boxed{} = k$ The variation constant is $\boxed{}$. The equation of variation is $y = \dfrac{\boxed{}}{x}$.	Find the variation constant and an equation of variation in which *y* varies inversely as *x* and $y = 75$ when $x = 0.1$.

EXAMPLE 2	YOUR TURN 2
Find the variation constant and an equation of variation in which y varies inversely as x and $y = 12$ when $x = 15$.	Find the variation constant and an equation of variation in which y varies inversely as x and $y = 32$ when $x = 9$.

We know that $(15, 12)$ is a solution of $y = k/x$.

$$y = \frac{k}{x} \qquad y \text{ varies inversely as } x.$$

$$\boxed{} = \frac{k}{\boxed{}} \qquad \text{Substituting}$$

$$15 \cdot 12 = k \qquad \text{Solving for } k$$

$$\boxed{} = k$$

The variation constant is $\boxed{}$. The equation of

variation is $y = \dfrac{\boxed{}}{x}$.

YOUR NOTES   Write your questions and additional notes.

Applications of Inverse Variation

ESSENTIALS

Example

- The cost per person to go on a ski trip for a day is inversely proportional to the number of people going on the trip. If it costs \$90 per person when 30 people go on the trip, how many are going on the ski trip if it costs \$135 per person?

  Let C = the cost per person and let N = the number of people going on the trip.

  Since inverse variation applies, we have $C = \dfrac{k}{N}$.

  Find the equation of variation.

  $$C = \frac{k}{N}$$

  $$90 = \frac{k}{30}$$

  $$30 \cdot 90 = k$$

  $$2700 = k$$

  The equation of variation is $C = \dfrac{2700}{k}$. When $C = 135$, we have

  $$135 = \frac{2700}{N}$$

  $$135N = 2700$$

  $$N = 20.$$

  It would cost \$135 per person if 20 people go on the ski trip.

⬛ **Textbook**   👤 **Instructor**   ▶ **Video**

GUIDED LEARNING

EXAMPLE 1	YOUR TURN 1
The number of minutes that a student should allow for each question on a test is inversely proportional to the number of questions on the test. If a student allows 3 minutes per question on a 20-question test, how many minutes per question should the student allow on a 25-question test? (The times allowed for the two tests are the same.)	The number of workers required to paint a house is inversely proportional to the time spent working. It takes 15 hours for 2 people to paint a house. How many workers are needed to paint the house in 6 hours?

Let M = the number of minutes per question and N = the number of questions on the test.

Since inverse variation applies, we have $M = \dfrac{k}{N}$.

Find the equation of variation.

$$M = \frac{k}{N}$$

$$\boxed{} = \frac{k}{\boxed{}} \qquad \text{Substituting 3 for } M \text{ and 20 for } N$$

$$20 \cdot 3 = k$$

$$\boxed{} = k$$

The equation of variation is $M = \dfrac{60}{N}$. When $N = 25$, we have

$$M = \frac{60}{\boxed{}}$$

$$M = \boxed{}.$$

A student should allow 2.4 minutes per question for a 25-question test.

YOUR NOTES   Write your questions and additional notes.

Other Kinds of Variation

ESSENTIALS

y varies directly as the *n*th power of *x* if there is some positive constant *k* such that $y = kx^n$.

y varies inversely as the *n*th power of *x* if there is some positive constant *k* such that $y = \dfrac{k}{x^n}$.

y varies **jointly** as *x* and *z*, if for some positive constant *k*, $y = kxz$.

Example

- Find an equation of variation in which *y* varies jointly as *w* and *x* and inversely as *z*, and $y = 100$ when $w = 2$, $x = 5$, and $z = 4$.

$$y = k \cdot \frac{wx}{z}$$

$$100 = k \cdot \frac{2 \cdot 5}{4}$$

$$100 = k \cdot \frac{5}{2}$$

$$40 = k$$

The equation of variation is $y = \dfrac{40wx}{z}$.

GUIDED LEARNING   📘 **Textbook**   👤 **Instructor**   ▶️ **Video**

EXAMPLE 1	YOUR TURN 1
Find an equation of variation in which *y* varies directly as the square of *x*, and $y = 50$ when $x = 10$.	Find an equation of variation in which *y* varies directly as the square of *x*, and $y = 36$ when $x = 3$.

$$y = kx^2$$

$\boxed{} = k \cdot \boxed{}^2$ Substituting

$50 = k \cdot \boxed{}$

$\boxed{} = k$

The equation of variation is $y = \boxed{}$.

EXAMPLE 2	YOUR TURN 2
Find an equation of variation in which W varies inversely as the square of d, and $W = 7$ when $d = 2$.	Find an equation of variation in which W varies inversely as the square of d, and $W = 8$ when $d = \dfrac{1}{2}$.

$$W = \frac{k}{d^2}$$

$$\boxed{} = \frac{k}{\left(\boxed{}\right)^2} \qquad \text{Substituting}$$

$$7 = \frac{k}{\boxed{}}$$

$$\boxed{} = k$$

The equation of variation is $y = \boxed{}$.

EXAMPLE 3	YOUR TURN 3
Find an equation of variation in which y varies jointly as x and z, and $y = 18$ when $x = 3$ and $z = 0.8$.	Find an equation of variation in which y varies jointly as x and z, and $y = 4$ when $x = 2$ and $z = \dfrac{1}{3}$.

$$y = kxz$$

$$\boxed{} = k \cdot 3(0.8) \qquad \text{Substituting}$$

$$18 = k(2.4)$$

$$\boxed{} = k$$

The equation of variation is $y = \boxed{}$.

YOUR NOTES   Write your questions and additional notes.

Other Applications of Variation

ESSENTIALS

Example

- The intensity I of a signal varies inversely as the square of the distance d from the transmitter. If the intensity is 20 watts per square meter (W/m²) at a distance of 3 km, what is the intensity at a distance of 5 km?

$$I = \frac{k}{d^2}$$

$$20 = \frac{k}{3^2}$$

$$20 = \frac{k}{9}$$

$$180 = k$$

The equation of variation is $I = \dfrac{180}{d^2}$.

To find the intensity at a distance of 5 km, substitute 5 for d.

$$I = \frac{180}{d^2} = \frac{180}{5^2} = \frac{180}{25} = 7.2$$

At a distance of 5 km, the intensity of the signal is 7.2 W/m².

EXAMPLE 1	YOUR TURN 1

EXAMPLE 1

The stopping distance d of a car after the brakes have been applied varies directly as the square of the speed r. If a car traveling 50 mph can stop in 250 ft, what is the stopping distance of a car traveling 30 mph?

First find k and the equation of variation.

$$d = kr^2$$

$$\boxed{} = k \cdot \boxed{}^2 \qquad \text{Substituting}$$

$$250 = k \cdot \boxed{}$$

$$\boxed{} = k$$

The equation of variation is $d = \boxed{} \cdot r^2$.

Now find d when $r = 30$ mph.

$$d = 0.1r^2$$

$$= 0.1 \cdot \boxed{}^2 \qquad \text{Substituting}$$

$$= 0.1 \cdot \boxed{}$$

$$= \boxed{}$$

A car traveling 30 mph can stop in $\boxed{}$ ft.

YOUR TURN 1

The intensity I of light from a light source varies inversely as the square of the distance r from the source. The intensity of light from a lamp is 80 watts per square meter (W/m^2) at a distance of 4 m from the lamp. What is the intensity at a distance of 10 m from the lamp?

YOUR NOTES   Write your questions and additional notes.

Practice Exercises

Readiness Check

Match each description of variation with the appropriate equation of variation.

1. *a* varies inversely as *b*
    a) $a = kb$

2. *a* varies directly as *b*
    b) $r = kt$

3. *r* is directly proportional to *t*
    c) $r = \dfrac{k}{t}$

4. *r* is inversely proportional to *t*
    d) $a = \dfrac{k}{b}$

Equations of Direct Variation

Find the variation constant and an equation of variation in which y varies directly as x and the following are true.

5. $y = 50$ when $x = 10$

6. $y = 9.6$ when $x = 1.2$

7. $y = 0.2$ when $x = 0.7$

8. $y = 400$ when $x = 800$

Applications of Direct Variation

9. The distance (in miles) traveled in a car is directly proportional to the average speed (in mph) of the car. If Cody travels 60 mi in a certain amount of time while driving at 55 mph, how many miles can he travel in that same amount of time at 44 mph?

Equations of Inverse Variation

Find the variation constant and an equation of variation in which y varies inversely as x and the following are true.

10. $y = 5$ when $x = 7$

11. $y = 45$ when $x = 0.2$

12. $y = 0.4$ when $x = 25$

13. $y = 22$ when $x = 15$

Applications of Inverse Variation

14. The time required to fill a man-made pond varies inversely as the rate of flow into the pond. A tanker can fill the pond in 4.5 hr at a rate of 800 L/min. How long would it take to fill the pond at a rate of 1200 L/min?

Other Kinds of Variations

Find an equation of variation in which the following are true.

15. y varies inversely as the square of x, and $y = 16$ when $x = 0.2$

16. y varies jointly as x and z and inversely as w, and $y = 60$ when $x = 15$, $z = 20$, and $w = 40$

Other Applications of Variation

17. The amount Q of water emptied by a pipe varies directly as the square of its diameter d. A pipe 2 in. in diameter will empty 300 gal of water over a fixed time period. If we assume the same kind of flow, how many gallons of water are emptied in the same amount of time by a pipe that is 4 in. in diameter?

Square Roots and Square-Root Functions

ESSENTIALS

The number c is a **square root** of a if $c^2 = a$.

Every positive real number has two real-number square roots.
The number 0 has just one square root, 0 itself.
Negative numbers do not have real-number square roots.

The **principal square root** of a nonnegative number is its nonnegative square root. The symbol $\sqrt{a}$ represents the principal square root of a. To name the negative square root of a, we can write $-\sqrt{a}$.

The symbol $\sqrt{}$ is called a **radical**.

An expression written with a radical is called a **radical expression**.
The expression written under the radical is the **radicand**.

Examples

- Find the two square roots of 49.

  The square roots are 7 and -7, because $7^2 = 49$ and $(-7)^2 = 49$.

- Simplify $\sqrt{49}$.

  The radical sign indicates the principal square root, so $\sqrt{49} = 7$.

GUIDED LEARNING		📖 **Textbook**		👤 **Instructor**		▶ **Video**

EXAMPLE 1	YOUR TURN 1
Simplify $\sqrt{16}$. $\sqrt{16} = \boxed{}$	Simplify $\sqrt{9}$.
EXAMPLE 2	**YOUR TURN 2**
Simplify $\sqrt{\dfrac{64}{100}}$. $\sqrt{\dfrac{64}{100}} = \dfrac{8}{\boxed{}} = \dfrac{4}{5}$	Simplify $\sqrt{\dfrac{25}{81}}$.
EXAMPLE 3	**YOUR TURN 3**
Simplify $-\sqrt{144}$. $-\sqrt{144} = -\boxed{}$	Simplify $-\sqrt{121}$.

EXAMPLE 4	YOUR TURN 4
Simplify $\sqrt{0.0081}$. $$\sqrt{0.0081} = \boxed{}$$	Simplify $\sqrt{0.0025}$.
EXAMPLE 5	**YOUR TURN 5**
Use a calculator to approximate $\sqrt{74}$ to three decimal places. Using a calculator with a 10-digit readout, we have $$\sqrt{74} \approx 8.602325267.$$ Rounding to three decimal places, we have $$\sqrt{74} \approx \boxed{}.$$	Use a calculator to approximate $\sqrt{83}$ to three decimal places.
EXAMPLE 6	**YOUR TURN 6**
Identify the radicand in $x\sqrt{2x^2 - 7}$. The radicand is the expression under the radical, or $\boxed{}$.	Identify the radicand in $4y^2\sqrt{y+8}$.
EXAMPLE 7	**YOUR TURN 7**
Find the indicated function value: $$f(x) = \sqrt{4x - 3}\,;\ f(2).$$ $f(2) = \sqrt{4(2) - 3}$ Substituting $$= \sqrt{\boxed{} - 3}$$ $$= \boxed{}$$	Find the indicated function value: $$f(x) = \sqrt{3x + 5}\,;\ f(3).$$

EXAMPLE 8	YOUR TURN 8

Find the domain of $f(x) = \sqrt{3x - 15}$.

The domain of f is the set of all x-values for which the radicand is nonnegative.

$3x - 15 \geq \boxed{}$ The radicand is nonnegative.

$3x \geq \boxed{}$

$x \geq \boxed{}$

The domain of $f = \left\{ x \,\middle|\, x \geq \boxed{} \right\}$, or $\left[\boxed{}, \infty \right)$.

YOUR TURN 8

Find the domain of $g(x) = \sqrt{2x + 5}$.

EXAMPLE 9	YOUR TURN 9

Graph: $f(x) = \sqrt{x - 3}$.

We choose some values for x and compute $f(x)$.

x	$f(x)$	$(x, f(x))$
3	$\boxed{}$	$(3, \boxed{})$
4	1	$\boxed{}$
5	1.4	$(5, 1.4)$
6	1.7	$(6, 1.7)$
7	$\boxed{}$	$\boxed{}$

We then plot the ordered pairs and connect them with a smooth curve.

YOUR TURN 9

Graph: $g(x) = 2 + \sqrt{x}$.

YOUR NOTES   Write your questions and additional notes.

Finding $\sqrt{a^2}$

ESSENTIALS

For any real number a, $\sqrt{a^2} = |a|$.

Example

- Simplify $\sqrt{16t^2}$. Assume the variable can represent any real number.

$$\sqrt{16t^2} = \sqrt{(4t)^2} = |4t|, \text{ or } 4|t|$$

GUIDED LEARNING	🔲 **Textbook** 👤 **Instructor** ▶ **Video**
EXAMPLE 1	**YOUR TURN 1**
Simplify $\sqrt{(-3x)^2}$. Assume that the variable can represent any real number. $$\sqrt{(-3x)^2} = \boxed{}$$ $$= |-3| \cdot |x|$$ $$= \boxed{}|x|$$	Simplify $\sqrt{(-9y)^2}$. Assume that the variable can represent any real number.
EXAMPLE 2	**YOUR TURN 2**
Simplify $\sqrt{(2x+1)^2}$. Assume that the variable can represent any real number. $$\sqrt{(2x+1)^2} = \boxed{}$$	Simplify $\sqrt{(x+2)^2}$. Assume that the variable can represent any real number.
EXAMPLE 3	**YOUR TURN 3**
Simplify $\sqrt{x^2+10x+25}$. Assume that the expression being squared is nonnegative. $$\sqrt{x^2+10x+25} = \sqrt{\left(\boxed{}\right)} = x+5$$	Simplify $\sqrt{x^2+4x+4}$. Assume that the expression being squared is nonnegative.

YOUR NOTES   Write your questions and additional notes.

Cube Roots

ESSENTIALS

The number c is the **cube root** of a if $c^3 = a$. In symbols, $\sqrt[3]{a}$ denotes the cube root of a.

Example

- Find the indicated function value: $f(x) = \sqrt[3]{x-1}$; $f(9)$.

$$f(9) = \sqrt[3]{9-1}$$
$$= \sqrt[3]{8}$$
$$= 2 \qquad \text{Since } 2 \cdot 2 \cdot 2 = 8$$

GUIDED LEARNING	**Textbook** **Instructor**	**Video**

EXAMPLE 1	YOUR TURN 1
Find $\sqrt[3]{64}$.	Find $\sqrt[3]{-8}$.
$\sqrt[3]{64} = \boxed{}$ because $\left(\boxed{}\right)^3 = 64$.	
EXAMPLE 2	YOUR TURN 2
Simplify $\sqrt[3]{-27x^3}$.	Simplify $\sqrt[3]{8x^3}$.
$\sqrt[3]{-27x^3} = \boxed{}$ because $\left(\boxed{}\right)^3 = -27x^3$.	
EXAMPLE 3	YOUR TURN 3
Find the indicated function value: $f(x) = \sqrt[3]{x+15}$; $f(-7)$.	Find the indicated function value: $f(x) = \sqrt[3]{2x-3}$; $f(15)$.
$f(-7) = \sqrt[3]{\boxed{} + 15}$ $= \sqrt[3]{\boxed{}}$ $= \boxed{}$	

YOUR NOTES   Write your questions and additional notes.

Odd and Even *k*th Roots

ESSENTIALS

In the expression $\sqrt[k]{a}$, we call k the **index**. When the index is 2, we do not write it.

If k is an *odd* natural number, then $\sqrt[k]{a^k} = a$.

If k is an *even* natural number, then $\sqrt[k]{a^k} = |a|$.

Examples

- Find $\sqrt[5]{-243}$.

  $\sqrt[5]{-243} = -3$ because $(-3)^5 = -243$.

- Simplify: $\sqrt[4]{x^4}$.

  $\sqrt[4]{x^4} = |x|$

	📖 **Textbook**	🧑 **Instructor**	▶ **Video**

GUIDED LEARNING

EXAMPLE 1	YOUR TURN 1
Find $\sqrt[4]{81}$. $\sqrt[4]{81} = \boxed{}$ because $\left(\boxed{}\right)^4 = 81$.	Find $\sqrt[4]{10{,}000}$.
EXAMPLE 2	YOUR TURN 2
Find $-\sqrt[7]{-128}$. $-\sqrt[7]{-128} = -\left(\boxed{}\right) = \boxed{}$	Find $-\sqrt[5]{32}$.
EXAMPLE 3	YOUR TURN 3
Find $\sqrt[4]{-16}$. Negative numbers do not have real *k*th roots when k is $\boxed{}$. Thus, $\sqrt[4]{-16}$ does not exist as a real number.	Find $\sqrt[6]{-64}$.
EXAMPLE 4	YOUR TURN 4
Simplify: $\sqrt[11]{(x-2)^{11}}$. $\sqrt[11]{(x-2)^{11}} = \boxed{}$	Simplify: $\sqrt[7]{(2x+1)^7}$.
EXAMPLE 5	YOUR TURN 5
Simplify: $\sqrt[4]{16x^4}$. Assume that the variable can represent any real number. $\sqrt[4]{16x^4} = |2x|$, or $2\boxed{}$	Simplify: $-\sqrt[4]{81x^4}$. Assume that the variable can represent any real number.

YOUR NOTES   Write your questions and additional notes.

Practice Exercises

Readiness Check

Determine whether each statement is true or false.

1. The index of a square root is 2.

2. The expression $\sqrt[4]{-100}$ does not exist as a real number.

3. The principal square root can be negative.

4. The domain of the function $f(x) = \sqrt{x}$ is $\{x | x \geq 0\}$.

Square Roots and Square-Root Functions

For each number, find all of its square roots.

5. 100

6. 256

Simplify.

7. $-\sqrt{\dfrac{100}{81}}$

8. $\sqrt{0.25}$

9. Use a calculator to approximate $\sqrt{153}$ to three decimal places.

10. Identify the radicand in $9x\sqrt{3x-7}$.

For each function, find the specified function value, if it exists.

11. $f(x) = \sqrt{x^2 + 5}$;
    $f(-2)$

12. $f(x)\sqrt{-4x}$;
    $f(5)$

13. $f(x) = \sqrt{(x+3)^2}$;
    $f(-4)$

Finding $\sqrt{a^2}$

Simplify, if possible. Assume variables can represent any real number.

14. $\sqrt{9x^2}$

15. $\sqrt{(-5x)^2}$

16. $\sqrt{y^2 - 12y + 36}$

Cube Roots

Simplify.

17. $\sqrt[3]{125}$

18. $-\sqrt[3]{-1}$

19. $\sqrt[3]{-216x^3}$

20. For $f(x) = \sqrt[3]{5 - 3x}$, find $f(2)$.

Odd and Even kth Roots

Find each of the following. Assume that letters can represent any real number.

21. $\sqrt[26]{x^{26}}$

22. $-\sqrt[5]{\dfrac{1}{32}}$

23. $\sqrt[7]{-1}$

24. $\sqrt[9]{a^9}$

25. $\sqrt[6]{(y+8)^6}$

26. $\sqrt[5]{(2x+5)^5}$

Rational Exponents

ESSENTIALS

$$a^{1/n} = \sqrt[n]{a}$$

For any natural numbers m and n $(n \neq 1)$ and any nonnegative real number a,

$a^{m/n}$ means $\sqrt[n]{a^m}$, or $\left(\sqrt[n]{a}\right)^m$.

Example

- Rewrite $64^{5/6}$ without rational exponents, and simplify, if possible.

$$64^{5/6} = \left(\sqrt[6]{64}\right)^5 = 2^5 = 32$$

	🔖 **Textbook**	👤 **Instructor**	▶ **Video**

GUIDED LEARNING

EXAMPLE 1	YOUR TURN 1
Rewrite without rational exponents, and simplify, if possible: $(-32)^{1/5}$. $(-32)^{1/5} = \sqrt[5]{-32} = \boxed{}$	Rewrite without rational exponents, and simplify, if possible: $49^{1/2}$.
EXAMPLE 2	YOUR TURN 2
Rewrite with rational exponents: $\sqrt[7]{5xy}$. $\sqrt[7]{5xy} = (5xy)^{\boxed{}}$	Rewrite with rational exponents: $\sqrt[5]{-3x^2y}$.
EXAMPLE 3	YOUR TURN 3
Rewrite without rational exponents, and simplify, if possible: $64^{2/3}$. $64^{2/3}$ means $\left(\sqrt[3]{64}\right)^2$ or, equivalently, $\sqrt[3]{64^2}$. We will use the form $\left(\sqrt[3]{64}\right)^2$. $64^{2/3} = \left(\sqrt[3]{64}\right)^2 = 4^2 = \boxed{}$	Rewrite without rational exponents, and simplify, if possible: $81^{3/2}$.

EXAMPLE 4	YOUR TURN 4
Rewrite with rational exponents: $\left(\sqrt[3]{15x}\right)^5$.	Rewrite with rational exponents: $\left(\sqrt[4]{9xy}\right)^5$.
$\left(\sqrt[3]{15x}\right)^5 = (15x)^{\boxed{}}$	

YOUR NOTES   Write your questions and additional notes.

Negative Rational Exponents

ESSENTIALS

For any rational number m/n and any positive real number a,

$$a^{-m/n} \text{ means } \frac{1}{a^{m/n}}.$$

Example

- Rewrite with positive exponents and simplify, if possible: $49^{-1/2}$.

$$49^{-1/2} = \frac{1}{49^{1/2}} = \frac{1}{\sqrt{49}} = \frac{1}{7}$$

GUIDED LEARNING	🔖 **Textbook**	🧑 **Instructor**	▶ **Video**

EXAMPLE 1	YOUR TURN 1
Rewrite with positive exponents, and simplify, if possible: $81^{-1/2}$. $$81^{-1/2} = \frac{1}{81^{1/2}} = \frac{1}{\sqrt{\Box}} = \frac{1}{\Box}$$	Rewrite with positive exponents, and simplify, if possible: $25^{-1/2}$.
EXAMPLE 2	YOUR TURN 2
Rewrite with positive exponents, and simplify, if possible: $27^{-2/3}$. $$27^{-2/3} = \frac{1}{27^{2/3}} = \frac{1}{\left(\sqrt[3]{\Box}\right)^2} = \frac{1}{\left(\Box\right)^2} = \frac{1}{9}$$	Rewrite with positive exponents, and simplify, if possible: $32^{-3/5}$.
EXAMPLE 3	YOUR TURN 3
Rewrite with positive exponents, and simplify, if possible: $(8xy)^{-4/5}$. $$(8xy)^{-4/5} = \frac{1}{(8xy)^{\Box}}$$	Rewrite with positive exponents, and simplify, if possible: $(11xyz)^{-2/3}$.
EXAMPLE 4	YOUR TURN 4
Rewrite with positive exponents, and simplify, if possible: $3x^{-2/3}y^{2/5}$. $$3x^{-2/3}y^{2/5} = 3 \cdot \frac{1}{x^{\Box}} \cdot y^{2/5} = \frac{\Box}{x^{2/3}}$$	Rewrite with positive exponents, and simplify, if possible: $7x^{-3/2}y^{4/5}$.

EXAMPLE 5	YOUR TURN 5
Rewrite with positive exponents, and simplify, if possible: $\left(\dfrac{5x}{7y}\right)^{-7/3}$. Recall that $\left(\dfrac{a}{b}\right)^{-n} = \left(\dfrac{b}{a}\right)^{n}$. $\left(\dfrac{5x}{7y}\right)^{-7/3} = \left(\dfrac{\square}{5x}\right)^{7/3}$	Rewrite with positive exponents, and simplify, if possible: $\left(\dfrac{2xy}{5ab}\right)^{-3/5}$.
EXAMPLE 6	**YOUR TURN 6**
Rewrite with positive exponents, and simplify, if possible: $\dfrac{1}{x^{-5/8}}$. Since $\dfrac{1}{a^{m/n}}$ means $a^{-m/n}$, $\dfrac{1}{x^{-5/8}} = x^{-(-5/8)} = x^{\boxed{}}$	Rewrite with positive exponents, and simplify, if possible: $\dfrac{1}{a^{-2/7}}$.

YOUR NOTES   Write your questions and additional notes.

Laws of Exponents

ESSENTIALS

For any real numbers a and b and any rational exponents m and n:

1. $a^m \cdot a^n = a^{m+n}$

2. $\dfrac{a^m}{a^n} = a^{m-n}$

3. $\left(a^m\right)^n = a^{mn}$

4. $(ab)^m = a^m b^m$

5. $\left(\dfrac{a}{b}\right)^n = \dfrac{a^n}{b^n}$

Example

- Use the laws of exponents to simplify $2^{1/5} \cdot 2^{2/5}$.

$$2^{1/5} \cdot 2^{2/5} = 2^{1/5+2/5} = 2^{3/5}$$

GUIDED LEARNING	🛈	**Textbook**	👤	**Instructor**	▶ **Video**

EXAMPLE 1	YOUR TURN 1
Use the laws of exponents to simplify $x^{1/7} \cdot x^{3/14}$. $$x^{1/7} \cdot x^{3/14} = x^{1/7+3/14}$$ $$= x^{2/14+3/14}$$ $$= \boxed{}$$	Use the laws of exponents to simplify $x^{1/5} \cdot x^{7/10}$.
EXAMPLE 2	YOUR TURN 2
Use the laws of exponents to simplify $\dfrac{a^{1/6}}{a^{1/2}}$. $$\dfrac{a^{1/6}}{a^{1/2}} = a^{1/6-1/2}$$ $$= a^{1/6-3/6}$$ $$= a^{\boxed{}}$$ $$= a^{-1/3}, \text{ or } \boxed{}$$	Use the laws of exponents to simplify $\dfrac{a^{1/4}}{a^{3/4}}$.

EXAMPLE 3	YOUR TURN 3
Use the laws of exponents to simplify $\left(1.2^{3/4}\right)^{1/3}$.	Use the laws of exponents to simplify $\left(3.4^{2/3}\right)^{3/4}$.
$$\left(1.2^{3/4}\right)^{1/3} = 1.2^{(3/4)(1/3)}$$ $$= 1.2^{3/12}$$ $$= \boxed{}$$	

EXAMPLE 4	YOUR TURN 4
Use the laws of exponents to simplify $\left(x^{-1/2}y^{3/4}\right)^{1/3}$.	Use the laws of exponents to simplify $\left(x^{-2/3}y^{1/2}\right)^{1/5}$.
$$\left(x^{-1/2}y^{3/4}\right)^{1/3} = x^{(-1/2)(1/3)}y^{(3/4)(1/3)}$$ $$= x^{-1/6}y^{3/12}$$ $$= x^{-1/6}y^{1/4}, \text{ or } \frac{y^{1/4}}{\boxed{}}$$	

YOUR NOTES   Write your questions and additional notes.

Simplifying Radical Expressions

ESSENTIALS

Simplifying Radical Expressions

1. Convert radical expressions to exponential expressions.
2. Use arithmetic and the laws of exponents to simplify.
3. Convert back to radical notation when appropriate.

Example

- Use rational exponents to simplify $\sqrt[7]{(3x)^{14}}$.

$$\sqrt[7]{(3x)^{14}} = (3x)^{14/7}$$
$$= (3x)^2$$
$$= 9x^2$$

GUIDED LEARNING	📘 **Textbook**　👤 **Instructor**　▶ **Video**	

EXAMPLE 1	YOUR TURN 1
Use rational exponents to simplify $\sqrt[6]{t^{24}}$. $\sqrt[6]{t^{24}} = t^{24/6}$ $= \boxed{}$	Use rational exponents to simplify $\sqrt[5]{t^{15}}$.
EXAMPLE 2	YOUR TURN 2
Use rational exponents to simplify $\sqrt[12]{x^9}$. $\sqrt[12]{x^9} = x^{9/\boxed{}}$ $= x^{3/\boxed{}}$ $= \sqrt[\boxed{}]{x^{\boxed{}}}$	Use rational exponents to simplify $\sqrt[10]{x^4}$.
EXAMPLE 3	YOUR TURN 3
Use rational exponents to simplify $\sqrt[8]{4}$. $\sqrt[8]{4} = 4^{1/\boxed{}}$ $= \left(2^2\right)^{1/8}$ $= 2^{1/\boxed{}}$ $= \sqrt[\boxed{}]{2}$	Use rational exponents to simplify $\sqrt[4]{9}$.

EXAMPLE 4	YOUR TURN 4
Use rational exponents to simplify $\sqrt[4]{x} \cdot \sqrt[6]{x}$. $$\sqrt[4]{x} \cdot \sqrt[6]{x} = x^{1/4} \cdot x^{1/6}$$ $$= x^{\Box/12} \cdot x^{\Box/12}$$ $$= x^{\Box/12}$$ $$= \sqrt[\Box]{x^{\Box}}$$	Use rational exponents to simplify $$\sqrt[6]{x} \cdot \sqrt[8]{x}.$$

EXAMPLE 5	YOUR TURN 5
Use rational exponents to simplify $\left(\sqrt[4]{x^2 y^4 z}\right)^{16}$. $$\left(\sqrt[4]{x^2 y^4 z}\right)^{16} = \left(x^2 y^4 z\right)^{16/4}$$ $$= \left(x^2 y^4 z\right)^4$$ $$= \left(x^2\right)^{\Box} \left(y^4\right)^{\Box} (z)^{\Box}$$ $$= \boxed{}$$	Use rational exponents to simplify $$\left(\sqrt[3]{x^5 y z^2}\right)^{15}.$$

EXAMPLE 6	YOUR TURN 6
Use rational exponents to write a single radical expression: $\sqrt[3]{\sqrt{x}}$. $$\sqrt[3]{\sqrt{x}} = \sqrt[3]{x^{1/2}}$$ $$= \left(x^{1/2}\right)^{\Box}$$ $$= x^{\Box}$$ $$= \sqrt[\Box]{x}$$	Use rational exponents to write a single radical expression: $\sqrt[5]{\sqrt[3]{x}}$.

YOUR NOTES   Write your questions and additional notes.

Practice Exercises

Readiness Check

Match each expression with its equivalent form.

1. $\left(2^4\right)^5$

2. $2^4 \cdot 2^5$

3. $\dfrac{2^4}{2^5}$

4. $\left(2 \cdot 4\right)^5$

     a) 2^{4+5}

     b) $2^5 \cdot 4^5$

     c) 2^{4-5}

     d) $2^{4 \cdot 5}$

Rational Exponents

Rewrite without rational exponents, and simplify, if possible.

5. $x^{2/3}$

6. $81^{1/2}$

7. $\left(x^2 y^2\right)^{1/3}$

8. $4^{5/2}$

Rewrite with rational exponents.

9. $\sqrt[3]{5}$

10. $\sqrt[4]{m^3}$

11. $\left(\sqrt{4xy}\right)^5$

12. $\left(\sqrt[3]{2x^2 y}\right)^8$

Negative Rational Exponents

Write an equivalent expression with positive exponents, and simplify, if possible.

13. $27^{-2/3}$

14. $x^2 y^{-1/3} z^4$

15. $\dfrac{12a}{c^{-4/7}}$

16. $\left(\dfrac{1}{27}\right)^{-2/3}$

17. $5a^{-3/2} b^{-1/5} c^{3/4}$

18. $\left(6xyz\right)^{-3/5}$

Laws of Exponents

Use the laws of exponents to simplify. Write the answers with positive exponents.

19. $7^{2/5} \cdot 7^{1/10}$

20. $\left(12^{3/4}\right)^{1/2}$

21. $\dfrac{2.1^{5/6}}{2.1^{-1/2}}$

22. $x^{2/3} \cdot x^{1/5}$

23. $\left(a^{-2/3}b^{1/7}\right)^{1/4}$

24. $\left(5^{-1/3}\right)^{3}$

Simplifying Radical Expressions

Use rational exponents to simplify. Write the answer in radical notation if appropriate.

25. $\sqrt[15]{x^3}$

26. $\left(\sqrt[6]{ab}\right)^{12}$

27. $\sqrt[8]{81}$

Use rational exponents to write a single radical expression.

28. $\sqrt[4]{\sqrt[3]{ab}}$

29. $\left(\sqrt[4]{x^2 y^3}\right)^{16}$

30. $\sqrt{2}\sqrt[4]{2}$

Multiplying and Simplifying Radical Expressions

ESSENTIALS

The Product Rule for Radicals

For any real numbers $\sqrt[k]{a}$ and $\sqrt[k]{b}$,

$$\sqrt[k]{a} \cdot \sqrt[k]{b} = \sqrt[k]{a \cdot b}, \text{ or } a^{1/k} \cdot b^{1/k} = (ab)^{1/k}.$$

Factoring Radical Expressions

For any nonnegative real numbers a and b and any index k,

$$\sqrt[k]{ab} = \sqrt[k]{a} \cdot \sqrt[k]{b}, \text{ or } (ab)^{1/k} = a^{1/k} \cdot b^{1/k}.$$

To simplify a radical expression by factoring:

1. Look for the largest factors of the radicand that are perfect kth powers (where k is the index).
2. Then take the kth root of the resulting factors.
3. A radical expression, with index k, is *simplified* when its radicand has no factors that are perfect kth powers.

Example

- Multiply and simplify: $\sqrt{12}\sqrt{8}$.

$$\sqrt{12}\sqrt{8} = \sqrt{12 \cdot 8}$$
$$= \sqrt{96}$$
$$= \sqrt{16 \cdot 6}$$
$$= \sqrt{16} \cdot \sqrt{6}$$
$$= 4\sqrt{6}$$

GUIDED LEARNING | 📖 **Textbook** | 👤 **Instructor** | ▶ **Video**

EXAMPLE 1	YOUR TURN 1
Multiply: $\sqrt{6}\sqrt{7}$.	Multiply: $\sqrt{5}\sqrt{13}$.
$\sqrt{6}\sqrt{7} = \sqrt{6 \cdot \boxed{}}$ $= \sqrt{\boxed{}}$	

EXAMPLE 2	YOUR TURN 2
Multiply: $\sqrt[5]{6x}\sqrt[5]{2y^2}$.	Multiply: $\sqrt[3]{4x^2}\sqrt[3]{5y}$.
$\sqrt[5]{6x}\sqrt[5]{2y^2} = \sqrt[5]{6x \cdot 2y^2}$ $= \boxed{}$	

EXAMPLE 3	YOUR TURN 3
Multiply: $\sqrt[4]{7}\sqrt{3}$.	Multiply: $\sqrt[3]{5}\sqrt{2}$.
The indices are different, so we must use rational exponents. $\sqrt[4]{7}\sqrt{3} = 7^{1/4} \cdot 3^{\boxed{}}$ $= 7^{1/4} \cdot 3^{\boxed{}/4}$ $= \left(7 \cdot 3^2\right)^{\boxed{}}$ $= 63^{1/4}$ $= \sqrt[\boxed{}]{63}$	

EXAMPLE 4	YOUR TURN 4
Simplify: $\sqrt{84}$.	Simplify: $\sqrt{125}$.
$\sqrt{84} = \sqrt{4 \cdot 21}$ $= \sqrt{\boxed{}} \cdot \sqrt{21}$ $= \boxed{}$	

EXAMPLE 5	YOUR TURN 5
Simplify: $\sqrt{24x^2y^3}$.	Simplify: $\sqrt{48xy^4}$.
$\sqrt{24x^2y^3} = \sqrt{4 \cdot 6 \cdot x^2 \cdot y^2 \cdot y^1}$ $= \sqrt{\boxed{}} \cdot \sqrt{6y}$ $= \boxed{}\sqrt{6y}$	

EXAMPLE 6	YOUR TURN 6
Simplify: $\sqrt[3]{24a^6b^{14}}$.	Simplify: $\sqrt[3]{32a^{10}b^7}$.
$\sqrt[3]{24a^6b^{14}} = \sqrt[3]{8 \cdot 3 \cdot a^6 \cdot b^{12} \cdot b^2}$	
$\qquad = \sqrt[3]{8}\,\boxed{}\,\sqrt[3]{b^{12}}\,\sqrt[3]{3b^2}$	
$\qquad = 2\,\boxed{}\,b^{12/3}\,\sqrt[3]{3b^2}$	
$\qquad = \boxed{}$	
EXAMPLE 7	YOUR TURN 7
Multiply and simplify: $\sqrt[3]{18}\,\sqrt[3]{3}$.	Multiply and simplify: $\sqrt[3]{12}\,\sqrt[3]{6}$.
$\sqrt[3]{18}\,\sqrt[3]{3} = \sqrt[3]{18 \cdot 3}$	
$\qquad = \sqrt[3]{2 \cdot 3 \cdot 3 \cdot \boxed{}}$	
$\qquad = \boxed{}\,\sqrt[3]{2}$	
EXAMPLE 8	YOUR TURN 8
Multiply and simplify: $2\sqrt{6} \cdot 5\sqrt{4}$.	Multiply and simplify: $-3\sqrt{5} \cdot 4\sqrt{8}$.
$2\sqrt{6} \cdot 5\sqrt{4} = 2 \cdot 5\sqrt{6 \cdot 4}$	
$\qquad = \boxed{}\,\sqrt{24}$	
$\qquad = 10\sqrt{\boxed{} \cdot 6}$	
$\qquad = 10\sqrt{4} \cdot \sqrt{6}$	
$\qquad = 10 \cdot 2\sqrt{6}$	
$\qquad = \boxed{}$	

EXAMPLE 9	YOUR TURN 9
Multiply and simplify: $\sqrt[4]{8x^5y^3}\,\sqrt[4]{6x^2y}$.	Multiply and simplify: $\sqrt[4]{9x^8y}\,\sqrt[4]{9x^3y}$.

$$\sqrt[4]{8x^5y^3}\,\sqrt[4]{6x^2y} = \sqrt[4]{8\cdot 6\cdot x^5\cdot x^2\cdot y^3\cdot y}$$

$$= \sqrt[4]{48x^7\,\boxed{}}$$

$$= \sqrt[4]{16x^4y^4\cdot 3x^3}$$

$$= \sqrt[4]{16x^4y^4}\,\sqrt[4]{3x^3}$$

$$= \boxed{}$$

YOUR NOTES   Write your questions and additional notes.

Dividing and Simplifying Radical Expressions

ESSENTIALS

The Quotient Rule for Radicals

For any nonnegative number a, any positive number b, and any index k,

$$\frac{\sqrt[k]{a}}{\sqrt[k]{b}} = \sqrt[k]{\frac{a}{b}}, \text{ or } \frac{a^{1/k}}{b^{1/k}} = \left(\frac{a}{b}\right)^{1/k}.$$

*k*th Roots of Quotients

For any nonnegative number a, any positive number b, and any index k,

$$\sqrt[k]{\frac{a}{b}} = \frac{\sqrt[k]{a}}{\sqrt[k]{b}}, \text{ or } \left(\frac{a}{b}\right)^{1/k} = \frac{a^{1/k}}{b^{1/k}}.$$

Example

- Simplify by taking the roots of the numerator and the denominator: $\sqrt{\dfrac{16}{y^2}}$.

$$\sqrt{\frac{16}{y^2}} = \frac{\sqrt{16}}{\sqrt{y^2}} = \frac{4}{y}$$

GUIDED LEARNING   📖 **Textbook**   👤 **Instructor**   ▶ **Video**

EXAMPLE 1	YOUR TURN 1
Divide and simplify: $\dfrac{\sqrt{32}}{\sqrt{2}}$.	Divide and simplify: $\dfrac{\sqrt{50}}{\sqrt{2}}$.
$\dfrac{\sqrt{32}}{\sqrt{2}} = \sqrt{\dfrac{32}{\square}} = \sqrt{16} = \square$	
EXAMPLE 2	**YOUR TURN 2**
Divide and simplify: $\dfrac{6\sqrt[3]{54}}{\sqrt[3]{2}}$.	Divide and simplify: $\dfrac{4\sqrt[3]{192}}{\sqrt[3]{3}}$.
$\dfrac{6\sqrt[3]{54}}{\sqrt[3]{2}} = 6\sqrt[3]{\dfrac{54}{2}}$ $= 6\sqrt[3]{\square}$ $= 6 \cdot 3$ $= \square$	

EXAMPLE 3	YOUR TURN 3
Simplify: $\sqrt[3]{\dfrac{27}{8}}$.	Simplify: $\sqrt[4]{\dfrac{16}{81}}$.
$\sqrt[3]{\dfrac{27}{8}} = \dfrac{\sqrt[3]{27}}{\boxed{}} = \dfrac{3}{\boxed{}}$	

EXAMPLE 4	YOUR TURN 4
Simplify: $\sqrt{\dfrac{50x^6}{y^{10}}}$.	Simplify: $\sqrt{\dfrac{3x^9}{25y^6}}$.
$\sqrt{\dfrac{50x^6}{y^{10}}} = \dfrac{\sqrt{50x^6}}{\sqrt{y^{10}}}$	
$\qquad = \dfrac{\sqrt{25x^6 \cdot \boxed{}}}{\sqrt{y^{10}}}$	
$\qquad = \dfrac{5\boxed{}\sqrt{2}}{y^5}$	

EXAMPLE 5	YOUR TURN 5
Divide and simplify: $\dfrac{\sqrt[4]{a^6 b^5}}{\sqrt[3]{a^2 b}}$.	Divide and simplify: $\dfrac{\sqrt[5]{a^7 b^8}}{\sqrt{ab^3}}$.
The indices are different, so we must use rational exponents.	

$$\frac{\sqrt[4]{a^6 b^5}}{\sqrt[3]{a^2 b}} = \frac{\left(a^6 b^5\right)^{1/4}}{\left(a^2 b\right)^{1/3}}$$

$$= \frac{a^{6/4} b^{\boxed{}}}{a^{2/3} b^{\boxed{}}}$$

$$= a^{6/4 - 2/3} b^{5/4 - 1/3}$$

$$= a^{18/12 - \boxed{}/12} b^{15/12 - \boxed{}/12}$$

$$= a^{\boxed{}/12} b^{\boxed{}/12}$$

$$= \sqrt[12]{a^{\boxed{}} b^{\boxed{}}}$$

YOUR NOTES   Write your questions and additional notes.

Practice Exercises

Readiness Check

Choose the word or expression that best completes each sentence.

1. When simplifying $\sqrt[4]{162x^5}$, the largest perfect fourth power is _____.

                                                             $81x^4$ / $27x^5$

2. When multiplying radicals with the same index, multiply the radicands and then _____, if possible.

        add / simplify

3. When simplifying $\sqrt[3]{54x^6}$, you must find the largest _____.

                                          perfect cube / perfect sixth power

4. In order to use the product and quotient rules for radicals, the indices must be _____.

    different / the same

Multiplying and Simplifying Radical Expressions

Simplify. Assume that no radicands were formed by raising negative numbers to even powers.

5. $\sqrt{50}$

6. $\sqrt[3]{40}$

7. $\sqrt{120x^5}$

8. $\sqrt[3]{-24x^9}$

9. $\sqrt{50ab^7}$

10. $\sqrt[4]{32x^{12}y^5}$

Multiply and simplify. Assume that no radicands were formed by raising negative numbers to even powers.

11. $\sqrt{6}\sqrt{10}$

12. $\sqrt{7a^3}\sqrt{14a^5}$

13. $\sqrt[3]{3a^2}\sqrt[3]{6a}$

14. $\sqrt[4]{x^2 y^3} \sqrt[4]{x^8 y^4}$ **15.** $\sqrt{12x^3 y} \sqrt[4]{20xy^5}$

Dividing and Simplifying Radical Expressions

Simplify. Assume that all expressions under radicals represent positive numbers.

16. $\dfrac{\sqrt{125}}{\sqrt{5}}$ **17.** $\dfrac{\sqrt{32x^5 y}}{\sqrt{16xy}}$

18. $\dfrac{\sqrt[3]{250x^4 y^7}}{\sqrt[3]{2x^2 y}}$ **19.** $\dfrac{\sqrt[4]{x^3 y^2}}{\sqrt[3]{xy}}$

Simplify.

20. $\sqrt{\dfrac{16}{25}}$ **21.** $\sqrt{\dfrac{49}{x^4}}$ **22.** $\sqrt[4]{\dfrac{3x^4 y^8}{z^4}}$

23. $\sqrt[3]{\dfrac{125a}{8b^3}}$ **24.** $\sqrt[3]{\dfrac{81x^5}{y^6}}$ **25.** $\sqrt[5]{\dfrac{a^9 b^2 c^6}{z^{15}}}$

Addition and Subtraction

ESSENTIALS

Like radicals are radical expressions that have the same indices and the same radicands.

Example

- Simplify $2\sqrt{3} + 5\sqrt{3}$ by collecting like radical terms, if possible.

$$2\sqrt{3} + 5\sqrt{3} = (2+5)\sqrt{3} = 7\sqrt{3}$$

GUIDED LEARNING	📖 **Textbook** 👤 **Instructor** ▶ **Video**

EXAMPLE 1	YOUR TURN 1
Simplify $4\sqrt{6} + 7\sqrt{6}$ by collecting like radical terms. $$4\sqrt{6} + 7\sqrt{6} = \left(4 + \boxed{}\right)\sqrt{6}$$ $$= 11\sqrt{6}$$	Simplify $2\sqrt{3} + 8\sqrt{3}$ by collecting like radical terms.

EXAMPLE 2	YOUR TURN 2
Simplify $10\sqrt[3]{5} - 4\sqrt[3]{3}$ by collecting like radical terms, if possible. $10\sqrt[3]{5} - 4\sqrt[3]{3}$ cannot be simplified because the radicands are different.	Simplify $8\sqrt[3]{2} - 3\sqrt[3]{7}$ by collecting like radical terms, if possible.

EXAMPLE 3	YOUR TURN 3
Simplify $9\sqrt[3]{3x} + 2\sqrt[3]{3x} - 10\sqrt[4]{3x}$ by collecting like radical terms, if possible. $$9\sqrt[3]{3x} + 2\sqrt[3]{3x} - 10\sqrt[4]{3x} = (9+2)\sqrt[3]{3x} - 10\sqrt[4]{3x}$$ $$= \boxed{} - 10\sqrt[4]{3x}$$	Simplify $6\sqrt[4]{2xy} + 3\sqrt[4]{2xy} - 5\sqrt[3]{2xy}$ by collecting like radical terms, if possible.

EXAMPLE 4	YOUR TURN 4
Simplify $3\sqrt{8} - \sqrt{32}$ by collecting like radical terms, if possible.	Simplify $5\sqrt{18} + \sqrt{50}$ by collecting like radical terms, if possible.

$3\sqrt{8} - \sqrt{32} = 3\sqrt{4 \cdot 2} - \sqrt{16 \cdot 2}$ $\left.\begin{array}{l} \\ = \boxed{} \cdot 2\sqrt{2} - 4\sqrt{2} \end{array}\right\}$ Simplifying the radicands

$= 6\sqrt{2} - 4\sqrt{2}$

$= (6 - 4)\sqrt{2}$

$= \boxed{}\sqrt{2}$

EXAMPLE 5	YOUR TURN 5
Simplify $\sqrt[3]{3x^3 y^5} + 5\sqrt[3]{3y^2}$ by collecting like radical terms, if possible.	Simplify $\sqrt[3]{5xy^9} + \sqrt[3]{5x}$ by collecting like radical terms, if possible.

$\sqrt[3]{3x^3 y^5} + 5\sqrt[3]{3y^2} = \sqrt[3]{x^3 y^3 \cdot 3y^2} + 5\sqrt[3]{3y^2}$

$= \sqrt[3]{x^3 y^3} \cdot \sqrt[3]{3y^2} + 5\sqrt[3]{3y^2}$

$= \boxed{} y\sqrt[3]{3y^2} + 5\sqrt[3]{3y^2}$

$= \left(\boxed{}\right)\sqrt[3]{3y^2}$

YOUR NOTES   Write your questions and additional notes.

More Multiplication

ESSENTIALS

Expressions of the form $\sqrt{a} + \sqrt{b}$ and $\sqrt{a} - \sqrt{b}$ are called **conjugates**. Their product is always an expression that contains no radicals.

To multiply radical expressions in which some factors contain more than one term, we use the procedures for multiplying polynomials.

Example

- Multiply: $\left(\sqrt{x} + 2\sqrt{3}\right)\left(\sqrt{x} + 5\sqrt{3}\right)$.

$$\left(\sqrt{x} + 2\sqrt{3}\right)\left(\sqrt{x} + 5\sqrt{3}\right) = \sqrt{x} \cdot \sqrt{x} + \sqrt{x} \cdot 5\sqrt{3} + 2\sqrt{3} \cdot \sqrt{x} + 2\sqrt{3} \cdot 5\sqrt{3} \quad \text{Using FOIL}$$

$$= x + 5\sqrt{3x} + 2\sqrt{3x} + 10 \cdot 3$$

$$= x + 7\sqrt{3x} + 30$$

GUIDED LEARNING	📖 **Textbook**	👤 **Instructor**	▶ **Video**

EXAMPLE 1	YOUR TURN 1
Multiply: $\sqrt{2}\left(x - \sqrt{3}\right)$. $\sqrt{2}\left(x - \sqrt{3}\right) = \sqrt{2} \cdot \boxed{} - \sqrt{2} \cdot \boxed{}$ $ = x\sqrt{2} - \boxed{}$	Multiply: $\sqrt{5}\left(\sqrt{3} + y\right)$.

EXAMPLE 2	YOUR TURN 2
Multiply: $\sqrt[3]{x}\left(\sqrt[3]{3x^2} + \sqrt[3]{24x^2}\right)$. $\sqrt[3]{x}\left(\sqrt[3]{3x^2} + \sqrt[3]{24x^2}\right) = \sqrt[3]{x} \cdot \sqrt[3]{3x^2} + \sqrt[3]{x} \cdot \sqrt[3]{24x^2}$ $ = \sqrt[3]{3x^3} + \boxed{}$ $ = \sqrt[3]{x^3}\sqrt[3]{3} + \sqrt[3]{8x^3}\sqrt[3]{3}$ $ = x\sqrt[3]{3} + \boxed{}\sqrt[3]{3} \quad \text{Simplifying}$ $ = \boxed{}\sqrt[3]{3}$	Multiply: $\sqrt[3]{y}\left(\sqrt[3]{4y^2} + \sqrt[3]{256y^2}\right)$.

EXAMPLE 3	YOUR TURN 3
Multiply: $\left(2+\sqrt{6}\right)^2$. $$\left(2+\sqrt{6}\right)^2 = \left(2+\sqrt{6}\right)\left(2+\sqrt{6}\right)$$ $$= (2)^2 + 2\sqrt{6} + 2\sqrt{6} + \left(\sqrt{6}\right)^2 \quad \text{FOIL}$$ $$= 4 + \boxed{}\sqrt{6} + 6$$ $$= \boxed{}$$	Multiply: $\left(3-\sqrt{7}\right)^2$.

EXAMPLE 4	YOUR TURN 4
Multiply: $\left(2\sqrt{3}+\sqrt{5}\right)\left(\sqrt{3}-4\sqrt{5}\right)$. $$\left(2\sqrt{3}+\sqrt{5}\right)\left(\sqrt{3}-4\sqrt{5}\right) = 2\left(\sqrt{3}\right)^2 - 8\sqrt{3}\cdot\sqrt{5} + \sqrt{5}\cdot\sqrt{3} - 4\left(\sqrt{5}\right)^2$$ $$= 2\cdot 3 - 8\sqrt{15} + \sqrt{15} - \boxed{}\cdot 5$$ $$= 6 - 8\sqrt{15} + \sqrt{15} - 20$$ $$= -14 - \boxed{}\sqrt{15}$$	Multiply: $$\left(4\sqrt{6}+\sqrt{2}\right)\left(\sqrt{6}-2\sqrt{2}\right).$$

EXAMPLE 5	YOUR TURN 5
Multiply: $\left(3+\sqrt{6}\right)\left(3-\sqrt{6}\right)$. $$\left(3+\sqrt{6}\right)\left(3-\sqrt{6}\right) = (3)^2 - 3\sqrt{6} + 3\sqrt{6} - \left(\sqrt{6}\right)^2$$ $$= 9 - 3\sqrt{6} + \boxed{} - 6$$ $$= 9 - \boxed{}$$ $$= \boxed{}$$	Multiply: $$\left(5-\sqrt{2}\right)\left(5+\sqrt{2}\right).$$

YOUR NOTES   Write your questions and additional notes.

Practice Exercises

Readiness Check

Determine whether each statement is true or false.

1. The conjugate of $\sqrt{a} + \sqrt{b}$ is $\sqrt{a} - \sqrt{b}$.

2. To add radical expressions, only the indices must be the same.

3. The simplified form of a product of two radical expressions always contains a radical.

4. The product of conjugates does not contain a radical.

Addition and Subtraction

Add or subtract. Then simplify by collecting like radical terms, if possible. Assume that no radicands were formed by raising negative numbers to even powers.

5. $4\sqrt{3} + 2\sqrt{3}$

6. $8\sqrt[3]{5} - 5\sqrt[3]{5}$

7. $2\sqrt{6} - 5\sqrt[4]{7} + 2\sqrt{6} + 9\sqrt[4]{7}$

8. $3\sqrt{20} - \sqrt{500}$

9. $\sqrt{49x^3} + 5\sqrt{x}$

10. $9\sqrt[3]{16} - 2\sqrt[3]{54}$

11. $\sqrt{9x + 36} - \sqrt{x + 4}$

12. $\sqrt[3]{8x^5} + \sqrt[3]{27x^5}$

More Multiplication

Multiply. Assume that no radicands were formed by raising negative numbers to even powers.

13. $\sqrt{5}\left(6 - 3\sqrt{5}\right)$

14. $\sqrt[3]{3y}\left(\sqrt[3]{9y^2} - \sqrt[3]{18y}\right)$

15. $\left(\sqrt{3} + x\right)^2$

16. $\left(\sqrt{a} + 4\right)\left(\sqrt{a} - 9\right)$

17. $\left(4\sqrt{5} + 3\right)\left(\sqrt{5} - 1\right)$

18. $\left(2\sqrt[3]{5} - 4\sqrt[3]{7}\right)\left(\sqrt[3]{5} - 6\sqrt[3]{7}\right)$

19. $\left(6 - \sqrt{3}\right)\left(6 + \sqrt{3}\right)$

20. $\left(\sqrt{x} + \sqrt{w}\right)\left(\sqrt{x} - \sqrt{w}\right)$

Rationalizing Denominators

ESSENTIALS

When we **rationalize the denominator** of a radical expression, we find an equivalent expression without a radical in the denominator.

Example

- Rationalize the denominator: $\dfrac{2\sqrt{5}}{3\sqrt{2}}$.

    Multiply by 1 using $\dfrac{\sqrt{2}}{\sqrt{2}}$ so that the radicand in the denominator will be a perfect square.

    $$\frac{2\sqrt{5}}{3\sqrt{2}} = \frac{2\sqrt{5}}{3\sqrt{2}} \cdot \frac{\sqrt{2}}{\sqrt{2}} = \frac{2\sqrt{5}\cdot\sqrt{2}}{3\sqrt{2}\cdot\sqrt{2}} = \frac{2\sqrt{10}}{3\sqrt{4}} = \frac{2\sqrt{10}}{3\cdot2} = \frac{\sqrt{10}}{3}$$

GUIDED LEARNING	📖 **Textbook** 👤 **Instructor** ▶ **Video**
EXAMPLE 1	YOUR TURN 1
Rationalize the denominator: $\sqrt{\dfrac{5}{3}}$. Multiply by 1 using $\dfrac{\sqrt{3}}{\sqrt{3}}$ so that the denominator of the radicand will be a perfect square. $\sqrt{\dfrac{5}{3}} = \dfrac{\sqrt{5}}{\sqrt{3}} \cdot \dfrac{\sqrt{\boxed{}}}{\sqrt{\boxed{}}} = \dfrac{\sqrt{5}\cdot\sqrt{3}}{\sqrt{3}\cdot\sqrt{3}} = \dfrac{\sqrt{15}}{\sqrt{\boxed{}}} = \dfrac{\sqrt{15}}{\boxed{}}$	Rationalize the denominator: $\sqrt{\dfrac{1}{2}}$.
EXAMPLE 2	YOUR TURN 2
Rationalize the denominator: $\sqrt[3]{\dfrac{3}{25}}$. Since $25 = 5\cdot5$, we need another factor of 5 so that the denominator of the radicand is a perfect cube. $\sqrt[3]{\dfrac{3}{25}} = \dfrac{\sqrt[3]{3}}{\sqrt[3]{25}} \cdot \dfrac{\sqrt[3]{\boxed{}}}{\sqrt[3]{\boxed{}}} = \dfrac{\sqrt[3]{15}}{\sqrt[3]{\boxed{}}} = \dfrac{\sqrt[3]{15}}{\boxed{}}$	Rationalize the denominator: $\sqrt[3]{\dfrac{9}{2}}$.

EXAMPLE 3	YOUR TURN 3
Rationalize the denominator: $\dfrac{\sqrt{3}}{3\sqrt{x^3}}$.	Rationalize the denominator: $\dfrac{\sqrt{5}}{2\sqrt{x}}$.

Multiply by 1 using $\dfrac{\sqrt{x}}{\sqrt{x}}$ so that the radicand in the denominator is a perfect square.

$$\frac{\sqrt{3}}{3\sqrt{x^3}} = \frac{\sqrt{3}}{3\sqrt{x^3}} \cdot \frac{\sqrt{x}}{\sqrt{x}} = \frac{\boxed{}}{3\sqrt{x^4}} = \boxed{}$$

EXAMPLE 4	YOUR TURN 4
Rationalize the denominator: $\dfrac{x\sqrt[3]{y}}{\sqrt[3]{9y^2z}}$.	Rationalize the denominator: $\dfrac{\sqrt[3]{xz}}{\sqrt[3]{16x^2yz^2}}$.

Multiply by $\dfrac{\sqrt[3]{3yz^2}}{\sqrt[3]{3yz^2}}$ so that the radicand in the denominator is a perfect cube.

$$\frac{x\sqrt[3]{y}}{\sqrt[3]{9y^2z}} = \frac{x\sqrt[3]{y}}{\sqrt[3]{9y^2z}} \cdot \frac{\sqrt[3]{3yz^2}}{\sqrt[3]{3yz^2}}$$

$$= \frac{x\sqrt[3]{\boxed{}}}{\sqrt[3]{27y^3z^3}}$$

$$= \frac{x\sqrt[3]{3y^2z^2}}{\boxed{}}$$

YOUR NOTES   Write your questions and additional notes.

Rationalizing When There Are Two Terms

ESSENTIALS

Pairs of radical expressions like $\sqrt{a} + \sqrt{b}$ and $\sqrt{a} - \sqrt{b}$ are called **conjugates**. The product of conjugates has no radicals.

Example

- Rationalize the denominator: $\dfrac{4}{6 + \sqrt{5}}$.

$$\frac{4}{6 + \sqrt{5}} = \frac{4}{6 + \sqrt{5}} \cdot \frac{6 - \sqrt{5}}{6 - \sqrt{5}} \qquad \text{Multiplying by 1, using the conjugate}$$
of $6 + \sqrt{5}$, which is $6 - \sqrt{5}$

$$= \frac{4\left(6 - \sqrt{5}\right)}{\left(6 + \sqrt{5}\right)\left(6 - \sqrt{5}\right)} \qquad \text{Multiplying numerators and denominators}$$

$$= \frac{4\left(6 - \sqrt{5}\right)}{6^2 - \left(\sqrt{5}\right)^2} \qquad \text{Using } \left(A + B\right)\left(A - B\right) = A^2 - B^2$$

$$= \frac{24 - 4\sqrt{5}}{36 - 5} = \frac{24 - 4\sqrt{5}}{31}$$

GUIDED LEARNING
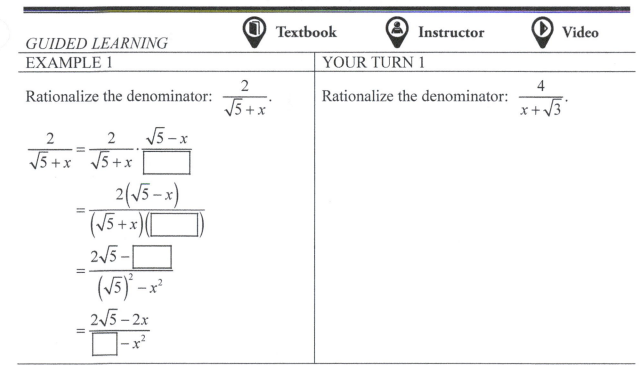

| 📖 **Textbook** | 👤 **Instructor** | ▶ **Video** |

EXAMPLE 1	YOUR TURN 1
Rationalize the denominator: $\dfrac{2}{\sqrt{5} + x}$.	Rationalize the denominator: $\dfrac{4}{x + \sqrt{3}}$.

$$\frac{2}{\sqrt{5} + x} = \frac{2}{\sqrt{5} + x} \cdot \frac{\sqrt{5} - x}{\boxed{}}$$

$$= \frac{2\left(\sqrt{5} - x\right)}{\left(\sqrt{5} + x\right)\left(\boxed{}\right)}$$

$$= \frac{2\sqrt{5} - \boxed{}}{\left(\sqrt{5}\right)^2 - x^2}$$

$$= \frac{2\sqrt{5} - 2x}{\boxed{} - x^2}$$

EXAMPLE 2	YOUR TURN 2
Rationalize the denominator: $\dfrac{3+\sqrt{3}}{\sqrt{7}-\sqrt{3}}$.	Rationalize the denominator: $\dfrac{4-\sqrt{11}}{\sqrt{2}+\sqrt{11}}$.

$$\frac{3+\sqrt{3}}{\sqrt{7}-\sqrt{3}} = \frac{3+\sqrt{3}}{\sqrt{7}-\sqrt{3}} \cdot \frac{\sqrt{7}+\sqrt{3}}{\sqrt{7}+\sqrt{3}}$$

$$= \frac{\left(3+\sqrt{3}\right)\left(\sqrt{7}+\sqrt{3}\right)}{\left(\sqrt{7}-\sqrt{3}\right)\left(\sqrt{7}+\sqrt{3}\right)}$$

$$= \frac{3\sqrt{7}+\boxed{}+\sqrt{3}\sqrt{7}+\left(\sqrt{3}\right)^2}{\left(\sqrt{7}\right)^2-\left(\sqrt{3}\right)^2}$$

$$= \frac{3\sqrt{7}+3\sqrt{3}+\sqrt{\boxed{}}+3}{7-3}$$

$$= \frac{3\sqrt{7}+3\sqrt{3}+\sqrt{21}+3}{\boxed{}}$$

YOUR NOTES   Write your questions and additional notes.

Practice Exercises

Readiness Check

Determine whether each statement is true or false.

1. The conjugate of $a - \sqrt{b}$ is $a + \sqrt{b}$.

2. To rationalize the denominator of $\dfrac{\sqrt[3]{5x}}{\sqrt[3]{3y}}$, we can multiply by 1 using $\dfrac{\sqrt[3]{3y}}{\sqrt[3]{3y}}$.

3. To rationalize the denominator of $\dfrac{2 + \sqrt{3}}{5 - \sqrt{7}}$, multiply by a form of 1 using $\dfrac{5 - \sqrt{7}}{5 - \sqrt{7}}$.

4. After a denominator is rationalized, it contains no radical expressions.

Rationalizing Denominators

Rationalize the denominator. Assume that no radicands were formed by raising negative numbers to even powers.

5. $\sqrt{\dfrac{1}{3}}$

6. $\dfrac{\sqrt{6}}{\sqrt{7}}$

7. $\dfrac{2\sqrt{5}}{3\sqrt{6}}$

8. $\sqrt[3]{\dfrac{3}{xy^2}}$

9. $\sqrt{\dfrac{7x}{12}}$

10. $\sqrt[4]{\dfrac{x^2}{16yz^4}}$

Rationalizing When There Are Two Terms

Rationalize the denominator. Assume that no radicands were formed by raising negative numbers to even powers.

11. $\dfrac{2}{5+\sqrt{3}}$

12. $\dfrac{4\sqrt{y}}{2-\sqrt{y}}$

13. $\dfrac{\sqrt{2}+\sqrt{3}}{\sqrt{2}-\sqrt{7}}$

14. $\dfrac{3-2\sqrt{x}}{2+3\sqrt{x}}$

The Principle of Powers

ESSENTIALS

A **radical equation** is an equation in which the variable appears in one or more radicands.

The Principle of Powers

If $a = b$ is true, then $a^n = b^n$ is true for any natural number n.

To solve an equation with one radical term:

1. Isolate the radical term on one side of the equation.
2. Use the principle of powers and then solve the resulting equation.
3. Check any possible solution in the original equation.

When we raise both sides of an equation to an even exponent, it is essential that we check the answer in the original equation.

Example

- Solve: $x = \sqrt{x+3} - 1$.

$$x = \sqrt{x+3} - 1$$

$$x + 1 = \sqrt{x+3} \qquad \text{Isolating the radical}$$

$$(x+1)^2 = \left(\sqrt{x+3}\right)^2 \qquad \text{Using the principle of powers}$$

$$x^2 + 2x + 1 = x + 3$$

$$x^2 + x - 2 = 0$$

$$(x+2)(x-1) = 0$$

$$x + 2 = 0 \quad or \quad x - 1 = 0$$

$$x = -2 \quad or \qquad x = 1$$

We check both −2 and 1 in the original equation. The number 1 checks but −2 does not, so the solution is 1.

GUIDED LEARNING	🔲 **Textbook**	👤 **Instructor**	▶ **Video**

EXAMPLE 1	YOUR TURN 1
Solve: $\sqrt{x} - 5 = 1$.	Solve: $\sqrt{x} + 3 = 8$.
$\sqrt{x} - 5 = 1$	
$\sqrt{x} = \boxed{}$ Isolating the radical	
$\left(\sqrt{x}\right)^2 = 6^2$ Using the principle of powers	
$x = \boxed{}$	
A check shows that 36 is the solution.	

EXAMPLE 2	YOUR TURN 2
Solve: $\sqrt{2y-1} = y-2$.	Solve: $\sqrt{3-x}+1 = x$.

$$\sqrt{2y-1} = y-2$$

$$\left(\sqrt{2y-1}\right)^2 = (y-2)^2$$

$$2y-1 = y^2 - \boxed{} + 4$$

$$0 = y^2 - \boxed{} + 5$$

$$0 = (y-1)\left(\boxed{}\right)$$

$$y = 1 \quad or \quad y = \boxed{}$$

Check: For 1:   For 5:

$$\begin{array}{c|c} \sqrt{2y-1} = y-2 \\ \hline \sqrt{2(1)-1} \;?\; 1-2 \\ \sqrt{2-1} \quad\Big|\; -1 \\ \sqrt{1} \\ 1 \;\Big|\; \text{FALSE} \end{array} \qquad \begin{array}{c|c} \sqrt{2y-1} = y-2 \\ \hline \sqrt{2(5)-1} \;?\; 5-2 \\ \sqrt{10-1} \quad\Big|\; 3 \\ \sqrt{9} \\ 3 \;\Big|\; \text{TRUE} \end{array}$$

The number 5 checks and 1 does not, so the solution is $\boxed{}$.

EXAMPLE 3	YOUR TURN 3
Solve: $\sqrt[3]{x-5}+2 = 1$.	Solve: $\sqrt[3]{x+4}+5 = 3$.

$$\sqrt[3]{x-5}+2 = 1$$

$$\sqrt[3]{x-5} = -1 \qquad \text{Isolating the radical}$$

$$\left(\sqrt[3]{x-5}\right)^3 = \boxed{} \qquad \text{Using the principle of powers}$$

$$x-5 = -1$$

$$x = \boxed{}$$

The number 4 checks and is the solution.

The solution is $\boxed{}$.

YOUR NOTES   Write your questions and additional notes.

Equations with Two Radical Terms

ESSENTIALS

To solve an equation with two or more radical terms:

1. Isolate one of the radical terms.
2. Use the principle of powers.
3. If a radical remains, perform steps (1) and (2) again.
4. Solve the resulting equation.
5. Check possible solutions in the original equation.

Example

- Solve: $\sqrt{x+2} - 2\sqrt{x-1} = 0$.

$$\sqrt{x+2} - 2\sqrt{x-1} = 0$$

$$\sqrt{x+2} = 2\sqrt{x-1} \qquad \text{Isolating a radical}$$

$$\left(\sqrt{x+2}\right)^2 = \left(2\sqrt{x-1}\right)^2 \qquad \text{Using the principle of powers}$$

$$x+2 = 4(x-1)$$

$$x+2 = 4x-4$$

$$6 = 3x$$

$$2 = x$$

Since 2 checks in the original equation, the solution is 2.

GUIDED LEARNING   📖 **Textbook**   👤 **Instructor**   ▶ **Video**

EXAMPLE 1	YOUR TURN 1
Solve: $\sqrt{4y-2} = \sqrt{3y+9}$.	Solve: $\sqrt{2y-7} = \sqrt{8-y}$.
$\sqrt{4y-2} = \sqrt{3y+9}$	
$\left(\sqrt{4y-2}\right)^2 = \left(\sqrt{3y+9}\right)^2$	
$4y - \boxed{} = 3y + \boxed{}$	
$\boxed{} - 2 = 9$	
$y = \boxed{}$	
A check shows that $\boxed{}$ is the solution.	

EXAMPLE 2	YOUR TURN 2
Solve: $\sqrt{a+1} = \sqrt{a-2} + 1$.	Solve: $\sqrt{a+8} - \sqrt{a} = 2$.

$$\sqrt{a+1} = \sqrt{a-2} + 1$$

$$\left(\sqrt{a+1}\right)^2 = \left(\sqrt{a-2} + 1\right)^2$$

$$a+1 = a-2+2\sqrt{a-2}+1$$

$$\boxed{} = 2\sqrt{a-2}$$

$$1 = \sqrt{a-2}$$

$$1^2 = \left(\boxed{}\right)^2$$

$$1 = a-2$$

$$\boxed{} = a$$

A check shows that $\boxed{}$ is the solution.

EXAMPLE 3	YOUR TURN 3
Solve: $\sqrt{3x+4} = 1 + \sqrt{2x+2}$.	Solve: $\sqrt{3-x} = 1 + \sqrt{4-2x}$.

$$\sqrt{3x+4} = 1 + \sqrt{2x+2}$$

$$\left(\sqrt{3x+4}\right)^2 = \left(1 + \sqrt{2x+2}\right)^2$$

$$3x + \boxed{} = 1 + 2\sqrt{2x+2} + 2x + \boxed{}$$

$$3x + 4 = \boxed{} + 2\sqrt{2x+2} + 2x$$

$$x + \boxed{} = 2\sqrt{2x+2}$$

$$\left(x+1\right)^2 = \left(2\sqrt{2x+2}\right)^2$$

$$x^2 + 2x + 1 = \boxed{}(2x+2)$$

$$x^2 + 2x + 1 = 8x + \boxed{}$$

$$x^2 - 6x - \boxed{} = 0$$

$$\left(x+1\right)\left(x - \boxed{}\right) = 0$$

$$x + 1 = 0 \quad or \quad x - 7 = 0$$

$$x = \boxed{} \quad or \quad x = \boxed{}$$

Since both numbers check, the solutions are $\boxed{}$ and $\boxed{}$.

YOUR NOTES   Write your questions and additional notes.

Applications

ESSENTIALS

Example

- At a height of h meters, one can see V kilometers to the horizon, where $V = 3.5\sqrt{h}$. A technician can see 70 km to the horizon from atop a tower. What is the altitude of the technician's eyes?

  1. **Familiarize**. Use the formula $V = 3.5\sqrt{h}$.

  2. **Translate**. Substitute 70 for V in the formula: $70 = 3.5\sqrt{h}$.

  3. **Solve**. Solve the equation for h.

  $$70 = 3.5\sqrt{h}$$
  $$20 = \sqrt{h}$$
  $$(20)^2 = (\sqrt{h})^2$$
  $$400 = h$$

  4. **Check**. Let $h = 400$ and calculate V: $V = 3.5\sqrt{400} = 3.5(20) = 70$. The answer checks.

  5. **State**. The altitude of the technician's eyes is 400 m.

GUIDED LEARNING   📔 **Textbook**   👤 **Instructor**   ▶ **Video**

EXAMPLE 1	YOUR TURN 1
At a height of h feet, one can see d miles to the horizon, where $d = \sqrt{1.5h}$. From an airplane window, Anne can see 90 miles to the horizon. What is the altitude of the airplane? 1. **Familiarize**. Use the formula ☐. 2. **Translate**. Substitute ☐ for d. $d = \sqrt{1.5h}$ ☐ $= \sqrt{1.5h}$ (continued)	The formula $r = 2\sqrt{5L}$ can be used to approximate the speed r, in miles per hour, of a car that has left a skid mark of length L, in feet. How far will a car skid at 30 mph?

3. **Solve**. Solve for h.

$$90 = \sqrt{1.5h}$$

$$(90)^2 = \left(\boxed{}\right)^2$$

$$8100 = \boxed{}$$

$$\frac{8100}{\boxed{}} = \frac{1.5h}{\boxed{}}$$

$$\boxed{} = h$$

4. **Check**. Let $h = \boxed{}$ and calculate d:

$$d = \sqrt{1.5h}$$

$$= \sqrt{1.5\left(\boxed{}\right)}$$

$$= \sqrt{\boxed{}}$$

$$= \boxed{}$$

The answer checks.

5. **State**. The altitude of the plane is $\boxed{}$ ft.

YOUR NOTES   Write your questions and additional notes.

Practice Exercises

Readiness Check

Fill in the word that best completes each sentence.

To solve an equation with two or more radical terms:

1. _____ one of the radical terms.

2. Use the principle of _____.

3. If a _____ remains, perform steps (1) and (2) again.

4. _____ the resulting equation.

5. Check possible solutions in the _____ equation.

The Principle of Powers

Solve.

6. $\sqrt{x} - 6 = 5$

7. $\sqrt[3]{2a+5} + 3 = 0$

8. $\sqrt{8x+1} = 5$

9. $3\sqrt{x-1} = x+1$

10. $\sqrt{2y+7} - y = 2$

11. $\sqrt[4]{y+16} + 4 = 0$

Equations with Two Radical Terms

Solve.

12. $\sqrt{2m-1} = \sqrt{1-2m}$

13. $\sqrt{x-12} = 2 - \sqrt{x}$

14. $\sqrt{2x-9} - \sqrt{x-1} = -1$

15. $\sqrt{x+1} + \sqrt{x+6} = 5$

16. $\sqrt[3]{6x-5} + 3 = -2$

17. $\sqrt{x+5} = \sqrt{2x-5}$

Applications

18. The speed v, in meters per second, of a wave on the surface of the ocean can be approximated by the formula $v = 3.1\sqrt{d}$, where d is the depth of the water, in meters. A wave is traveling 12.4 m/sec. What is the water depth?

Applications

ESSENTIALS

The Pythagorean Theorem

In any right triangle, if a and b are the lengths of the legs and c is the length of the hypotenuse, then $a^2 + b^2 = c^2$.

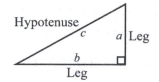

Example

- One leg of a right triangle is 6 m and the hypotenuse is 12 m. Find the length of the other leg. Give an exact answer and an approximation to three decimal places.

$$a^2 + b^2 = c^2$$
$$6^2 + b^2 = 12^2$$
$$36 + b^2 = 144$$
$$b^2 = 108$$
$$b = \sqrt{108} \qquad \text{The length must be positive.}$$
$$b = \sqrt{36 \cdot 3}$$
$$b = 6\sqrt{3} \approx 10.392$$

The exact answer is $6\sqrt{3}$ m, and the approximation is 10.392 m.

GUIDED LEARNING   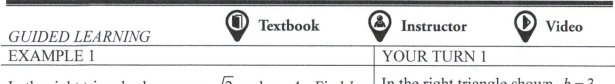 **Textbook**   **Instructor**   **Video**

EXAMPLE 1	YOUR TURN 1

In the right triangle shown, $a = \sqrt{2}$ and $c = 4$. Find b. Give an exact answer and an approximation to three decimal places.

$$a^2 + b^2 = c^2$$
$$\left(\boxed{}\right)^2 + b^2 = 4^2$$
$$\boxed{} + b^2 = 16$$
$$b^2 = \boxed{}$$
$$b = \sqrt{14} \approx \boxed{}$$

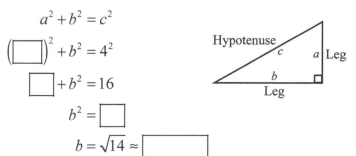

In the right triangle shown, $b = 3$ and $c = \sqrt{11}$. Find a. Give an exact answer and an approximation to three decimal places.

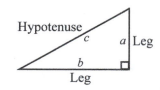

EXAMPLE 2	YOUR TURN 2
How long is a guy wire reaching from the top of a 36-ft pole to a point on the ground 12 ft from the pole? Give an exact answer and an approximation to three decimal places.	A ladder is leaning against a house. The bottom of the ladder is 3 ft from the house. The top of the ladder touches the house at a point 20 ft above the ground. How long is the ladder? Give an exact answer and an approximation to three decimal places.

Let c = the length (in feet) of the guy wire.

$$12^2 + 36^2 = c^2$$
$$144 + 1296 = c^2$$
$$1440 = c^2$$

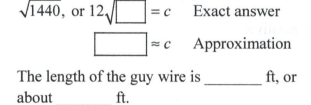

$$\sqrt{\boxed{}} = c$$

(Use the positive square root because length is not negative.)

$\sqrt{1440}$, or $12\sqrt{\boxed{}} = c$ Exact answer

$\boxed{} \approx c$ Approximation

The length of the guy wire is _____ ft, or about _____ ft.

YOUR NOTES   Write your questions and additional notes.

Practice Exercises

Readiness Check

Determine whether each statement is true or false.

1. The longest side of a right triangle is the hypotenuse.

2. The Pythagorean theorem is true for any triangle.

3. When solving for a length, we do not consider negative numbers since length cannot be negative.

4. If the lengths of the legs of a right triangle are rational numbers, then the length of the hypotenuse is an irrational number.

Applications

In Exercises 5-10, give an exact answer and, where appropriate, an approximation to three decimal places.

5. In the right triangle shown, $a = 6$ and $b = 7$. Find c.

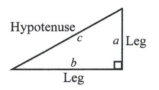

6. In the right triangle shown, $a = 6$ and $c = 10$. Find b.

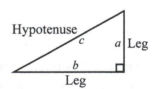

7. A right triangle's hypotenuse is 10 cm, and one leg is $3\sqrt{2}$ cm. Find the length of the other leg.

8. One leg in a right triangle is 2 m, and the hypotenuse measures $\sqrt{6}$ m. Find the length of the other leg.

9. A zipline ride extends 400 ft along the ground in a wooded area. The riders start the ride from a platform 30 ft high. How long is the zipline?

10. Mary walks through a rectangular park on her way to work. If the park is 50 m long and 45 m wide, how far does Mary walk when she walks across the park diagonally?

Increasing, Decreasing, and Constant Functions

ESSENTIALS

A function f is said to be **increasing** on an *open* interval I, if for all a and b in that interval, $a < b$ implies $f(a) < f(b)$.

A function f is said to be **decreasing** on an *open* interval I, if for all a and b in that interval, $a < b$ implies $f(a) > f(b)$.

A function f is said to be **constant** on an *open* interval I, if for all a and b in that interval, $f(a) = f(b)$.

In other words, on a given interval, if the graph of a function rises from left to right, it is said to be **increasing** on that interval. If the graph drops from left to right, it is said to be **decreasing**. If the function values stay the same on the interval, the function is said to be **constant**.

Example

- Determine the intervals on which the function is
  (a) increasing;   (b) decreasing;   (c) constant.

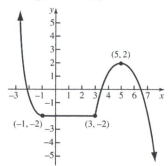

a) As x-values increase from 3 to 5, y-values increase from -2 to 2. Thus the function is increasing on $(3, 5)$.

b) As x-values increase from $-\infty$ to -1, y-values decrease. Also, as x-values increase from 5 to ∞, y-values decrease. Thus the function is decreasing on $(-\infty, -1)$ and $(5, \infty)$.

c) As x-values increase from -1 to 3, the y-values stay the same, -2. The function is constant on $(-1, 3)$.

GUIDED LEARNING   📖 **Textbook**   👤 **Instructor**   ▶ **Video**

EXAMPLE 1	YOUR TURN 1
Determine the intervals on which the function is (a) increasing; (b) decreasing; (c) constant.	Determine the intervals on which the function is (a) increasing; (b) decreasing; (c) constant.

a) As x-values increase from ⬚ to -3, y-values increase. Also, as x-values increase from ⬚ to 4, y-values increase. Thus the function is increasing on $\left(-\infty, \boxed{}\right)$ and $\left(\boxed{}, 4\right)$.

b) As x-values increase from 0 to 2, y-values _____ . Thus the function
    increase / decrease
    is decreasing on $\left(0, \boxed{}\right)$.

c) As x-values increase from -3 to ⬚ , the y-values stay the same, 2. Thus the function is constant on $\left(\boxed{}, 0\right)$.

YOUR NOTES   Write your questions and additional notes.

Relative Maximum and Minimum Values

ESSENTIALS

Suppose that *f* is a function for which $f(c)$ exists for some *c* in the domain of *f*. Then:

$f(c)$ is a **relative maximum** (plural, **maxima**) if there exists an *open* interval *I* containing *c* such that $f(c) > f(x)$, for all *x* in *I* where $x \neq c$; and

$f(c)$ is a **relative minimum** (plural **minima**) if there exists an *open* interval *I* containing *c* such that $f(c) < f(x)$, for all *x* in *I* where $x \neq c$.

Simply stated, $f(c)$ is a *relative maximum* if $(c, f(c))$ is the highest point in some *open* interval, and $f(c)$ is a *relative minimum* if $(c, f(c))$ is the lowest point in some *open* interval.

Example

- Using the graph shown below, determine any relative maxima or minima of the function $f(x) = x^3 - 2x^2 - 4.5x + 3$ and the intervals on which the function is increasing or decreasing.

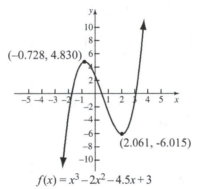

$$f(x) = x^3 - 2x^2 - 4.5x + 3$$

We see that the relative maximum value of the function is 4.830. It occurs when $x = -0.728$. We also see that the relative minimum is -6.015 at $x = 2.061$.

The graph starts rising, or increasing, from the left and stops increasing at the relative maximum. Then it decreases to the relative minimum and then rises again. Thus the function is increasing on the intervals $(-\infty, -0.728)$ and $(2.061, \infty)$ and is decreasing on the interval $(-0.728, 2.061)$.

GUIDED LEARNING   **Textbook**   **Instructor**   **Video**

EXAMPLE 1	YOUR TURN 1
Using the graph, determine any relative maxima or minima of the function $f(x) = -2x^3 - 6x^2 - 2.5x + 2$ and the intervals on which the function is increasing or decreasing.	Using the graph, determine any relative maxima or minima of the function $f(x) = 2x^3 + 4x^2 - 1.5x + 1$ and the intervals on which the function is increasing or decreasing.

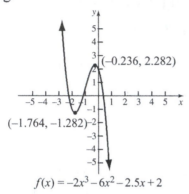

$f(x) = -2x^3 - 6x^2 - 2.5x + 2$

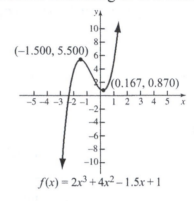

$f(x) = 2x^3 + 4x^2 - 1.5x + 1$

We see that the relative maximum value of the function is [____] at $x =$ [____] . We also see that the relative minimum value is [____] at

$x =$ [____] .

The graph starts falling or decreasing from the left and stops decreasing at the relative

_____ .
 maximum / minimum

Then it increases to the relative

_____ and then falls again. Then
 maximum / minimum

the function is increasing on the interval

$\left(-1.764, \boxed{}\right)$ and is decreasing on

$\left(-\infty, \boxed{}\right)$ and $\left(-0.236, \boxed{}\right)$.

YOUR NOTES   Write your questions and additional notes.

Applications of Functions

ESSENTIALS

Many real-world situations can be modeled by functions.

Example

Jenna's restaurant supply store has 20 ft of dividers with which to set off a rectangular area for the storage of overstock. If a corner of the store is used for the storage area, the partition needs to form only two sides of a rectangle.

a) Express the floor area of the storage space as a function of the length of the partition.

Let $x =$ the length of the partition, in feet.

$$A(x) = x(20 - x)$$
$$A(x) = 20x - x^2$$

b) Find the domain of the function.

$$A(x) = 20x - x^2$$

The domain could be $(-\infty, \infty)$ but, since only 20 ft of dividers are available, a more realistic domain is $(0, 20)$.

c) Using the graph shown below, determine the dimensions that maximize the area of the floor.

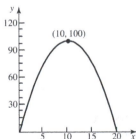

Relative maximum is 100 when $x = 10$.

Maximum area $= 100$ ft^2 when $x = 10$ ft.

Length: 10 ft

Width: $20 - 10 = 10$ ft

Dimensions: 10 ft $\times$ 10 ft

GUIDED LEARNING   **Textbook**   **Instructor**   **Video**

EXAMPLE 1	YOUR TURN 1
A dentist has 24 ft of dividers with which to set off a rectangular area for a play space in her waiting room. If a corner of the waiting room is used for the play space, the partition needs to form only two sides of the rectangle.	Joseph has 36 ft of fencing with which to enclose a garden plot in his yard. If an already fenced corner of the yard is used for the plot, the new fencing needs to form only two sides of the plot.

a) Express the floor area of the play space as a function of the length of the partition.

b) Find the domain of the function.

c) Using the graph shown below, determine the dimensions that maximize the floor area.

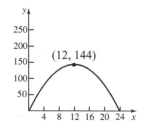

a) Express the area of the garden plot as a function of the length of the plot.

b) Find the domain of the function.

c) Using the graph shown below, determine the dimensions that maximize the area of the plot.

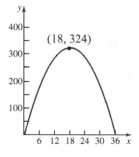

a) Let x = the length of the rectangle, in feet.

Then $\boxed{} - x$ = the width.

$$A(x) = x(24 - x) = 24x - \boxed{}$$

b) Because the rectangle's length and width must be positive and only 24 ft of dividers are available, we restrict the domain of A to $\{x \mid 0 < x < 24\}$, or $(0, 24)$.

c) From the graph we see that the maximum value of the area appears to be 144 when $x = 12$. Thus the dimensions that maximize the area are

Length $= x = 12$ ft and

Width $= 24 - x = 24 - \boxed{} = 12$ ft.

YOUR NOTES   Write your questions and additional notes.

Functions Defined Piecewise

ESSENTIALS

Some functions are defined **piecewise** using different output formulas for different pieces, or parts, of the domain.

A piecewise function with importance in calculus and computer programming is the **greatest integer function** denoted $f(x) = [\![x]\!]$, or $f(x) = \text{int}(x)$, and defined as follows:

$$f(x) = [\![x]\!] = \text{ the greatest integer less than or equal to } x.$$

Example

- For the function defined as

$$f(x) = \begin{cases} x+1, & \text{for } x < -2, \\ 5, & \text{for } -2 \le x \le 3, \\ x^2, & \text{for } x > 3, \end{cases}$$

find $f(-5), f(0), f(3),$ and $f(10)$.

Since $-5 < -2$, we use the formula $f(x) = x+1$:

$$f(-5) = -5 + 1 = -4.$$

Since $-2 \le 0 \le 3$, we use the formula $f(x) = 5$:

$$f(0) = 5.$$

Since $-2 \le 3 \le 3$, we use the formula $f(x) = 5$ again:

$$f(3) = 5.$$

Since $10 > 3$, we use the formula $f(x) = x^2$:

$$f(10) = 10^2 = 100.$$

GUIDED LEARNING

EXAMPLE 1	YOUR TURN 1

For the function defined as

$$g(x) = \begin{cases} x^2 - 1, & \text{for } x \leq -1, \\ -x, & \text{for } -1 \leq x \leq 3 \\ 6, & \text{for } x > 3, \end{cases}$$

find $g(-2)$, $g(1)$, $g(3)$, and $g(4)$.

Since $-2 \leq -1$, we use the formula $g(x) = x^2 - 1$:

$$g(-2) = (-2)^2 - 1 = 4 - 1 = \boxed{}.$$

Since $-1 < 1 \leq 3$, we use the formula $g(x) = -x$:

$$g(1) = -1.$$

Since $-1 < 3 \leq 3$, we use the formula $g(x) = -x$:

$$g(3) = \boxed{}.$$

Since $4 > 3$, we use the formula $g(x) = 6$:

$$g(4) = \boxed{}.$$

For the function defined as

$$f(x) = \begin{cases} 2x + 3, & \text{for } x < -2, \\ 1, & \text{for } -2 \leq x \leq 1, \\ x^2, & \text{for } x > 1, \end{cases}$$

find $f(-5)$, $f(-2)$, $f(0)$, and $f(3)$.

EXAMPLE 2	YOUR TURN 2

EXAMPLE 2

Graph the function defined as

$$g(x) = \begin{cases} 2x, & \text{for } x < -1, \\ x-3, & \text{for } x \geq -1. \end{cases}$$

We create the graph in two pieces, or parts. First we graph $g(x) = 2x$ only for inputs less than -1. We find some ordered pairs that are solutions of this piece of the function.

x $(x < -1)$	$g(x) = 2x$
-3	-6
-2	☐
$-\dfrac{3}{2}$	-3

We graph $g(x) = x - 3$ only for inputs greater than or equal to ☐. We find some ordered pairs that are solutions of this piece of the function.

x $(x \geq -1)$	$g(x) = x - 3$
-1	☐
1	-2
4	☐

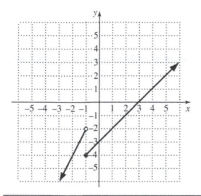

YOUR TURN 2

Graph the function defined as

$$f(x) = \begin{cases} -x+2, & \text{for } x \leq 3, \\ x-5, & \text{for } x > 3. \end{cases}$$

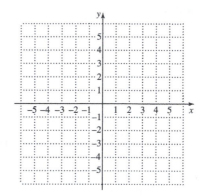

EXAMPLE 3	YOUR TURN 3

Graph the greatest integer function $f(x) = [\![x]\!]$.

The greatest integer function can be thought of as a piecewise function with an infinite number of statements.

$$f(x) = \begin{cases} \vdots \\ -3, & \text{for } -3 \le x < -2, \\ -2, & \text{for } -2 \le x < -1, \\ \boxed{}, & \text{for } -1 \le x < 0, \\ 0, & \text{for } 0 \le x < 1, \\ \boxed{}, & \text{for } 1 \le x < 2, \\ 2, & \text{for } 2 \le x < 3, \\ \boxed{}, & \text{for } 3 \le x < 4, \\ \vdots \end{cases}$$

Graph $f(x) = [\![x]\!] - 1$.

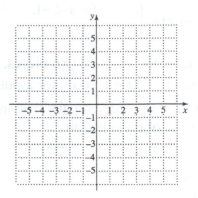

YOUR NOTES   Write your questions and additional notes.

Practice Exercises

Readiness Check

Determine whether each statement is true or false.

1. If the graph of a function drops from left to right on an interval, the function is increasing on the interval.

2. For a function f, $f(c)$ is a relative maximum if $(c, f(c))$ is this highest point in some open interval.

3. The greatest integer function is a piecewise function.

4. For the greatest integer function $f(x) = [\![x]\!]$, $f(-5.3) = -5$.

Increasing, Decreasing, and Constant Functions

5. Determine the intervals on which the function is

   (a) increasing; (b) decreasing; (c) constant.

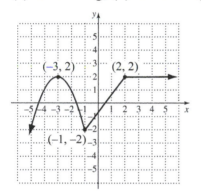

Relative Maximum and Minimum Values

6. Using the graph, determine any maxima or minima of the function
   $f(x) = -x^3 - 1.1x^2 + 5x + 2$ and the intervals on which the function is increasing or decreasing.

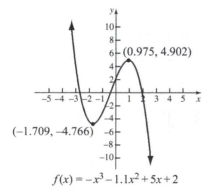

$$f(x) = -x^3 - 1.1x^2 + 5x + 2$$

Applications of Functions

7. Mason is designing a triangular pennant so that the height of the triangle, in inches, is 6 more than twice the base. Express the area of the pennant as a function of the base.

8. A rancher has 360 yd of fencing with which to enclose two adjacent rectangular corrals. A river forms one side of the corrals. Suppose the width of each corral is x yards.

    a) Express the total area of the two corrals as a function of x.

    b) Find the domain of the function.

    c) Using the graph, determine the dimensions that yield the maximum area.

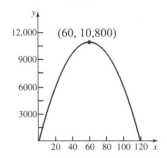

Functions Defined Piecewise

Simplify.

9. For the function defined as

$$f(x) = \begin{cases} 2x - 3, & \text{for } x \le -3, \\ 1, & \text{for } -3 < x < 0, \\ x^2 + 2, & \text{for } x \ge 0, \end{cases}$$

    find $f(-3)$, $f(-1)$, $f(0)$, and $f(4)$.

Graph each of the following.

10. $g(x) = \begin{cases} -x, & \text{for } x \le -4, \\ x + 1, & \text{for } x > -4 \end{cases}$

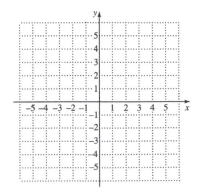

11. $f(x) = 2[\![x]\!]$

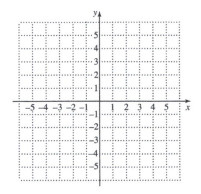

Section 7.1 Symmetry

> **Algebraic Tests of Symmetry**
>
> *x-axis*: If replacing y with $-y$ produces an equivalent equation, then the graph is *symmetric with respect to the x-axis*.
>
> *y-axis*: If replacing x with $-x$ produces an equivalent equation, then the graph is *symmetric with respect to the y-axis*.
>
> *Origin*: If replacing x with $-x$ and y with $-y$ produces an equivalent equation, then the graph is *symmetric with respect to the origin*.

Example 1 Test $y = x^2 + 2$ for symmetry with respect to the x-axis, the y-axis, and the origin.

x-axis:

$$y = x^2 + 2$$
$$\downarrow$$
$$-y = x^2 + 2 \qquad\qquad \text{Replacing } y \text{ with } -y$$
$$y = -x^2 - \boxed{} \qquad\qquad \text{Not equivalent to } y = x^2 + 2$$

The graph _____ symmetric with respect to the x-axis.
is / is not

y-axis:

$$y = x^2 + 2$$
$$\downarrow$$
$$y = (-x)^2 + 2 \qquad\qquad \text{Replacing } x \text{ with } -x$$
$$y = \boxed{} + 2 \qquad\qquad \text{Obtaining the original equation}$$

The graph _____ symmetric with respect to the y-axis.
is / is not

Origin:

$$y = x^2 + 2$$
$$\downarrow \quad \downarrow$$
$$-y = (-x)^2 + 2 \qquad\qquad \text{Replacing } y \text{ with } -y \text{ and } x \text{ with } -x$$
$$\boxed{} = x^2 + 2$$
$$\boxed{} = -x^2 - 2 \qquad\qquad \text{Not equivalent to } y = x^2 + 2$$

The graph _____ symmetric with respect to the origin.
is / is not

Example 2   Test $x^2 + y^4 = 5$ for symmetry with respect to the *x*-axis, the *y*-axis, and the origin.

x-axis:

$$x^2 + y^4 = 5$$

$$x^2 + (-y)^4 = 5 \qquad \text{Replacing } y \text{ with } -y$$

$$x^2 + \boxed{} = 5 \qquad \text{Obtaining the original equation}$$

The graph _____ symmetric with respect to the *x*-axis.
is / is not

y-axis:

$$x^2 + y^4 = 5$$

$$(-x)^2 + y^4 = 5 \qquad \text{Replacing } x \text{ with } -x$$

$$x^2 + y^4 = \boxed{} \qquad \text{Obtaining the original equation}$$

The graph _____ symmetric with respect to the *y*-axis.
is / is not

Origin:

$$x^2 + y^4 = 5$$

$$(-x)^2 + (-y)^4 = 5 \qquad \text{Replacing } x \text{ with } -x \text{ and } y \text{ with } -y$$

$$\boxed{} + y^4 = 5 \qquad \text{Obtaining the original equation}$$

The graph _____ symmetric with respect to the origin.
is / is not

Even Functions and Odd Functions

If the graph of a function f is symmetric with respect to the y-axis, we say that it is an **even function**. That is, for each x in the domain of f, $f(x) = f(-x)$.

If the graph of a function f is symmetric with respect to the origin, we say that it is an **odd function**. That is, for each x in the domain of f, $f(-x) = -f(x)$.

Example 3

a) Determine whether $f(x) = 5x^7 - 6x^3 - 2x$ is odd, even, or neither.

$$f(-x) = 5(-x)^7 - 6(-x)^3 - 2(-x)$$
$$= -5x^7 + 6x^3 + 2x$$
$$f(-x) \neq f(x)$$

Thus, $f(x)$ _____ even.
$\qquad$ is / is not

$$-f(x) = -\left(5x^7 - 6x^3 - 2x\right)$$
$$= -5x^7 + 6x^3 + 2x$$
$$f(-x) = -f(x)$$

Thus, $f(x)$ _____ odd.
$\qquad$ is / is not

b) Determine if $h(x) = 5x^6 - 3x^2 - 7$ is even, odd, or neither.

$$h(-x) = 5(-x)^6 - 3(-x)^2 - 7$$
$$= 5x^6 - 3x^2 - \boxed{}$$
$$h(-x) = h(x)$$

Thus, $h(x)$ is an _____ function.
$\qquad$ even / odd

The function cannot be both even and odd, so we stop here.

Section 7.2 Transformations

Vertical Translation and Horizontal Translation

For $b > 0$:

the graph of $y = f(x) + b$ is the graph of $y = f(x)$ shifted *up b* units;

the graph of $y = f(x) - b$ is the graph of $y = f(x)$ shifted *down b* units.

For $d > 0$:

the graph of $y = f(x - d)$ is the graph of $y = f(x)$ shifted *to the right d* units;

the graph of $y = f(x + d)$ is the graph of $y = f(x)$ shifted *to the left d* units.

Example 1   Graph each of the following. Before doing so, describe how each graph can be obtained from one of the basic graphs shown earlier in this section.

a) $g(x) = x^2 - 6$

We compare this with the graph of $f(x) = x^2$.

$g(x) = f(x) - 6$, so the graph of $g(x)$ is the graph of $f(x)$ translated

_____ 6 units.
up / down

We compare some points on the graphs of *f* and *g*.

$$\frac{f}{\quad} \qquad \frac{g}{\quad}$$

$(-3, 9) \quad (-3, 3)$

$(0, 0) \quad \left(0, \boxed{}\right)$

$(2, 4) \quad \left(2, \boxed{}\right)$

The *y*-coordinate of a point on the graph of $g(x)$ is 6 _____ than the
more / less

corresponding point on the graph of $f(x)$.

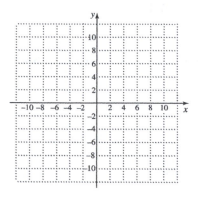

b) $g(x) = |x - 4|$

We compare this with the graph of $f(x) = |x|$.

$g(x) = f(x - 4)$, so the graph of $g(x)$ is the graph of $f(x)$ shifted

_____ 4 units.
right / left

We compare some points on the graphs of *f* and *g*.

$$
\begin{array}{cc}
\underline{f} & \underline{g} \\
(-4, 4) & (0, 4) \\
(0, 0) & (\boxed{}, 0) \\
(6, 6) & (10, \boxed{})
\end{array}
$$

The *x*-coordinate of a point on the graph of $g(x)$ is 4 _____ than the
more / less

corresponding point on the graph of $f(x)$.

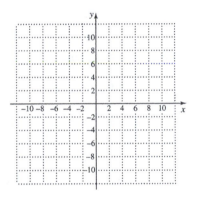

c) $g(x) = \sqrt{x + 2}$

We compare this with the graph of $f(x) = \sqrt{x}$.

$g(x) = f(x + 2)$, so the graph of $g(x)$ is the graph of $f(x)$ shifted

_____ 2 units, as we can see from the graphs of the functions.
right / left

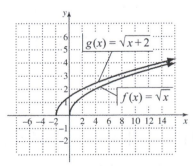

d) $h(x) = \sqrt{x+2} - 3$

We compare this with the graph of $f(x) = \sqrt{x}$.

In part (c) we saw that the graph of $g(x) = \sqrt{x+2}$ is the graph of $f(x)$ shifted left ☐ units.

$h(x) = g(x) - 3$, so the graph of $h(x)$ is the graph of $g(x)$ shifted _____ 3 units.
up / down

Together, the graph of $h(x)$ is the graph of $f(x) = \sqrt{x}$ shifted left ☐ units and down ☐ units.

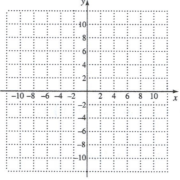

Reflections

The graph of $y = -f(x)$ is the **reflection** of the graph of $y = f(x)$ across the x-axis.

The graph of $y = f(-x)$ is the **reflection** of the graph of $y = f(x)$ across the y-axis.

If a point (x, y) is on the graph of $y = f(x)$, then $(x, -y)$ is on the graph of $y = -f(x)$, and $(-x, y)$ is on the graph of $y = f(-x)$.

Example 2   Graph each of the following. Before doing so, describe how each graph can be obtained from the graph of $f(x) = x^3 - 4x^2$.

a) $g(x) = (-x)^3 - 4(-x)^2$

$f(-x) = (-x)^3 - 4(-x)^2 = g(x)$, so $g(x)$ is a reflection of $f(x)$ across the

_____ axis.
x- / y-

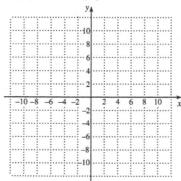

If (x, y) is on the graph of f, then $\left(\boxed{}, y\right)$ is on the graph of g. For example, if $(2, -8)$ is on the graph of f, then $(-2, -8)$ is on the graph of g.

b) $h(x) = 4x^2 - x^3$

$-f(x) = -(x^3 - 4x^2) = -x^3 + 4x^2 = 4x^2 - x^3 = h(x)$, so the graph of $h(x)$ is the

reflection of the graph of $f(x)$ across the _____ axis.

x- / y-

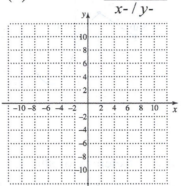

If (x, y) is on the graph of f, then $\left(x, \boxed{}\right)$ is on the graph of h. For example, if $(2, -8)$

is on the graph of f, then $(2, 8)$ is on the graph of h.

Vertical Stretching and Shrinking and Horizontal Stretching and Shrinking

The graph of $y = af(x)$ can be obtained from the graph of $y = f(x)$ by

    stretching vertically for $|a| > 1$, or

    shrinking vertically for $0 < |a| < 1$.

For $a < 0$, the graph is also reflected across the x-axis. (The y-coordinates of the graph of $y = af(x)$ can be obtained by multiplying the y-coordinates of $y = f(x)$ by a.)

The graph of $y = f(cx)$ can be obtained from the graph of $y = f(x)$ by

    shrinking horizontally for $|c| > 1$, or

    stretching horizontally for $0 < |c| < 1$.

For $c < 0$, the graph is also reflected across the y-axis. (The x-coordinates of the graph of $y = f(cx)$ can be obtained by dividing the x-coordinates of the graph of $y = f(x)$ by c.)

Example 3   Shown is a graph of $y = f(x)$ for some function f. No formula for f is given. Graph each of the following.

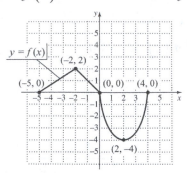

a) $g(x) = 2f(x)$

The graph of $y = af(x)$ is a vertical stretching or shrinking of the graph of $y = f(x)$. Here $|a| = |2| > 1$, so this is a stretching. Each y-value on $f(x)$ will be multiplied by 2 to obtain the graph of $g(x)$.

Points on f	Corresponding points on g
$(-5, 0)$	$(-5, 0)$
$(-2, 2)$	$\left(-2, \boxed{}\right)$
$(0, 0)$	$(0, 0)$
$(2, -4)$	$\left(2, \boxed{}\right)$
$(4, 0)$	$(4, 0)$

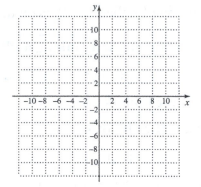

b) $h(x) = \dfrac{1}{2} f(x)$

The graph of $y = af(x)$ is a vertical stretching or shrinking of the graph of $y = f(x)$. Here $|a| = \left|\dfrac{1}{2}\right| < 1$, so this is a shrinking. Each y-value on $f(x)$ will be multiplied by $\dfrac{1}{2}$ to obtain the graph of $h(x)$.

Points on f	Corresponding points on h
$(-5, 0)$	$\left(-5, \boxed{}\right)$
$(-2, 2)$	$(-2, 1)$
$(0, 0)$	$\left(0, \boxed{}\right)$
$(2, -4)$	$\left(2, \boxed{}\right)$
$(4, 0)$	$(4, 0)$

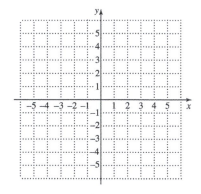

c) $r(x) = f(2x)$

The graph of $y = f(cx)$ is a horizontal stretching or shrinking of the graph of $y = f(x)$.
Here $|c| = |2| > 1$, so this is a shrinking. Each x-coordinate on $f(x)$ will be divided by 2
to obtain the graph of r.

Points on f	Corresponding points on r
$(-5, 0)$	$(-2.5, 0)$
$(-2, 2)$	$(\boxed{}, 2)$
$(0, 0)$	$(\boxed{}, 0)$
$(2, -4)$	$(1, -4)$
$(4, 0)$	$(\boxed{}, 0)$

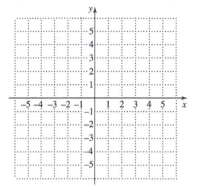

d) $s(x) = f\left(\dfrac{1}{2}x\right)$

The graph of $y = f(cx)$ is a horizontal stretching or shrinking of the graph of $y = f(x)$.
Here $|c| = \left|\dfrac{1}{2}\right| < 1$, so this is a stretching. Each x-coordinate on $f(x)$ will be multiplied by
2 to obtain the graph of s.

Points on f	Corresponding points on s
$(-5, 0)$	$(\boxed{}, 0)$
$(-2, 2)$	$(-4, 2)$
$(0, 0)$	$(\boxed{}, 0)$
$(2, -4)$	$(\boxed{}, -4)$
$(4, 0)$	$(8, 0)$

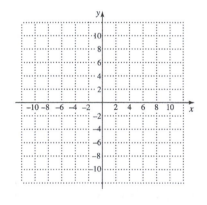

e) $t(x) = f\left(-\dfrac{1}{2}x\right)$

The graph of $y = f(cx)$ is a horizontal stretching or shrinking of the graph of $y = f(x)$.

Here $c = -\dfrac{1}{2}$. Because $c < 0$, we will also have the reflection of f across the y-axis. We

have $|c| = \left|-\dfrac{1}{2}\right| = \boxed{} < 1$, so this is a horizontal stretching along with the reflection

across the y-axis. We will multiply each x-coordinate of f by 2 and then change its sign.

Points on f	Corresponding points on t
$(-5, 0)$	$(10, 0)$
$(-2, 2)$	$(\boxed{}, 2)$
$(0, 0)$	$(0, 0)$
$(2, -4)$	$(\boxed{}, -4)$
$(4, 0)$	$(\boxed{}, 0)$

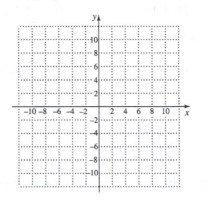

Example 4   Use the graph of $y = f(x)$ shown below to graph $y = -2f(x-3)+1$.

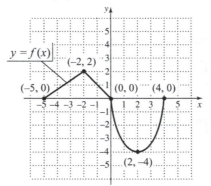

First consider $y = f(x-3)$. This is a shift of the given graph $\boxed{}$ units to

the _____ .
    right / left

Points on $y = f(x)$	Corresponding points on $y = f(x-3)$
$(-5, 0)$	$(-2, 0)$
$(-2, 2)$	$(\boxed{}, 2)$
$(0, 0)$	$(3, 0)$
$(2, -4)$	$(\boxed{}, -4)$
$(4, 0)$	$(7, 0)$

Now sketch the graph of $y = f(x-3)$.

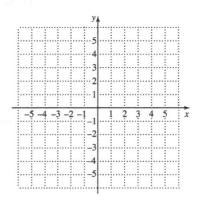

Next consider $y = 2f(x-3)$. This will stretch the graph of $y = f(x-3)$ vertically by a factor of 2.

Points on $y = f(x-3)$	Corresponding points on $y = 2f(x-3)$
$(-2, 0)$	$(-2, 0)$
$(3, 0)$	$(3, 0)$
$(7, 0)$	$(7, \boxed{})$
$(1, 2)$	$(1, \boxed{})$
$(5, -4)$	$(5, -8)$

Now sketch the graph of $y = 2f(x-3)$.

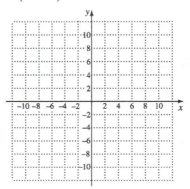

Next consider $y = -2f(x-3)$. This is the reflection of the graph of $y = 2f(x-3)$ across the _____ axis.

x- / y-

Points on $y = 2f(x-3)$	Corresponding points on $y = -2f(x-3)$
$(-2, 0)$	$(-2, 0)$
$(3, 0)$	$\left(3, \boxed{}\right)$
$(7, 0)$	$(7, 0)$
$(1, 4)$	$\left(1, \boxed{}\right)$
$(5, -8)$	$\left(5, \boxed{}\right)$

Now sketch the graph of $y = -2f(x-3)$.

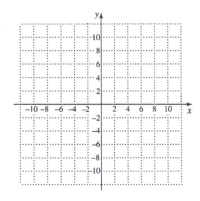

Finally, consider $y = -2f(x-3)+1$. This is a shift of the graph of $y = -2f(x-3)$

_____ 1 unit.
up / down

Points on $y = -2f(x-3)$	Corresponding points on $y = -2f(x-3)+1$
$(-2, 0)$	$(-2, 1)$
$(1, -4)$	$\left(1, \boxed{}\right)$
$(3, 0)$	$\left(3, \boxed{}\right)$
$(5, 8)$	$(5, 9)$
$(7, 0)$	$\left(7, \boxed{}\right)$

Now we can sketch the graph of $y = -2f(x-3)+1$.

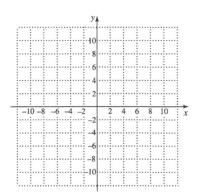

Section 7.3 The Complex Numbers

The Number *i*

The number *i* is defined such that

$$i = \sqrt{-1} \text{ and } i^2 = -1.$$

Example 1   Express each number in terms of *i*.

a) $\sqrt{-7} = \sqrt{-1 \cdot 7}$

$\phantom{\sqrt{-7}} = \boxed{} \cdot \sqrt{7}$

$\phantom{\sqrt{-7}} = \boxed{} \sqrt{7}, \text{ or } \sqrt{7}i$

b) $\sqrt{-16} = \sqrt{-1 \cdot 16}$

$\phantom{\sqrt{-16}} = \boxed{} \cdot \sqrt{16}$

$\phantom{\sqrt{-16}} = i \cdot \boxed{}$

$\phantom{\sqrt{-16}} = 4i$

c) $-\sqrt{-13} = -\sqrt{-1 \cdot 13}$

$\phantom{-\sqrt{-13}} = -\sqrt{-1} \cdot \sqrt{13}$

$\phantom{-\sqrt{-13}} = -\boxed{} \sqrt{13}, \text{ or } -\sqrt{13}i$

d) $-\sqrt{-64} = -\sqrt{-1 \cdot 64}$

$\phantom{-\sqrt{-64}} = -\boxed{} \cdot \sqrt{64}$

$\phantom{-\sqrt{-64}} = -i \cdot \boxed{}$

$\phantom{-\sqrt{-64}} = -8i$

e) $\sqrt{-48} = \sqrt{-1 \cdot 48}$

$\phantom{\sqrt{-48}} = \sqrt{-1} \cdot \sqrt{48}$

$\phantom{\sqrt{-48}} = \boxed{} \cdot \sqrt{48}$

$\phantom{\sqrt{-48}} = i \cdot \sqrt{16 \cdot 3}$

$\phantom{\sqrt{-48}} = i \cdot \boxed{} \cdot \sqrt{3}$

$\phantom{\sqrt{-48}} = i \cdot 4 \cdot \sqrt{3}$

$\phantom{\sqrt{-48}} = \boxed{}$

Complex Numbers

A **complex number** is a number of the form $a + bi$, where a and b are real numbers. The number a is said to be the **real part** of $a + bi$ and the number b is said to be the **imaginary part** of $a + bi$.

Example 2   Add or subtract and simplify each of the following.

a) $(8 + 6i) + (3 + 2i)$

$$(8 + 6i) + (3 + 2i) = \left(8 + \boxed{}\right) + \left(6i + \boxed{}\right)$$
$$= 11 + \boxed{}$$

b) $(4 + 5i) - (6 - 3i)$

$$(4 + 5i) - (6 - 3i) = (4 - 6) + \left(5i + \boxed{}\right)$$
$$= \boxed{} + 8i$$

Example 3   Multiply and simplify each of the following.

a) $\sqrt{-16} \cdot \sqrt{-25} = \boxed{} \cdot \sqrt{16} \cdot \boxed{} \cdot \sqrt{25}$

$$= \boxed{} \cdot 4 \cdot i \cdot \boxed{}$$
$$= 20i^2 \qquad\qquad i = \sqrt{-1}$$
$$= 20\left(\boxed{}\right) \qquad\quad i^2 = -1$$
$$= -20$$

b) $(1 + 2i)(1 + 3i) = 1 + 3i + 2i + \boxed{}$

$$= 1 + 3i + 2i + 6\left(\boxed{}\right)$$
$$= 1 + \boxed{} - 6$$
$$= -5 + 5i$$

c) $(3 - 7i)^2 = 9 - \boxed{} + 49i^2$

$$= 9 - 42i + 49(-1)$$
$$= 9 - 42i - 49$$
$$= \boxed{} - 42i$$

Example 4   Simplify each of the following.

a) $i^{37} = \boxed{} \cdot i$

$= \left(i^2\right)^{\boxed{}} \cdot i$

$= \left(\boxed{}\right)^{18} \cdot i$

$= 1 \cdot i$

$= i$

b) $i^{58} = \left(i^2\right)^{\boxed{}}$

$= (-1)^{29}$

$= \boxed{}$

c) $i^{75} = \boxed{} \cdot i$

$= \left(i^2\right)^{\boxed{}} \cdot i$

$= (-1)^{37} \cdot i$

$= -1 \cdot i$

$= \boxed{}$

d) $i^{80} = \left(i^2\right)^{\boxed{}}$

$= (-1)^{40}$

$= \boxed{}$

Conjugate of a Complex Number

The **conjugate** of a complex number $a + bi$ is $a - bi$. The numbers $a + bi$ and $a - bi$ are **complex conjugates**.

Example 5   Multiply each of the following.

a) $(5 + 7i)(5 - 7i)$ $(A + B)(A - B) = A^2 - B^2$

$= 5^2 - \left(\boxed{}\right)^2$

$= 25 - 49i^2$

$= 25 - 49\left(\boxed{}\right)$

$= 25 + 49$

$= \boxed{}$

b) $(8i)(-8i) = -64i^2$

$= -64\left(\boxed{}\right)$

$= 64$

Example 6   Divide $2 - 5i$ by $1 - 6i$.

$$\frac{2-5i}{1-6i} = \frac{2-5i}{1-6i} \cdot \frac{\boxed{}}{\boxed{}} = \frac{(2-5i)(1+6i)}{(1-6i)(1+6i)}$$

$$= \frac{2+12i-5i-\boxed{}}{1-\left(\boxed{}\right)^2}$$

$$= \frac{2+12i-5i+\boxed{}}{1-36i^2} = \frac{\boxed{}+7i}{1+36}$$

$$= \frac{32+7i}{\boxed{}} = \frac{32}{37} + \frac{7}{37}i$$

Section 7.4 Quadratic Equations, Functions, Zeros, and Models

Quadratic Equations

A **quadratic equation** is an equation that can be written in the form

$ax^2 + bx + c = 0$, $a \neq 0$, where a, b, and c are real numbers.

Quadratic Functions

A **quadratic function** f is a function that can be written in the form

$f(x) = ax^2 + bx + c$, $a \neq 0$, where a, b, and c are real numbers.

The **zeros** of a quadratic function $f(x) = ax^2 + bx + c$ are the *solutions* of the associated quadratic equation $ax^2 + bx + c = 0$.

Equation-Solving Principles

The Principle of Zero Products: If $ab = 0$ is true, then $a = 0$ *or* $b = 0$, and if $a = 0$ *or* $b = 0$, then $ab = 0$.

The Principle of Square Roots: If $x^2 = k$, then $x = \sqrt{k}$ *or* $x = -\sqrt{k}$.

Example 1   Solve: $2x^2 - x = 3$.

$$2x^2 - x = 3$$
$$2x^2 - x - 3 = 0$$
$$(2x - 3)(\boxed{}) = 0$$

$2x - 3 = 0$ *or* $x + 1 = 0$
$2x = 3$ *or* $x = -1$
$x = \boxed{}$ *or* $x = -1$

The solutions are $\boxed{}$ and $\boxed{}$.

Check:
For $x = -1$:

$$2x^2 - x = 3$$
$$2(\boxed{})^2 - (\boxed{}) \;?\; 3$$
$$2 + 1$$
$$3 \;|\; 3 \quad \text{TRUE}$$

For $x = \dfrac{3}{2}$:

$$2x^2 - x = 3$$
$$2\left(\frac{3}{2}\right)^2 - \frac{3}{2} \;?\; 3$$
$$2\left(\frac{9}{4}\right) - \frac{3}{2}$$
$$\boxed{} - \frac{3}{2}$$
$$3 \;|\; 3 \quad \text{TRUE}$$

Example 2   Solve: $2x^2 - 10 = 0$.

$$2x^2 - 10 = 0$$
$$2x^2 = 10$$
$$x^2 = 5$$
$$x = \boxed{}, \text{ or } \sqrt{5} \text{ and } -\sqrt{5}$$

Check:

$$
\begin{array}{c|c}
\multicolumn{2}{c}{2x^2 - 10 = 0} \\
\hline
2\left(\boxed{}\right)^2 - 10 \; ? \; 0 & \\
10 - 10 & \\
0 & 0 \quad \text{TRUE}
\end{array}
$$

The solutions are $\sqrt{5}$ and $-\sqrt{5}$, or $\boxed{}$.

Example 3   Find the zeros of $f(x) = x^2 - 6x - 10$ by completing the square.

$$x^2 - 6x - 10 = 0$$
$$x^2 - 6x \quad = 10$$
$$x^2 - 6x + \boxed{} = 10 + \boxed{} \qquad \frac{1}{2}\left(\boxed{}\right) = -3; \; (-3)^2 = \boxed{}$$
$$\left(\boxed{}\right)^2 = 19$$
$$x - 3 = \boxed{}$$
$$x = 3 \pm \sqrt{19}$$

$$x = 3 + \sqrt{19} \quad or \quad x = 3 - \sqrt{19}$$
$$x \approx 7.359 \quad or \quad x \approx -1.359$$

Zeros of $f(x) = x^2 - 6x - 10$ are

$$3 + \sqrt{19} \text{ and } \boxed{},$$

or approximately

$\boxed{}$ and -1.359.

Example 4 Solve: $2x^2 - 1 = 3x$.

$$2x^2 - 1 = 3x$$

$$2x^2 - \boxed{} - 1 = 0$$

$$2x^2 - 3x \qquad = 1$$

$$x^2 - \frac{3}{2}x \qquad = \boxed{}$$

$$x^2 - \frac{3}{2}x + \boxed{} = \frac{1}{2} + \boxed{} \qquad \text{Completing the square}$$

$$\left(\boxed{}\right)^2 = \frac{17}{16}$$

$$x - \frac{3}{4} = \pm\sqrt{\frac{17}{16}} = \pm\frac{\sqrt{17}}{\sqrt{16}} = \pm\frac{\sqrt{17}}{\boxed{}}$$

$$x = \frac{3}{4} \pm \frac{\sqrt{17}}{4}$$

$$x = \frac{3 \pm \sqrt{17}}{4}$$

The solutions are $\dfrac{3 + \sqrt{17}}{4}$ and $\dfrac{3 - \sqrt{17}}{4}$, or $\boxed{}$.

The Quadratic Formula

The solutions of $ax^2 + bx + c = 0$, $a \neq 0$, are given by

$$x = \frac{-b \pm \sqrt{b^2 - 4ac}}{2a}.$$

Example 5 Solve $3x^2 + 2x = 7$. Find exact solutions and approximate solutions rounded to three decimal places.

$$3x^2 + 2x = 7$$

$$3x^2 + 2x - 7 = 0$$

$$a = 3 \quad b = \boxed{} \quad c = \boxed{}$$

$$x = \frac{-b \pm \sqrt{b^2 - 4ac}}{2a}$$

$$x = \frac{-\boxed{} \pm \sqrt{2^2 - 4 \cdot 3 \cdot (\boxed{})}}{2(3)}$$

$$= \frac{-2 \pm \sqrt{4+84}}{6} = \frac{-2 \pm \sqrt{\boxed{}}}{6}$$

$$= \frac{-2 \pm \sqrt{4 \cdot \boxed{}}}{6} = \frac{-2 \pm 2\sqrt{22}}{6}$$

$$= \frac{2(-1 \pm \sqrt{22})}{2 \cdot 3} = \frac{\boxed{} \pm \sqrt{22}}{3}$$

Exact solutions: $\dfrac{-1-\sqrt{22}}{3}$ and $\dfrac{-1+\sqrt{22}}{3}$

Approximate solutions: -1.897 and $\boxed{}$

Example 6   Solve: $x^2 + 5x + 8 = 0$.

$a = \boxed{}$ $b = 5$ $c = 8$

$$x = \frac{-b \pm \sqrt{b^2 - 4ac}}{2a}$$

$$x = \frac{-\boxed{} \pm \sqrt{5^2 - 4 \cdot \boxed{} \cdot \boxed{}}}{2 \cdot 1}$$

$$x = \frac{-5 \pm \sqrt{\boxed{}}}{2}$$

$$x = \frac{-5 \pm \sqrt{7}\,\boxed{}}{2}$$

$$x = -\frac{5}{2} + \frac{\sqrt{7}}{2}i \text{ or } x = -\frac{5}{2} - \frac{\sqrt{7}}{2}i$$

Discriminant

For $ax^2 + bx + c = 0$, where a, b, and c are real numbers, $a \neq 0$:

$b^2 - 4ac = 0 \rightarrow$ One real-number solution;

$b^2 - 4ac > 0 \rightarrow$ Two different real-number soluions;

$b^2 - 4ac < 0 \rightarrow$ Two different imaginary-number solutions, complex conjugates.

Example 7 Solve: $x^4 - 5x^2 + 4 = 0$.

Let $u = \boxed{}$.

$u^2 - 5\boxed{} + 4 = 0$

$(u-1)\left(\boxed{}\right) = 0$

$u - 1 = 0$ or $\boxed{} = 0$

$u = 1$ or $u = 4$

$x^2 = \boxed{}$ or $x^2 = 4$

$x = \pm 1$ or $x = \boxed{}$

The solutions are -1, 1, -2, and 2.

Example 8 Solve: $t^{2/3} - 2t^{1/3} - 3 = 0$.

Let $u = t^{1/3}$.

$\left(t^{1/3}\right)^{\boxed{}} - 2t^{1/3} - 3 = 0$

$u^2 - 2\boxed{} - 3 = 0$

$\left(\boxed{}\right)(u-3) = 0$

$u + 1 = 0$ or $u - 3 = 0$

$u = \boxed{}$ or $u = 3$

$\boxed{} = -1$ or $t^{1/3} = 3$

$\left(t^{1/3}\right)^3 = (-1)^3$ or $\left(t^{1/3}\right)^3 = 3^3$

$\boxed{} = -1$ $t = 27$

The solutions are $\boxed{}$ and $\boxed{}$.

Example 9   *Museums in China.* The number of museums in China increased from approximately 2000 in the year 2000 to over 3500 by the end of 2012. In 2012, a record 451 new museums opened. For comparison, in the United States, only 20–40 new museums were opened per year from 2000 to 2008. The function

$$h(x) = 30.992x^2 + 4.108x + 2294.594$$

can be used to estimate the number of museums in China, x years after 2005.

a) Estimate the number of museums that will be in China in 2017 if the number of new museums that open per year continues at the same rate.

$$h(12) = 30.992\left(\boxed{}\right)^2 + 4.108\left(\boxed{}\right) + 2294.594$$

$$h(12) \approx 6807$$

In the year 2017, there are approximately $\boxed{}$ museums in China.

b) In what year was the number of museums in China 2600?

$$\boxed{} = 30.992x^2 + 4.108x + 2294.594$$

$$0 = 30.992x^2 + 4.108x - \boxed{}$$

$$a = 30.992 \qquad b = 4.108 \qquad c = -305.406$$

$$x = \frac{-b \pm \sqrt{b^2 - 4ac}}{2a}$$

$$x = \frac{-\boxed{} \pm \sqrt{(4.108)^2 - 4(30.992)(-305.406)}}{2(30.992)}$$

$$x = \frac{-4.108 \pm \sqrt{37,877.44667}}{61.984}$$

$$x = \boxed{} \quad or \quad x = \boxed{}$$

We are looking for a year after 2005, so we use the positive solution 3.074. Thus, there were about 2600 museums in China 3 years after 2005, or in $\boxed{}$.

Example 10　*Sales of New Homes.*　Sales of new homes have increased in recent years. The function

$$h(x) = 22.1x^2 - 72.2x + 371.9$$

can be used to estimate the sales of new homes, in thousands, in the United States, where x is the number of years after 2009 (*Source*: IHS Global Insight). In what year were the number of sales of new homes about 563,400, or 563.4 thousands?

$$\boxed{} = 22.1x^2 - 72.2x + 371.9$$

$$0 = 22.1x^2 - 72.2x - \boxed{}$$

$$a = 22.1 \quad b = -72.2 \quad c = -191.5$$

$$x = \frac{-b \pm \sqrt{b^2 - 4ac}}{2a}$$

$$x = \frac{-\left(\boxed{}\right) \pm \sqrt{(-72.2)^2 - 4(22.1)(-191.5)}}{2(22.1)}$$

$$x = \frac{72.2 \pm \sqrt{22,141.44}}{44.2}$$

$$x = \boxed{} \quad or \quad x = \boxed{}$$

We are looking for a year after 2009, so we use the positive solution. Thus, there were about 563,400 sales of new homes 5 years after 2009, or in $\boxed{}$.

Example 11   *Train Speeds*. Two trains leave a station at the same time. One train travels due west, and the other travels due south. The train traveling west travels 20 km/h faster than the train traveling south. After 2 hr, the trains are 200 km apart. Find the speed of each train.

We use $d = rt$ to find the distance each train traveled in 2 hr.

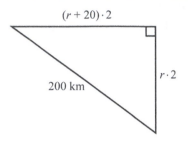

We use the Pythagorean theorem.

$$\left[2(r+20)\right]^2 + \left(\boxed{}\right)^2 = 200^2$$

$$\frac{\boxed{}}{4} + \frac{4r^2}{4} = \frac{40,000}{4}$$

$$(r+20)^2 + r^2 = 10,000$$

$$r^2 + 40r + \boxed{} + r^2 = 10,000$$

$$\frac{2r^2}{2} + \frac{40r}{2} - \frac{9600}{2} = 0$$

$$r^2 + 20r - 4800 = 0$$

$$\left(\boxed{}\right)(r - 60) = 0$$

$$r + 80 = 0 \quad or \quad r - 60 = 0$$

$$r = -80 \quad or \quad r = \boxed{}$$

The negative value of r doesn't make sense in the problem. We check 60.

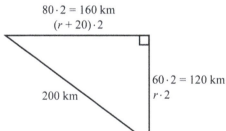

Check:

$$160^2 + 120^2 = 200^2$$

$$40,000 = 40,000$$

The answer checks.

The speed of the train heading south is $\boxed{}$ km/h, and the speed of the train heading west is $\boxed{}$ km/h.

Section 7.5 Analyzing Graphs of Quadratic Functions

Graphing Quadratic Functions

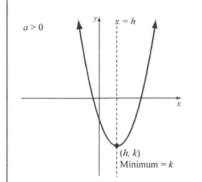

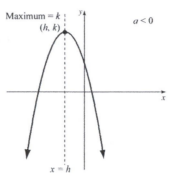

The graph of the function

$f(x) = a(x-h)^2 + k$ is a

parabola that
- opens up if $a > 0$ and down if $a < 0$;
- has (h, k) as the vertex;
- has $x = h$ as the axis of symmetry;
- has k as a minimum value (output) if $a > 0$;
- has k as a maximum value if $a < 0$.

Example 1   Find the vertex, the axis of symmetry, and the maximum or minimum value of $f(x) = x^2 + 10x + 23$. Then graph the function.

$$f(x) = a(x-h)^2 + k$$

$$f(x) = \left(x^2 + 10x + \boxed{}\right) - \boxed{} + 23$$

$$= (x+5)^2 - 2$$

$$= \left[x - \left(\boxed{}\right)\right]^2 + \left(\boxed{}\right)$$

Vertex: $(-5, -2)$

Axis of symmetry: $x = \boxed{}$

Minimum: -2

Some points on the graph:

$(-5, -2)$

$(-4, -1)$

$(-2, 7)$

$(-7, 2)$

$(-8, 7)$

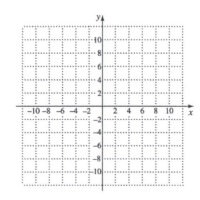

Example 2   Find the vertex, the axis of symmetry, and the maximum or minimum value of $g(x) = \dfrac{x^2}{2} - 4x + 8$. Then graph the function.

$$g(x) = a(x-h)^2 + k$$

$$g(x) = \frac{x^2}{2} - 4x + 8 = \frac{1}{2}\left(x^2 - 8x\right) + 8$$

$$= \frac{1}{2}\left(x^2 - 8x + \boxed{} - \boxed{}\right) + 8$$

$$= \frac{1}{2}\left(x^2 - 8x + 16\right) - \frac{1}{2}\cdot 16 + 8$$

$$= \frac{1}{2}\left(\boxed{}\right)^2 + \boxed{}, \text{ or } \frac{1}{2}(x-4)^2$$

Vertex: $(4, 0)$

Axis of symmetry: $x = \boxed{}$

Minimum: $\boxed{}$

Some points on the graph:

$(4, 0)$

$(0, 8)$

$(2, 2)$

$(6, 2)$

$(8, 8)$

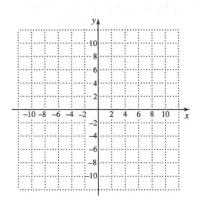

Example 3   Find the vertex, the axis of symmetry, and the maximum or minimum value of $f(x) = -2x^2 + 10x - \dfrac{23}{2}$. Then graph the function.

$$f(x) = -2x^2 + 10x - \frac{23}{2}$$

$$= -2\left(\boxed{}\right) - \frac{23}{2}$$

$$= -2\left(x^2 - 5x + \frac{25}{4} - \frac{25}{4}\right) - \frac{23}{2} \qquad \left(-\frac{5}{2}\right)^2 = \frac{25}{4}$$

$$= -2\left(x^2 - 5x + \frac{25}{4}\right) + \boxed{} - \frac{23}{2}$$

$$f(x) = -2\left(\boxed{}\right)^2 + \boxed{}$$

Vertex: $\left(\dfrac{5}{2}, 1\right)$

Axis of symmetry: $x = \boxed{}$

Maximum value is $\boxed{}$ when $x = \dfrac{5}{2}$.

Some points on the graph:

$\left(\dfrac{5}{2}, 1\right)$

$\left(2, \dfrac{1}{2}\right)$

$\left(3, \dfrac{1}{2}\right)$

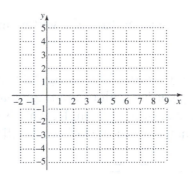

The Vertex of a Parabola

The vertex of the graph of $f(x) = ax^2 + bx + c$ is

$$\left(-\dfrac{b}{2a},\ f\left(-\dfrac{b}{2a}\right)\right).$$

We calculate the x-coordinate. Then we substitute to find the y-coordinate.

Example 4 For the function $f(x) = -x^2 + 14x - 47$:

a) Find the vertex.

$a = -1 \qquad b = 14 \qquad c = -47$

Vertex: $\left(\dfrac{-14}{2(-1)},\ \boxed{}\right) = (7,\ \ )$

$f(7) = -(7)^2 + 14 \cdot 7 - 47$

$ = -49 + 98 - 47 = \boxed{}$

The vertex is $(7, 2)$.

b) Determine whether there is a maximum or a minimum value and find that value.

Since the coefficient of x^2 is negative, the function has a $\underline{\hspace{4cm}}$ value at

$\hspace{6cm}$ maximum / minimum

the vertex.

Maximum value: $\boxed{}$

c) Find the range.

Vertex: $(7, 2)$

Maximum value: 2

Range: $\left(-\infty, \boxed{}\,\right]$

d) On what intervals is the function increasing? decreasing?

Increasing as we approach the vertex from the left: $\left(-\infty, \boxed{}\,\right)$

Decreasing as we move away from the vertex on the right: $\left(\boxed{}, \infty\right)$

Example 5   *Maximizing Area.* A landscaper has enough stone to enclose a rectangular koi pond next to an existing garden wall of the Englemans' house with 24 ft of stone wall. If the garden wall forms one side of the rectangle, what is the maximum area that the landscaper can enclose? What dimensions of the koi pond will yield this area?

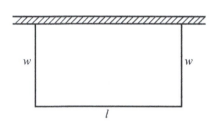

$2w + l = \boxed{}$

$l = 24 - 2w$

$A = l \cdot w$

$A = \left(\boxed{}\right) \cdot w$

$A = 24w - 2w^2$

$A = -2w^2 + 24w \qquad\qquad A(w) = aw^2 + bw + c$

Maximum occurs at the vertex:

$w = \dfrac{-b}{2a} = \dfrac{\boxed{}}{2(-2)} = \dfrac{-24}{-4} = \boxed{}$

$l = 24 - 2w = 24 - 2\left(\boxed{}\right) = 24 - 12 = \boxed{}$

$A = 12 \text{ ft} \cdot 6 \text{ ft} = 72 \text{ ft}^2$

The maximum possible area is $\boxed{}$ ft^2 when the koi pond is $\boxed{}$ ft wide and $\boxed{}$ ft long.

Example 6 *Height of a Rocket.* A model rocket is launched with an initial velocity of 100 ft/sec from the top of a hill that is 20 ft high. Its height, in feet, t seconds after it has been launched is given by the function $s(t) = -16t^2 + 100t + 20$. Determine the time at which the rocket reaches its maximum height and find the maximum height.

1. & 2.   Familiarize and Translate.

$$s(t) = -16t^2 + 100t + 20$$

3. Carry out.

$$a = -16 \qquad b = 100 \qquad c = 20$$

$$t = -\frac{b}{2a} = -\frac{\boxed{}}{2(-16)} = -\frac{100}{-32}$$

$$= \boxed{}$$

$$s(3.125) = -16\left(\boxed{}\right)^2 + 100\left(\boxed{}\right) + 20$$

$$= \boxed{}$$

4. Check. We can check the answer by completing the square on the function. If we did that, we would get

$$s(t) = -16(t - 3.125)^2 + 176.25.$$

This confirms that the vertex is $(3.125, 176.25)$.

5. State. The rocket reaches a maximum height of $\boxed{}$ ft $\boxed{}$ sec after it has been launched.

Example 7   Jared drops a screwdriver from the top of an elevator shaft. Exactly 5 sec later, he hears the sound of the screwdriver hitting the bottom of the shaft. The speed of sound is 1100 ft/sec. How tall is the elevator shaft?

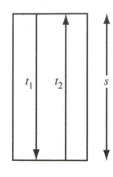

$$t_1 + t_2 = \boxed{}$$

$$s = 16 \cdot t_1^2$$

$$\frac{s}{16} = t_1^2 \quad \rightarrow \quad t_1 = \sqrt{\frac{s}{16}} = \frac{\sqrt{s}}{\boxed{}}$$

$$d = rt$$

$$s = \boxed{} \cdot t_2$$

$$\frac{s}{1100} = t_2$$

$$t_1 + t_2 = 5$$

$$\frac{\sqrt{s}}{4} + \boxed{} = 5$$

$$275\sqrt{s} + s = \boxed{}$$

$$s + 275\sqrt{s} - 5500 = 0$$

Let $u = \sqrt{s}$.

$$u^2 + 275u - 5500 = 0$$

$$u = \frac{-\boxed{} \pm \sqrt{275^2 - 4(1)(-5500)}}{2 \cdot 1}$$

$$u = \frac{-275 \pm \sqrt{97,625}}{2}$$

$$u = \frac{-275 + \sqrt{97,625}}{2} \qquad \text{We want only the positive solution.}$$

$$u \approx \boxed{}$$

$$\sqrt{s} \approx 18.725$$

$$s \approx 350.6 \text{ ft}$$

The height of the elevator shaft is about $\boxed{}$ ft.

Section 8.1 Polynomial Functions and Models

Polynomial Function

A **polynomial function** P is given by

$$P(x) = a_n x^n + a_{n-1} x^{n-1} + a_{n-2} x^{n-2} + \cdots + a_1 x + a_0,$$

where the coefficients a_n, a_{n-1}, ..., a_1, a_0 are real numbers and the exponents are whole numbers.

The Leading-Term Test

If $a_n x^n$ is the leading term of a polynomial function, then the behavior of the graph as $x \to \infty$ or as $x \to -\infty$ can be described in one of the four following ways:

n	$a_n > 0$	$a_n < 0$
Even		
Odd		

The ⋙ portion of the graph is not determined by this test.

Example 1 Use the leading-term test to match the functions below with one of the graphs on the next page.

a) $f(x) = 3x^4 - 2x^3 + 3$

Exponent: <u>even / odd</u> (circle one)
Coefficient: <u>positive / negative</u> (circle one)
Graph: <u>A, B, C, or D</u> (circle one)

b) $f(x) = -5x^3 - x^2 + 4x + 2$

Exponent: <u>even / odd</u> (circle one)
Coefficient: <u>positive / negative</u> (circle one)
Graph: <u>A, B, C, D</u> (circle one)

c) $f(x) = x^5 + \dfrac{1}{4}x + 1$

Exponent: <u>even / odd</u> (circle one)
Coefficient: <u>positive / negative</u> (circle one)
Graph: <u>A, B, C, D</u> (circle one)

d) $f(x) = -x^6 + x^5 - 4x^3$

Exponent: <u>even / odd</u> (circle one)
Coefficient: <u>positive / negative</u> (circle one)
Graph: <u>A, B, C, D</u> (circle one)

A. 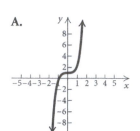   **B.** 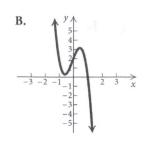   **C.** 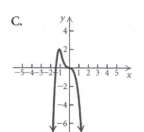   **D.**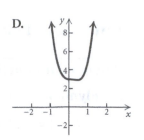

Example 2   Consider $P(x) = x^3 + x^2 - 17x + 15$. Determine whether each of the numbers 2 and -5 is a zero of $P(x)$.

If $P(c) = 0$, then c is a zero of the polynomial.

Let us first evaluate $P(2)$.

$P(2) = 2^3 + 2^2 - 17 \cdot 2 + 15$

$P(2) = 8 + \boxed{} - 34 + 15$

$P(2) = \boxed{}$

Since $P(2) \neq 0$, 2 <u>is / is not</u> (circle one) a zero of this function.

Let us now evaluate $P(-5)$.

$P(-5) = (-5)^3 + (-5)^2 - 17 \cdot (-5) + 15$

$P(-5) = -125 + 25 + 85 + 15$

$P(-5) = 0$

Since $P(-5) = 0$, -5 <u>is / is not</u> (circle one) a zero of this function.

Example 3   Find the zeros of

$$f(x) = 5(x-2)(x-2)(x-2)(x+1)$$
$$= 5(x-2)^3(x+1).$$

According to the principle of zero products, either

$\quad x - 2 = 0 \quad or \quad x + 1 = 0$

$\qquad x = 2 \quad or \qquad x = -1 \quad$ So the zeros of $f(x)$ are $\boxed{}$ and $\boxed{}$.

Example 4   Find the zeros of

$$g(x) = -(x-1)(x-1)(x+2)(x+2)$$
$$= -(x-1)^2(x+2)^2.$$

To find the zeros, we set $g(x) = 0$, and use the principle of zero products.

$0 = -(x-1)^2(x+2)^2$

$x - 1 = 0 \quad or \quad x + 2 = 0$

$\quad x = 1 \quad or \qquad x = -2 \quad$ The zeros are $\boxed{}$ and $\boxed{}$.

Example 5   Find the zeros of $f(x) = x^3 - 2x^2 - 9x + 18$.

$$f(x) = x^3 - 2x^2 - 9x + 18$$
$$= x^2\left(x - \boxed{}\right) - 9\left(\boxed{} - 2\right)$$
$$= (x - 2)(x^2 - 9)$$
$$= (x - 2)(x + 3)\left(x - \boxed{}\right)$$

To find the zeros, we solve the equation $f(x) = 0$ using the principle of zero products.

$$0 = (x - 2)(x + 3)(x - 3)$$

$x - 2 = 0$ *or* $x + 3 = 0$ *or* $x - 3 = 0$

$x = 2$ *or* $x = -3$ *or* $x = 3$

The solutions of $f(x) = 0$ are 2, $\boxed{}$, and 3.

The zeros of $f(x)$ are $\boxed{}$, -3, and $\boxed{}$.

Example 6   Find the zeros of $f(x) = x^4 + 4x^2 - 45$.

To find the zeros, we solve the equation $f(x) = 0$.

$$0 = x^4 + 4x^2 - 45$$
$$0 = \left(x^2 - \boxed{}\right)(x^2 + 9)$$

Using the principle of zero products, this becomes

$x^2 - 5 = 0$ *or* $x^2 + 9 = 0$

$x^2 = 5$ *or* $x^2 = -9$

$x = \pm\sqrt{5}$ *or* $x = \pm\sqrt{-9}$

$x = \pm 3i$

The solutions of $f(x) = 0$ are $-\sqrt{5}$, $\sqrt{5}$, $\boxed{}$, and $3i$.

The zeros of $f(x)$ are $-\sqrt{5}$, $\boxed{}$, $-3i$, and $\boxed{}$.

Example 7   The polynomial function

$$M(t) = 0.5t^4 + 3.45t^3 - 96.65t^2 + 347.7t$$

can be used to estimate the number of milligrams of the pain relief medication ibuprofen in the bloodstream t hours after 400 mg of the medication has been taken. Find the number of milligrams in the bloodstream at $t = 0$, 0.5, 1, 1.5, and so on, up to 6 hr. Round the function values to the nearest tenth.

Using a calculator, we compute the function values:

$M(0) = 0$, $M(3.5) = 255.9$,

$M(0.5) = 150.2$, $M(4) = \boxed{}$,

$M(1) = 255$, $M(4.5) = 126.9$,

$M(1.5) = 318.3$, $M(5) = 66$,

$M(2) = \boxed{}$, $M(5.5) = 20.2$,

$M(2.5) = 338.6$, $M(6) = 0$.

$M(3) = 306.9$,

Section 8.2 Graphing Polynomial Functions

> **To Graph a Polynomial Function:**
> 1. Use the leading-term test to determine the end behavior.
> 2. Find the zeros of the function by solving $f(x) = 0$. Any real zeros are the first coordinates of the x-intercepts.
> 3. Use the x-intercepts (zeros) to divide the x-axis into intervals and choose a test point in each interval to determine the sign of all function values in that interval.
> 4. Find $f(0)$. This gives the y-intercept of the function.
> 5. If necessary, find additional function values to determine the general shape of the graph and then draw the graph.
> 6. As a partial check, use the facts that the graph has at most n x-intercepts and at most $n-1$ turning points. Multiplicity of zeros can also be considered in order to check where the graph crosses or is tangent to the x-axis.

Example 1   Graph the polynomial function $h(x) = -2x^4 + 3x^3$.

1. Use the leading-term test.
    Degree: 4, even; coefficient: $-2 < 0$
    As $x \to \infty$ and as $x \to -\infty$, $h(x) \to \boxed{}$.

2. Find the zeros. Solve $h(x) = 0$.

    $$-2x^4 + 3x^3 = 0$$
    $$-x^3(2x-3) = 0$$

    $\boxed{} = 0$ *or* $2x - 3 = 0$

    $x = 0$ *or* $x = \boxed{}$

3.

    Choose a test value from each interval and find $h(x)$.

Interval	$(-\infty, 0)$	$\left(0, \dfrac{3}{2}\right)$	$\left(\dfrac{3}{2}, \infty\right)$
Test value	-1	1	2
Function value, $h(x)$	-5	1	-8
Sign of $h(x)$	$-$	$+$	$-$
Location of points on graph	Above/Below x-axis (circle one)	Above/Below x-axis (circle one)	Above/Below x-axis (circle one)

Three points on the graph are $\left(-1, \boxed{}\right)$, $(1, 1)$, and $\left(\boxed{}, -8\right)$.

4. To determine the y-intercept, find $h(0)$.

$$h(x) = -2x^4 + 3x^3$$
$$h(0) = -2 \cdot 0^4 + 3 \cdot 0^3 = \boxed{}$$

y-intercept: $(0, 0)$

5. Find additional function values and draw the graph.

x	$h(x)$
-1.5	-20.25
-0.5	-0.5
0.5	$\boxed{}$
2.5	-31.25

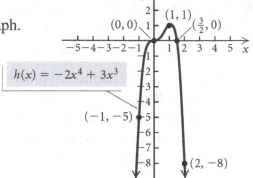

Example 2   Graph the polynomial function $f(x) = 2x^3 + x^2 - 8x - 4$.

1. Use the leading-term test.

   Degree: 3, odd; coefficient: $2 > 0$

   As $x \to \infty$, $f(x) \to \infty$; as $x \to -\infty$, $f(x) \to -\infty$.

2. Find the zeros. Solve $f(x) = 0$.

$$2x^3 + x^2 - 8x - 4 = 0$$
$$x^2(2x + 1) - 4(2x + 1) = 0$$
$$(2x + 1)\left(x^2 - \boxed{}\right) = 0$$
$$(2x + 1)(x + 2)(x - 2) = 0$$

$\boxed{} = 0$ *or* $x + 2 = 0$ *or* $x - 2 = 0$

$x = -\dfrac{1}{2}$ *or* $x = \boxed{}$ *or* $x = 2$

3.

$$-2 \quad -\frac{1}{2} \quad 2$$

$$-5\ -4\ -3\ -2\ -1\ \ 0\ \ 1\ \ 2\ \ 3\ \ 4\ \ 5 \quad x$$

Choose a test value from each interval and find $f(x)$.

Interval	$(-\infty, -2)$	$\left(\boxed{}, -\frac{1}{2}\right)$	$\left(-\frac{1}{2}, 2\right)$	$\left(\boxed{}, \infty\right)$
Test value	-3	-1	1	3
Function value, $f(x)$	-25	3	-9	35
Sign of $f(x)$	$-$	$+$	$-$	$+$
Location of points on graph	Above/Below x-axis (circle one)	Above/Below x-axis (circle one)	Above/Below x-axis (circle one)	Above/Below x-axis (circle one)

Four points on the graph are $(-3, -25)$, $\left(-1, \boxed{}\right)$, $(1, -9)$, and $\left(\boxed{}, 35\right)$.

4. To determine the y-intercept, find $f(0)$.

$$f(x) = 2x^3 + x^2 - 8x - 4$$

$$f(0) = 2 \cdot 0^3 + 0^2 - 8 \cdot 0 - 4 = \boxed{}$$

y-intercept: $\left(0, \boxed{}\right)$

5. Find additional function values and draw the graph.

x	$f(x)$
-2.5	-9
-1.5	3.5
0.5	-7.5
1.5	-7

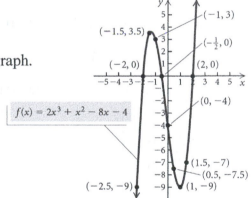

Example 3   Graph the polynomial function

$$g(x) = x^4 - 7x^3 + 12x^2 + 4x - 16$$

$$= (x+1)(x-2)^2(x-4).$$

1. Use the leading-term test.

Degree: 4, even; coefficient: $1 > 0$

As $x \to \infty$ and as $x \to -\infty$, $g(x) \to \infty$.

2. Find the zeros. Solve $g(x) = 0$.

$$(x+1)(x-2)^2(x-4) = 0$$

$$x+1=0 \quad or \quad (x-2)^2 =0 \quad or \quad x-4=0$$
$$x=-1 \quad or \qquad x=2 \quad or \qquad x=4$$

3.   ![number line from -5 to 5 with points at -1, 2, 4 marked]
−5 −4 −3 −2 −1 0 1 2 3 4 5 x

Choose a test value from each interval and find $g(x)$.

Interval	$(-\infty,-1)$	$(\boxed{}, 2)$	$(2, 4)$	$(\boxed{}, \infty)$
Test value	−1.25	1	3	4.25
Function value, $g(x)$	≈ 13.9	−6	−4	≈ 6.6
Sign of $g(x)$	+	−	−	+
Location of points on graph	Above/Below x-axis (circle one)	Above/Below x-axis (circle one)	Above/Below x-axis (circle one)	Above/Below x-axis (circle one)

4. To determine the y-intercept, find $g(0)$.

$$g(x)= x^4 -7x^3 +12x^2 +4x -16$$
$$g(0)= 0^4 -7\cdot0^3 +12\cdot0^2 +4\cdot0 -16 = \boxed{}$$

y-intercept: $\left(\boxed{},-16\right)$

5. Find additional function values and draw the graph.

x	$g(x)$
−0.5	−14.1
0.5	−11.8
1.5	−1.6
2.5	−1.3
3.5	−5.1

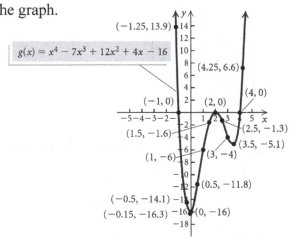

The Intermediate Value Theorem

For any polynomial function $P(x)$ with real coefficients, suppose that for $a \neq b$, $P(a)$ and $P(b)$ are of opposite signs. Then the function has a real zero between a and b.

The intermediate value theorem *cannot* be used to determine whether there is a real zero between a and b when $P(a)$ and $P(b)$ have the *same* sign.

Example 4

a) Using the intermediate value theorem, determine, if possible, whether
$f(x) = x^3 + x^2 - 6x$ has a real zero between -4 and -2.

$$f(-4) = (-4)^3 + (-4)^2 - 6(-4) = \boxed{}$$

$$f(-2) = (-2)^3 + (-2)^2 - 6(-2) = \boxed{}$$

Because $f(-4)$ and $f(-2)$ have opposite signs, $f(x)$ <u>has / does not have</u> (circle one) a
zero between -4 and -2.

b) Using the intermediate value theorem, determine, if possible, whether
$f(x) = x^3 + x^2 - 6x$ has a real zero between -1 and 3.

$$f(-1) = (-1)^3 + (-1)^2 - 6(-1) = \boxed{}$$

$$f(3) = (3)^3 + (3)^2 - 6(3) = \boxed{}$$

Because $f(-1)$ and $f(3)$ have the same sign, it <u>is / is not</u> (circle one) possible to
determine with the intermediate value theorem if there is a zero between -1 and 3.

c) Using the intermediate value theorem, determine, if possible, whether $g(x) = \dfrac{1}{3}x^4 - x^3$

has a real zero between $-\dfrac{1}{2}$ and $\dfrac{1}{2}$.

$$g\left(-\frac{1}{2}\right) = \frac{1}{3}\left(-\frac{1}{2}\right)^4 - \left(-\frac{1}{2}\right)^3 = \boxed{}$$

$$g\left(\frac{1}{2}\right) = \frac{1}{3}\left(\frac{1}{2}\right)^4 - \left(\frac{1}{2}\right)^3 = \boxed{}$$

Because $g\left(-\dfrac{1}{2}\right)$ and $g\left(\dfrac{1}{2}\right)$ have opposite signs, $g(x)$ <u>has / does not have</u> (circle one) a

zero between $-\dfrac{1}{2}$ and $\dfrac{1}{2}$.

d) Using the intermediate value theorem, determine, if possible, whether $g(x) = \dfrac{1}{3}x^4 - x^3$

has a real zero between 1 and 2.

$$g(1) = \frac{1}{3}(1)^4 - (1)^3 = \boxed{}$$

$$g(2) = \frac{1}{3}(2)^4 - (2)^3 = \boxed{}$$

Because $g(1)$ and $g(2)$ have the same sign, it <u>is / is not</u> (circle one) possible to
determine with the intermediate value theorem if there is a zero between 1 and 2.

Section 8.3   Polynomial Division; The Remainder Theorem and the Factor Theorem

Example 1   Divide to determine whether $x+1$ and $x-3$ are factors of x^3+2x^2-5x-6.

First Divisor: $x+1$

$$
\begin{array}{r}
\boxed{} \\
x+1\,\overline{)\,x^3\ +\ 2x^2\ -\ 5x\ -\ 6} \\
\underline{x^3\ +\ \boxed{}} \\
x^2\ -\ 5x \\
\underline{x^2\ +\ x} \\
\boxed{}\ -\ 6 \\
\underline{-6x\ -\ 6} \\
\boxed{}\quad \leftarrow \text{Remainder}
\end{array}
$$

Since the remainder is 0, we know that $x+1$ <u>is / is not</u> (circle one) a factor of x^3+2x^2-5x-6.

Second Divisor: $x-3$

$$
\begin{array}{r}
\boxed{} \\
x-3\,\overline{)\,x^3\ +\ 2x^2\ -\ 5x\ -\ 6} \\
\underline{x^3\ -\ 3x^2} \\
5x^2\ -\ 5x \\
\underline{\boxed{}\ -\ 15x} \\
10x\ -\ \boxed{} \\
\underline{10x\ -\ 30} \\
\boxed{}\quad \leftarrow \text{Remainder}
\end{array}
$$

Since the remainder is 24, we know that $x-3$ <u>is / is not</u> (circle one) a factor of x^3+2x^2-5x-6.

Example 2   Use synthetic division to find the quotient and the remainder:

$$\left(2x^3 + 7x^2 - 5\right) \div \left(x + 3\right).$$

First, rewrite the dividend including the missing x-term, and write the divisor in the form $(x - c)$:

$$\left(2x^3 + 7x^2 + 0x - 5\right) \div \left(x - (-3)\right)$$

Now perform synthetic division:

$$
\begin{array}{r|rrrr}
-3 & 2 & 7 & 0 & -5 \\
   &   & -6 & \boxed{} & 9 \\
\hline
   & 2 & \boxed{} & -3 & \boxed{} \\
\end{array}
$$

The quotient is $2x^2 + x - 3$. The remainder is $\boxed{}$.

The Remainder Theorem

If a number c is substituted for x in the polynomial $f(x)$, then the result $f(c)$ is the remainder that would be obtained by dividing $f(x)$ by $x - c$.

That is, if $f(x) = (x - c) \cdot Q(x) + R$, then $f(c) = R$.

Example 3   Given that $f(x) = 2x^5 - 3x^4 + x^3 - 2x^2 + x - 8$, find $f(10)$.

$f(10)$ is the remainder when $f(x)$ is divided by $x - 10$.

We can use synthetic division to find $f(10)$.

$$
\begin{array}{r|rrrrrr}
10 & 2 & -3 & 1 & -2 & 1 & -8 \\
   &   & 20 & \boxed{} & 1710 & 17{,}080 & 170{,}810 \\
\hline
   & 2 & 17 & 171 & \boxed{} & 17{,}081 & \boxed{} \\
\end{array}
$$

Thus, $f(10) = \boxed{}$.

Example 4   Determine whether 5 is a zero of $g(x)$, where $g(x) = x^4 - 26x^2 + 25$.

If 5 is a zero of $g(x)$, then $g(5) = 0$. If we divide $g(x)$ by $x - 5$ and get a remainder of 0, then $g(5) = 0$ and 5 is a zero of the function.

We use synthetic division to find $g(5)$, remembering to write zeros for the missing terms in $g(x)$.

$$
\begin{array}{r|rrrrr}
5 & 1 & 0 & -26 & 0 & 25 \\
  &   & 5 & \boxed{} & -5 & -25 \\
\hline
  & \boxed{} & 5 & -1 & -5 & \boxed{}
\end{array}
$$

$g(5) = \boxed{}$, so 5 is a zero of $g(x)$.

Example 5   Determine whether i is a zero of $f(x)$, where $f(x) = x^3 - 3x^2 + x - 3$.

If i is a zero of $f(x)$, then $f(i) = 0$. If we divide $f(x)$ by $x - i$ and get a remainder of 0, then $f(i) = 0$ and i is a zero of the function.

We use synthetic division to find $f(i)$.

$$
\begin{array}{r|rrrr}
i & 1 & -3 & 1 & -3 \\
  &   & \boxed{} & -1-3i & 3 \\
\hline
  & 1 & -3+i & \boxed{} & \boxed{}
\end{array}
$$

$f(i) = \boxed{}$, so i is a zero of $f(x)$.

The Factor Theorem

For a polynomial $f(x)$, if $f(c) = 0$, then $x - c$ is a factor of $f(x)$.

Example 6 Let $f(x) = x^3 - 3x^2 - 6x + 8$. Factor $f(x)$ and solve the equation $f(x) = 0$.

According to the factor theorem, if $f(c) = 0$, then $x - c$ is a factor of $f(x)$. At this point, we make some arbitrary choices for c and determine if they are zeros.

First try $c = -1$. This means we are dividing by $x + 1$.

$$
\begin{array}{r|rrrr}
-1 & 1 & -3 & -6 & 8 \\
   &   & -1 & \square & \square \\
\hline
   & 1 & -4 & -2 & \square
\end{array}
$$

$10 \neq 0$, so -1 is not a zero of $f(x)$. That means that $x + 1$ is not a factor of $f(x)$.

Now try $c = 1$. This tests whether $x - 1$ is a factor of $f(x)$.

$$
\begin{array}{r|rrrr}
1 & 1 & -3 & -6 & 8 \\
  &   & 1 & -2 & -8 \\
\hline
  & 1 & \square & -8 & \square
\end{array}
$$

This tells us that one of the factors of $f(x)$ is $x - 1$. Another factor is the quotient that we found when dividing. We have

$$
\begin{aligned}
f(x) &= (x-1)(x^2 - 2x - 8) \\
     &= (x-1)(x-4)(x+2).
\end{aligned}
$$

Finally, we solve the equation $f(x) = 0$.

$$(x-1)(x-4)(x+2) = 0$$

$x - 1 = 0$ *or* $x - 4 = 0$ *or* $x + 2 = 0$

$x = \square$ *or* $x = \square$ *or* $x = \square$

The solutions are $\square$, $\square$, and $\square$.

Section 8.4 Theorems about Zeros of Polynomial Functions

In this section, we will need to use the following relationship between the factors of $f(x)$ and the zeros of $f(x)$.

The Factor Theorem

For a polynomial $f(x)$, if $f(c) = 0$, then $x - c$ is a factor of $f(x)$.

The Fundamental Theorem of Algebra

Every polynomial function of degree n, with $n \geq 1$, has at least one zero in the set of complex numbers.

We will also need to know one of the results of the fundamental theorem of algebra:

Every polynomial function f of degree n, with $n \geq 1$, can be factored into n linear factors (not necessarily unique); that is,

$$f(x) = a_n(x - c_1)(x - c_2) \cdots (x - c_n).$$

Example 1   Find a polynomial function of degree 3, having the zeros 1, $3i$, and $-3i$.

If these are the zeros, then the factors must be

$(x - 1),\ (x - 3i),\ (x - (-3i))$, or

$(x - 1),\ (x - 3i),\ (x + 3i)$.   Writing $(x - (-3i))$ as $(x + 3i)$

So the function must have the form

$$f(x) = a_n(x - 1)(x - 3i)(x + 3i).$$

Let $a_n = 1$.

$$f(x) = (x - 1)(x - 3i)(x + 3i)$$

$$= (x - 1)(x^2 + 9)$$

$$= \boxed{} + 9x - x^2 - 9$$

$$= x^3 - \boxed{} + 9x - 9$$

Example 2   Find a polynomial function of degree 5 with -1 as a zero of multiplicity 3, 4 as a zero of multiplicity 1, and 0 as a zero of multiplicity 1.

If these are the zeros, then the function in factored form must be:

$$f(x) = a_n \left(x + \boxed{} \right)^3 \left(x - \boxed{} \right)(x - 0)$$

Let $a_n = 1$.

$$f(x) = (x+1)^3 (x-4)(x-0)$$
$$= x^5 - \boxed{} - 9x^3 - \boxed{} - 4x$$

Nonreal Zeros: $a + bi$ and $a - bi,\ b \neq 0$

If a complex number $a + bi$, $b \neq 0$, is a zero of a polynomial function $f(x)$ with *real* coefficients, then its conjugate, $a - bi$, is also a zero. For example, if $2 + 7i$ is a zero of a polynomial function $f(x)$, with real coefficients, then its conjugate, $2 - 7i$, is also a zero. (Nonreal zeros occur in conjugate pairs.)

Irrational Zeros: $a + c\sqrt{b}$ and $a - c\sqrt{b}$, b Is Not a Perfect Square

If $a + c\sqrt{b}$, where a, b, and c are rational and b is not a perfect square, is a zero of a polynomial function $f(x)$ with *rational* coefficients, then its conjugate, $a - c\sqrt{b}$, is also a zero. For example, if $-3 + 5\sqrt{2}$ is a zero of a polynomial function $f(x)$ with rational coefficients, then its conjugate, $-3 - 5\sqrt{2}$, is also a zero. (Irrational zeros occur in conjugate pairs.)

Example 3   Suppose that a polynomial function of degree 6 with rational coefficients has $-2 + 5i$, $-2i$, and $1 - \sqrt{3}$ as three of its zeros. Find the other zeros.

Since the function has rational coefficients, the other zeros are the conjugates of the given zeros. They are

$$-2 - 5i,\ \boxed{},\ \text{and } 1 + \sqrt{3}.$$

Example 4   Find a polynomial function of lowest degree with rational coefficients that has $-\sqrt{3}$, and $1+i$ as two of its zeros.

Two other zeros must be ☐ and ☐. (We will not include additional zeros, because we want the function of lowest degree.)

$$f(x) = \left[x - \left(-\sqrt{3}\right)\right]\left[x - \left(\boxed{}\right)\right]\left[x - (1+i)\right]\left[x - \left(\boxed{}\right)\right]$$

$$= \left(x + \sqrt{3}\right)\left(x - \sqrt{3}\right)\left((x-1) - i\right)\left((x-1) + \boxed{}\right)$$

$$= \left(x^2 - \boxed{}\right)\left((x-1)^2 - i^2\right)$$

$$= \left(x^2 - 3\right)\left(x^2 - 2x + 1^2 + 1\right)$$

$$= \left(x^2 - 3\right)\left(x^2 - 2x + 2\right)$$

$$= x^4 - 2x^3 - \boxed{} + 6x - \boxed{}$$

The Rational Zeros Theorem

Let $P(x) = a_n x^n + a_{n-1} x^{n-1} + \cdots + a_1 x + a_0$, where all the coefficients are integers. Consider a rational number denoted by p/q, where p and q are relatively prime (having no common factor besides -1 and 1). If p/q is a zero of $P(x)$, then p is a factor of a_0 and q is a factor of a_n.

Example 5 Given $f(x) = 3x^4 - 11x^3 + 10x - 4$:

a) Find the rational zeros and then the other zeros; that is, solve $f(x) = 0$.

b) Factor $f(x)$ into linear factors.

First, notice the degree of the polynomial is 4, so there are at most 4 distinct zeros.

a) The rational zeros theorem says that if a rational number p/q is a zero of $f(x)$, then p must be a factor of -4 and q must be a factor of 3.

$$\frac{\text{Possibilities for } p}{\text{Possibilities for } q} : \frac{\pm 1, \ \pm 2, \ \pm 4}{\pm 1, \ \pm 3}$$

Possibilities for $\dfrac{p}{q}$: $1, -1, 2, -2, 4, -4, \dfrac{1}{3}, -\dfrac{1}{3}, \dfrac{2}{3}, -\dfrac{2}{3}, \dfrac{4}{3}, -\dfrac{4}{3}$

To find which are zeros, we use synthetic division.

Try 1:

$$
\begin{array}{r|rrrrr}
1 & 3 & -11 & 0 & 10 & -4 \\
  &   & 3 & \boxed{} & -8 & 2 \\
\hline
  & \boxed{} & -8 & -8 & 2 & \boxed{} \\
\end{array}
$$

Since $f(1) = \boxed{}$, 1 is not a zero.

Try -1:

$$
\begin{array}{r|rrrrr}
-1 & 3 & -11 & 0 & 10 & -4 \\
   &   & \boxed{} & 14 & -14 & \boxed{} \\
\hline
   & 3 & -14 & \boxed{} & -4 & 0 \\
\end{array}
$$

Since $f(-1) = \boxed{}$, -1 is a zero.

Using the results of the synthetic division, we can rewrite $f(x)$ as

$$f(x) = (x+1)(3x^3 - 14x^2 + 14x - 4).$$

Now consider $3x^3 - 14x^2 + 14x - 4$. Continue using synthetic division to find other zeros.

Try -1 again since it might have multiplicity 2.

$$
\begin{array}{r|rrrr}
-1 & 3 & -14 & 14 & -4 \\
   &   & \boxed{} & 17 & \boxed{} \\
\hline
   & 3 & -17 & \boxed{} & -35 \\
\end{array}
$$

Since $f(-1) = -35$, -1 does not have multiplicity 2.

Synthetic division shows that $2,\ -2,\ 4,$ and -4 are not zeros.

Now try $\dfrac{2}{3}$:

$$\begin{array}{r|rrrr}
2/3 & 3 & -14 & 14 & -4 \\
& & 2 & \boxed{} & 4 \\
\hline
& \boxed{} & -12 & 6 & \boxed{}
\end{array}$$

Since $f\left(\dfrac{2}{3}\right)=\boxed{}$, $\dfrac{2}{3}$ is a zero of $f(x)$ and $\left(x-\dfrac{2}{3}\right)$ is a factor of $f(x)$.

We can rewrite $f(x)$ as

$$f(x)=(x+1)\left(x-\frac{2}{3}\right)\left(3x^2-12x+6\right)$$

$$=(x+1)\left(x-\frac{2}{3}\right)3\left(x^2-4x+2\right)$$

and the last factor can be used to find the two remaining zeros by solving the equation $x^2-4x+2=0$ using the quadratic formula.

$$x=\frac{-b\pm\sqrt{b^2-4ac}}{2a}$$

$$x=\frac{-(-4)\pm\sqrt{(-4)^2-4\cdot1\cdot2}}{2\cdot1}$$

$$x=\frac{4\pm\sqrt{\boxed{}}}{2}=\frac{4\pm2\sqrt2}{2}=\frac{2\left(2\pm\sqrt2\right)}{2}=2\pm\sqrt2$$

So the rational zeros are $\boxed{}$, $\dfrac{2}{3}$, $2+\sqrt2,\ 2-\sqrt2$.

b) The complete factorization of $f(x)$ is

$$f(x)=3(x+1)\left(x-\frac{2}{3}\right)\left[x-\left(2-\sqrt2\right)\right]\left[x-\left(2+\sqrt2\right)\right],\ \text{or}$$

$$f(x)=(x+1)\left(3x-\boxed{}\right)\left[x-\left(2-\sqrt2\right)\right]\left[x-\left(2+\sqrt2\right)\right].$$

Example 6   Given $f(x) = 2x^5 - x^4 - 4x^3 + 2x^2 - 30x + 15$:

a) Find the rational zeros and then the other zeros; that is, solve $f(x) = 0$.

b) Factor $f(x)$ into linear factors.

First, notice the degree of the polynomial is 5, so there are at most 5 distinct zeros.

a) The rational zeros theorem says that if a rational number p/q is a zero of $f(x)$, then p must be a factor of 15 and q must be a factor of 2.

$$\frac{Possibilities\ for\ p}{Possibilities\ for\ q}: \frac{\pm 1,\ \pm 3,\ \pm 5,\ \pm 15}{\pm 1,\ \pm 2}$$

$Possibilities\ for\ \dfrac{p}{q}: 1, -1, 3, -3, 5, -5, 15, -15, \dfrac{1}{2}, -\dfrac{1}{2}, \dfrac{3}{2}, -\dfrac{3}{2}, \dfrac{5}{2}, -\dfrac{5}{2}, \dfrac{15}{2}, -\dfrac{15}{2}$

To find which are zeros, we use synthetic division.

Try 1:

$$
\begin{array}{r|rrrrrr}
1 & 2 & -1 & -4 & 2 & -30 & 15 \\
  &   & 2 & 1 & -3 & -1 & -31 \\
\hline
  & 2 & 1 & -3 & -1 & -31 & \boxed{}
\end{array}
$$

Since $f(1) = \boxed{}$, 1 is not a zero.

Try -1:

$$
\begin{array}{r|rrrrrr}
-1 & 2 & -1 & -4 & 2 & -30 & 15 \\
   &   & -2 & 3 & 1 & -3 & 33 \\
\hline
   & 2 & -3 & -1 & 3 & -33 & \boxed{}
\end{array}
$$

Since $f(-1) = \boxed{}$, -1 is not a zero.

Try $\dfrac{1}{2}$:

$$
\begin{array}{r|rrrrrr}
1/2 & 2 & -1 & -4 & 2 & -30 & 15 \\
    &   & 1 & 0 & -2 & 0 & -15 \\
\hline
    & 2 & \boxed{} & -4 & 0 & -30 & \boxed{}
\end{array}
$$

Since $f\left(\dfrac{1}{2}\right) = \boxed{}$, $\dfrac{1}{2}$ is a zero of $f(x)$ and $\left(x - \dfrac{1}{2}\right)$ is a factor of $f(x)$.

We can rewrite $f(x)$ as

$$f(x) = \left(x - \frac{1}{2}\right)\left(2x^4 - 4x^2 - 30\right)$$

$$= \left(x - \frac{1}{2}\right)2\left(x^4 - 2x^2 - 15\right) = \left(x - \frac{1}{2}\right)2\left(x^2 - 5\right)\left(x^2 + 3\right).$$

We now solve the equation $f(x) = 0$ using the principle of zero products.

$$\left(x - \frac{1}{2}\right)2\left(x^2 - 5\right)\left(x^2 + 3\right) = 0$$

$x - \dfrac{1}{2} = 0 \qquad or \qquad x^2 - 5 = 0 \qquad or \qquad x^2 + 3 = 0$

$x = \dfrac{1}{2} \qquad or \qquad x^2 = 5 \qquad or \qquad x^2 = -3$

$x = \boxed{} \qquad or \qquad x = \boxed{} \qquad or \qquad x = \boxed{}$

The rational zero is $\dfrac{1}{2}$. The other zeros are $\sqrt{5}$, $-\sqrt{5}$, $\sqrt{3}i$, and $-\sqrt{3}i$.

b) The complete factorization of $f(x)$ is

$$f(x) = 2\left(x - \boxed{}\right)\left(x + \sqrt{5}\right)\left(x - \sqrt{5}\right)\left(x + \sqrt{3}i\right)\left(x - \boxed{}\right), \text{ or}$$

$$f(x) = (2x - 1)\left(x + \sqrt{5}\right)\left(x - \boxed{}\right)\left(x + \sqrt{3}i\right)\left(x - \sqrt{3}i\right).$$

Descartes' Rule of Signs

Let $P(x)$, written in descending order or ascending order, be a polynomial function with real coefficients and a nonzero constant term. The number of positive real zeros of $P(x)$ is either:

1. The same as the number of variations of sign in $P(x)$, or

2. Less than the number of variations of sign in $P(x)$ by a positive even integer.

The number of negative real zeros of $P(x)$ is either:

3. The same as the number of variations of sign in $P(-x)$, or

4. Less than the number of variations of sign in $P(-x)$ by a positive even integer.

A zero of multiplicity m must be counted m times.

Example 7   Determine the number of variations of sign in the polynomial function $P(x) = 2x^5 - 3x^2 + x + 4$.

There are $\boxed{}$ variations of sign.

Example 8   Given the function $P(x) = 2x^5 - 5x^2 - 3x + 6$, what does Descartes' rule of signs tell you about the number of positive real zeros and the number of negative real zeros?

There are $\boxed{}$ variations of sign in $P(x)$.

Thus the number of positive real zeros is $\boxed{}$ or $2 - 2 = \boxed{}$.

Now find $P(-x)$:

$$P(-x) = 2(-x)^5 - 5(-x)^2 - 3(-x) + 6$$

$$= -2x^5 \boxed{} 5x^2 + 3x + 6$$

There is $\boxed{}$ variation of sign in $P(-x)$.

Thus there is $\boxed{}$ negative real zero.

Total Number of Zeros	5	
Positive Real	2	0
Negative Real	1	1
Nonreal	2	4

Example 9 Given the function $P(x) = 5x^4 - 3x^3 + 7x^2 - 12x + 4$, what does Descartes' rule of signs tell you about the number of positive real zeros and the number of negative real zeros?

There are $\boxed{}$ variations of sign in $P(x)$.

Thus $P(x)$ has $\boxed{}$ or $4 - 2 = \boxed{}$ or $4 - 4 = \boxed{}$ positive real zeros.

Now find $P(-x)$:

$$P(-x) = 5(-x)^4 - 3(-x)^3 + 7(-x)^2 - 12(-x) + 4$$

$$= 5x^4 + 3x^3 + 7x^2 + 12x + 4$$

There are $\boxed{}$ variations of sign in $P(-x)$.

Thus there are $\boxed{}$ negative real zeros.

Total Number of Zeros	4		
Positive Real	4	2	0
Negative Real	0	0	0
Nonreal	0	2	4

Example 10 Given the function $P(x) = 6x^6 - 2x^2 - 5x$, what does Descartes' rule of signs tell you about the number of positive real zeros and the number of negative real zeros?

$$P(x) = 6x^6 - 2x^2 - 5x = x(6x^5 - 2x - 5)$$

Note that $x = 0$ is a zero of this function.

Let $Q(x) = 6x^5 - 2x - 5$.

There is $\boxed{}$ variation of sign in $Q(x)$.

Thus there is $\boxed{}$ positive real zero.

Now find $Q(-x)$:

$$Q(-x) = 6(-x)^5 - 2(-x) - 5 = -6x^5 + 2x - 5$$

There are $\boxed{}$ variations of sign in $Q(-x)$.

Thus there are $\boxed{}$ or $2 - 2 = \boxed{}$ negative real zeros.

Total Number of Zeros	6	
0 as a Zero	1	1
Positive Real	1	1
Negative Real	2	0
Nonreal	2	4

Section 8.5 Rational Functions

Rational Function

A **rational function** is a function f that is a quotient of two polynomials. That is,

$$f(x) = \frac{p(x)}{q(x)}$$

where $p(x)$ and $q(x)$ are polynomials and where $q(x)$ is not the zero polynomial. The domain of f consists of all inputs x for which $q(x) \neq 0$.

Example 1 Consider $f(x) = \dfrac{1}{x-3}$. Find the domain and graph f.

The domain is all the x-values which do not make the denominator zero.

$x - 3 = 0$ when $x = \boxed{}$.

Domain: $\left\{ x \mid x \neq \boxed{} \right\}$, or $(-\infty, 3) \cup (3, \infty)$

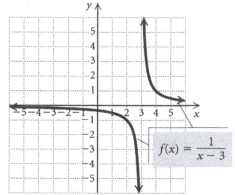

$f(x) = \dfrac{1}{x-3}$

Example 2 Determine the domain of each of the functions.

Function	Domain
$f(x) = \dfrac{1}{x}$	$\left\{ x \mid x \neq \boxed{} \right\}$, or $(-\infty, 0) \cup (0, \infty)$
$f(x) = \dfrac{1}{x^2}$	$\left\{ x \mid x \neq 0 \right\}$, or $\left(-\infty, \boxed{} \right) \cup (0, \infty)$
$f(x) = \dfrac{x-3}{x^2+x-2} = \dfrac{x-3}{(x+2)(x-1)}$	$\left\{ x \mid x \neq -2 \text{ and } x \neq \boxed{} \right\}$, or $(-\infty, -2) \cup (-2, 1) \cup (1, \infty)$
$f(x) = \dfrac{2x+5}{2x-6} = \dfrac{2x+5}{2(x-3)}$	$\left\{ x \mid x \neq \boxed{} \right\}$, or $(-\infty, 3) \cup (3, \infty)$
$f(x) = \dfrac{x^2+2x-3}{x^2-x-2} = \dfrac{x^2+2x-3}{(x+1)(x-2)}$	$\left\{ x \mid x \neq -1 \text{ and } x \neq 2 \right\}$, or $(-\infty, -1) \cup (-1, 2) \cup (2, \infty)$
$f(x) = \dfrac{-x^2}{x+1}$	$\left\{ x \mid x \neq -1 \right\}$, or $\left(-\infty, -1 \right) \cup \left(\boxed{}, \infty \right)$

Determining Vertical Asymptotes

For a rational function $f(x) = p(x)/q(x)$, where $p(x)$ and $q(x)$ are polynomials with *no common factors other than constants*, if a is a zero of the denominator, then the line $x = a$ is a vertical asymptote for the graph of the function.

Example 3 Determine the vertical asymptotes for the graph of each of the following functions.

a) $f(x) = \dfrac{2x - 11}{x^2 + 2x - 8}$

First, factor the denominator.

$$f(x) = \frac{2x - 11}{(x + 4)(x - \boxed{})}$$

The numerator and the denominator have no common factors.

Zeros of the denominator: -4, $\boxed{}$

Vertical asymptotes: $x = \boxed{}$ and $x = 2$

b) $h(x) = \dfrac{x^2 - 4x}{x^3 - x}$

First, factor the numerator and the denominator.

$$h(x) = \frac{x(x - \boxed{})}{x(x^2 - 1)} = \frac{x(x - \boxed{})}{x(x + 1)(x - 1)}$$

Zeros of the numerator: 0, $\boxed{}$

Zeros of the denominator: 0, -1, and $\boxed{}$

Vertical asymptotes: $x = -1$ and $x = 1$

c) $g(x) = \dfrac{x - 2}{x^3 - 5x}$

First, factor the denominator: $g(x) = \dfrac{x - 2}{x(x^2 - 5)}$.

The numerator and the denominator have no common factors.

Solving $x(x^2 - 5) = 0$, we have $x = \boxed{}$ or $x = \pm\sqrt{5}$.

Zeros of the denominator: 0, $-\sqrt{5}$, and $\boxed{}$

Vertical asymptotes: $x = \boxed{}$, $x = -\sqrt{5}$, and $x = \sqrt{5}$

Example 4   Find the horizontal asymptote: $f(x) = \dfrac{-7x^4 - 10x^2 + 1}{11x^4 + x - 2}$.

When the numerator and the denominator of a rational function have the same degree, the horizontal asymptote is given by $y = a/b$ where a and b are the leading coefficients of the numerator and the denominator, respectively.

Thus, the horizontal asymptote is $y = \dfrac{-7}{\boxed{}} = -\dfrac{\boxed{}}{11}$, or $y = -0.\overline{63}$.

Example 5   Find the horizontal asymptote: $f(x) = \dfrac{2x + 3}{x^3 - 2x^2 + 4}$.

Let's call the numerator $p(x)$ and the denominator $q(x)$. Notice $p(x)$ has degree 1 and $q(x)$ has degree 3. As x increases, a degree 3 polynomial will increase much faster than a degree 1 polynomial.

So, as $x \to \infty$, $\dfrac{p(x)}{q(x)} \to 0$, and as $x \to -\infty$, $\dfrac{p(x)}{q(x)} \to 0$.

Thus, the horizontal asymptote is $y = \boxed{}$.

This is the case anytime the degree of the numerator is less than the degree of the denominator.

Determining a Horizontal Asymptote

- When the numerator and the denominator of a rational function have the same degree, the line $y = a/b$ is the horizontal asymptote, where a and b are the leading coefficients of the numerator and the denominator, respectively.

- When the degree of the numerator of a rational function is less than the degree of the denominator, the x-axis, or $y = 0$, is the horizontal asymptote.

- When the degree of the numerator of a rational function is greater than the degree of the denominator, there is no horizontal asymptote.

Example 6   Graph $g(x) = \dfrac{2x^2 + 1}{x^2}$. Include and label all asymptotes.

The denominator is zero when $x = \boxed{}$, so the vertical asymptote is $x = 0$.

The degree of the numerator is the same as the degree of the denominator, so the ratio of the leading coefficients gives us the horizontal asymptote: $y = \boxed{}$.

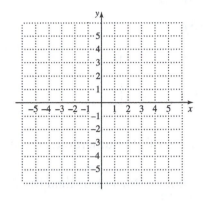

Find and plot some points on the graph. We have:

$$(-2, 2.25), \ (-1, 3), \ (-0.5, 6),$$
$$(0.5, 6), \ (1.5, 2.\overline{4}), \ (2, 2.25)$$

> Sometimes a line that is neither horizontal nor vertical is an asymptote. Such a line is called an **oblique asymptote**, or a **slant asymptote**.

Example 7   Find all the asymptotes of $f(x) = \dfrac{2x^2 - 3x - 1}{x - 2}$.

The denominator is zero when $x = \boxed{}$, so the vertical asymptote is $x = 2$.

There are no horizontal asymptotes, because the degree of the numerator is greater than that of the denominator.

When the degree of the numerator is 1 more than the degree of the denominator, we divide to find an equivalent expression.

$$f(x) = \frac{2x^2 - 3x - 1}{x - 2} = 2x + 1 + \frac{1}{x - 2}$$

As $x \to \infty$ or as $x \to -\infty$, $\dfrac{1}{x - 2} \to 0$, so $f(x) \to 2x + 1$.

The line $y = \boxed{}$ is an oblique asymptote.

Example 8   Graph: $f(x) = \dfrac{2x+3}{3x^2 + 7x - 6}$.

First, find the zeros of the denominator.

$$3x^2 + 7x - 6 = 0$$
$$(3x - 2)(x + 3) = 0$$

$$3x - 2 = 0 \quad or \quad x + 3 = 0$$
$$x = \frac{2}{3} \quad or \qquad x = \boxed{}$$

Neither zero is a zero of the numerator, so the vertical asymptotes are $x = \dfrac{2}{3}$ and $x = -3$.

The degree of the numerator is less than the degree of the denominator, so the horizontal asymptote is $y = \boxed{}$.

The zero of the numerator tells us the x-intercept.

$$2x + 3 = 0$$
$$2x = -3$$

$$x = \boxed{}$$

Thus the x-intercept is $\left(-\dfrac{3}{2}, 0\right)$.

To find the y-intercept, find $f(0)$.

$$f(0) = \frac{2 \cdot 0 + 3}{3 \cdot 0^2 + 7 \cdot 0 - 6}$$
$$= \frac{3}{-6} = -\frac{1}{2}$$

The y-intercept is $\left(0, -\dfrac{1}{2}\right)$.

Finally, choose some other x-values and find their corresponding function values to find enough points to determine the shape of the graph.

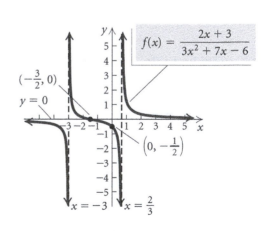

Example 9   Graph: $g(x) = \dfrac{x^2 - 1}{x^2 + x - 6}$.

First, find the zeros of the denominator.

$$x^2 + x - 6 = 0$$

$$(x + 3)(x - 2) = 0$$

$$x = -3 \quad or \quad x = 2$$

Neither zero is a zero of the numerator, so the vertical asymptotes are $x = -3$ and $x = \boxed{}$.

The degree of the numerator is the same as the degree of the denominator, so the horizontal asymptote is $y = \dfrac{1}{1} = 1$.

The zeros of the numerator tell us the *x*-intercepts.

$$x^2 - 1 = 0$$

$$(x + 1)(x - 1) = 0$$

$$x = -1 \quad or \quad x = \boxed{}$$

The *x*-intercepts are $(-1, 0)$ and $(1, 0)$.

To find the *y*-intercept, find $g(0)$.

$$g(0) = \frac{0^2 - 1}{0^2 + 0 - 6} = \frac{-1}{-6} = \frac{1}{6}$$

The *y*-intercept is $\left(0, \dfrac{1}{6}\right)$.

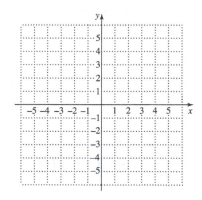

Finally, choose some other *x*-values and find their corresponding function values to find enough points to determine the shape of the graph.

Example 10   Graph: $g(x) = \dfrac{x - 2}{x^2 - x - 2}$.

Factoring the denominator, we have

$$g(x) = \frac{x - 2}{(x - 2)(x + 1)}.$$

The denominator is equal to 0 when $x = 2$ or $x = -1$.

Domain: $\{x \mid x \neq -1 \ and \ x \neq 2\}$

2 is also a zero of the numerator, so the only vertical asymptote is $x = -1$.

The degree of the numerator is less than the degree of the denominator, so the horizontal asymptote is $y = \boxed{}$.

The function has no zeros because the only zero of the numerator, 2, is not in the domain of the function.

$$g(x) = \frac{x-2}{x^2-x-2} = \frac{x-2}{(x-2)(x+1)} = \frac{1}{x+1}, \ x \neq 2$$

The graph will be the graph of

$g(x) = \dfrac{1}{x+1}$ with a hole at $x = 2$.

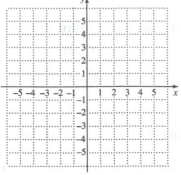

Example 11   Graph: $f(x) = \dfrac{-2x^2-x+15}{x^2-x-12}$.

Factoring: $f(x) = \dfrac{(-2x+5)(x+3)}{(x-4)(x+3)}$

Domain: $\left\{ x \,\middle|\, x \neq \boxed{} \text{ and } x \neq \boxed{} \right\}$, or $(-\infty, -3) \cup (-3, 4) \cup (4, \infty)$

The only zero of the denominator that is not a zero of the numerator is 4, so the only vertical asymptote is $x = 4$.

The degree of the numerator is equal to the degree of the denominator, so the horizontal asymptote is $y = \dfrac{-2}{1} = -2$.

The zeros of the numerator are $\dfrac{5}{2}$ and -3, but -3 is not in the domain of the function, so the only x-intercept is $\left(\dfrac{5}{2}, 0 \right)$.

To find the y-intercept, find $f(0)$.

$$f(0) = \frac{15}{-12} = -\frac{5}{4}$$

The y-intercept is $\left(0, -\dfrac{5}{4} \right)$.

Simplifying: $f(x) = \dfrac{(-2x+5)(x+3)}{(x-4)(x+3)} = \dfrac{-2x+5}{x-4}$, $x \neq -3$ *and* $x \neq 4$

The graph will be the graph of $f(x) = \dfrac{-2x+5}{x-4}$ with a hole at $x = -3$.

Determine the second coordinate of the hole:

$$f(-3) = \frac{-2(-3)+5}{-3-4} = \frac{11}{\boxed{}} = -\frac{11}{7}$$

The hole is at $\left(-3, \boxed{}\right)$, indicated by an open

circle on the graph.

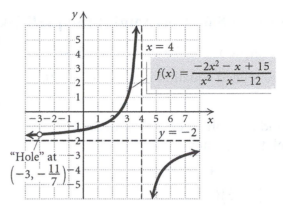

$f(x) = \dfrac{-2x^2 - x + 15}{x^2 - x - 12}$

"Hole" at $\left(-3, -\dfrac{11}{7}\right)$

Example 12 *Temperature During an Illness.* A person's temperature T, in degrees Fahrenheit, during an illness is given by the function

$$T(t) = \frac{4t}{t^2 + 1} + 98.6,$$

where time t is given in hours since the onset of the illness.

a) Find the temperature at $t = 0, 1, 2, 5, 12,$ and 24.

$$T(0) = 98.6$$

$$T(1) = 100.6$$

$$T(2) = \boxed{}$$

$$T(5) = 99.369$$

$$T(12) = 98.931$$

$$T(24) = \boxed{}$$

b) Find the horizontal asymptote of the graph of $T(t)$. Complete $T(t) \rightarrow \boxed{}$ as $t \rightarrow \infty$.

$$T(t) = \frac{4t}{t^2 + 1} + 98.6$$

$$= \frac{4t + 98.6t^2 + 98.6}{t^2 + 1} \qquad \text{Writing with a common denominator}$$

$$= \frac{98.6t^2 + 4t + 98.6}{t^2 + 1}$$

Since the degree of the numerator is the same as the degree of the denominator, the horizontal asymptote is the ratio of the leading coefficients: $y = \dfrac{98.6}{1}$, or $y = \boxed{}$.

So, $T(t) \rightarrow \boxed{}$ as $t \rightarrow \infty$.

c) Give the meaning of the answer to part (b) in terms of the application.
   As time goes on, a person's temperature returns to normal, $98.6°\,F$.

Section 8.6   Polynomial Inequalities and Rational Inequalities

Example 1   Solve: $x^2 - 4x - 5 > 0$.

The related equation is $x^2 - 4x - 5 = 0$.

Graph $f(x) = x^2 - 4x - 5$.

Find the *x*-intercepts by solving the related equation.

$$x^2 - 4x - 5 = 0$$

$$(x - 5)(x + 1) = 0$$

$$x - 5 = 0 \quad or \quad x + 1 = 0$$

$$x = \boxed{} \quad or \quad x = \boxed{}$$

x-intercepts: $(5, 0)$ and $(-1, 0)$

The intercepts divide the *x*-axis into three intervals.

$f(x) = x^2 - 4x - 5$

Interval	$(-\infty, -1)$	$(-1, 5)$	$(5, \infty)$
Test value	$f(-2) = 7$	$f(0) = -5$	$f(7) = 16$
Sign of $f(x)$	+	−	+

The solution set is $\left(-\infty, \boxed{}\right) \cup \left(\boxed{}, \infty\right)$, or $\left\{x \mid x < -1 \ or \ x \boxed{} 5\right\}$.

Example 2   Solve: $x^2 + 3x - 5 \le x + 3$.

$$x^2 + 3x - 5 - x - 3 \le 0$$

$$x^2 + 2x - 8 \le 0$$

Graph $f(x) = x^2 + 2x - 8$.

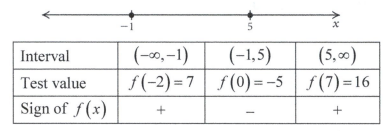

Find the *x*-intercepts by solving the related equation.

$$x^2 + 2x - 8 = 0$$

$$\left(x + \boxed{}\right)\left(x - \boxed{}\right) = 0$$

$$x + 4 = 0 \quad or \quad x - 2 = 0$$

$$x = -4 \quad or \quad x = 2$$

x-intercepts: $\left(\boxed{}, 0\right), \left(\boxed{}, 0\right)$

The solution set is $\left[-4, \boxed{}\right]$, or $\left\{x \mid \boxed{} \le x \le 2\right\}$.

Example 3   Solve: $x^3 - x > 0$.

Find the zeros of the related equation.

$$x^3 - x = 0$$
$$x\left(x^2 - \boxed{}\right) = 0$$
$$x(x+1)(x-1) = 0$$

$$x = 0 \quad or \quad x+1 = 0 \qquad or \quad \boxed{} = 0$$
$$x = 0 \quad or \qquad x = \boxed{} \quad or \qquad x = 1$$

x-intercepts: $(0,0)$, $(-1,0)$, $\left(\boxed{}, 0\right)$

Interval	$(-\infty, -1)$	$(-1, 0)$	$(0,1)$	$(1, \infty)$
Test value	$f(-2) = -6$	$f(-0.5) = 0.375$	$f(0.5) = -0.375$	$f(2) = 6$
Sign of $f(x)$	$-$	$+$	$-$	$+$

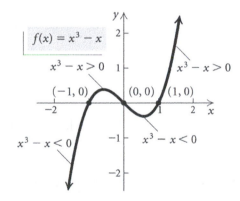

The solution set is $\left(-1, \boxed{}\right) \cup \left(\boxed{}, \infty\right)$, or $\left\{x \mid -1 < x < 0 \ or \ x > \boxed{}\right\}$.

Example 4   Solve: $3x^4 + 10x \le 11x^3 + 4$.

$$3x^4 - 11x^3 + 10x - 4 \le 0$$

Find the zeros of the related equation $3x^4 - 11x^3 + 10x - 4 = 0$.

This equation was solved in Example 5 in Section 8.4. The exact solutions are

$$-1, \ 2 - \sqrt{2}, \ \frac{2}{3}, \ \text{and } 2 + \sqrt{2}.$$

The approximate solutions are

$$-1, \ 0.586, \ 0.667, \ \text{and } 3.414.$$

Interval	Test Value	Sign of $f(x)$
$(-\infty, \ -1)$	$f(-2)=112$	+
$\left(-1, \ 2-\sqrt{2}\right)$	$f(0)=-4$	−
$\left(2-\sqrt{2}, \ \dfrac{2}{3}\right)$	$f(0.6)=0.0128$	+
$\left(\dfrac{2}{3}, \ 2+\sqrt{2}\right)$	$f(1)=-2$	−
$\left(2+\sqrt{2}, \ \infty\right)$	$f(4)=100$	+

The solution set is $\left[\ \boxed{}, 2-\sqrt{2}\right] \cup \left[\dfrac{2}{3}, 2+\sqrt{2}\right]$, or

$$\left\{x \middle| -1 \le x \le 2-\sqrt{2} \ or \ \dfrac{2}{3} \le x \le \boxed{}\right\}.$$

Example 5   Solve: $\dfrac{3x}{x+6}<0.$

Related function:

$$f(x)=\dfrac{3x}{x+6}.$$

To find the critical values, first find the values of x for which the function is not defined.

$$x+6=0 \text{ when } x=\boxed{}.$$

The values of x for which the related function equals 0 are also critical values.

$$f(x)=0$$

$$\dfrac{3x}{x+6}=0$$

$$\left(x+\boxed{}\right)\cdot\dfrac{3x}{x+6}=0\cdot\left(x+\boxed{}\right)$$

$$3x=0$$

$$x=\boxed{}$$

Interval	$(-\infty,-6)$	$(-6,0)$	$(0,\infty)$
Test value	$f(-8)=12$	$f(-2)=-\dfrac{3}{2}$	$f(3)=1$
Sign of $f(x)$	+	−	+

The solution set is $\left(-6, \boxed{}\right)$, or $\left\{x \middle| \boxed{} < x < 0\right\}.$

Example 6   Solve: $\dfrac{x+1}{2x-4} \le 1$.

$$\frac{x+1}{2x-4} - 1 \le 0$$

Related function:

$$f(x) = \frac{x+1}{2x-4} - 1.$$

Critical values are the values of x for which $f(x)$ is not defined or 0. $f(x)$ is not defined when $2x - 4 = 0$, or $x = 2$.

Now solve $f(x) = 0$.

$$\frac{x+1}{2x-4} - 1 = 0$$
$$\frac{x+1}{2x-4} = 1$$
$$x+1 = 2x - 4$$
$$1 = x - 4$$
$$\boxed{} = x$$

The critical values are 2 and 5.

Interval	Test Value	Sign of $f(x)$
$(-\infty, 2)$	$f(0) = -\dfrac{5}{4}$	$\boxed{}$
$(2, 5)$	$f(3) = 1$	$\boxed{}$
$(5, \infty)$	$f(6) = -\dfrac{1}{8}$	$\boxed{}$

The critical value 2 is not in the solution set. The critical value 5 is in the solution set.

The solution set is $\left(-\infty, \boxed{} \right) \cup \left[\boxed{}, \infty \right)$.

Example 7   Solve: $\dfrac{x-3}{x+4} \geq \dfrac{x+2}{x-5}$.

$$\dfrac{x-3}{x+4} - \dfrac{x+2}{x-5} \geq 0$$

Related function:

$$f(x) = \dfrac{x-3}{x+4} - \dfrac{x+2}{x-5}.$$

Find the critical values.

$f(x)$ is not defined for $x = -4$ and $x = \boxed{}$.

Now solve $f(x) = 0$.

$$\dfrac{x-3}{x+4} - \dfrac{x+2}{x-5} = 0$$

$$(x+4) \cdot (x-5) \cdot \left[\dfrac{x-3}{x+4} - \dfrac{x+2}{x-5} \right] = 0 \cdot (x+4) \cdot (x-5)$$

$$(x-5) \cdot \left(x - \boxed{} \right) - (x+4) \cdot \left(x + \boxed{} \right) = 0$$

$$x^2 - 8x + 15 - \left(x^2 + \boxed{}\, x + 8 \right) = 0$$

$$x^2 - 8x + 15 - x^2 - 6x - 8 = 0$$

$$-14x + 7 = 0$$

$$-14x = -7$$

$$x = \dfrac{7}{14} = \boxed{}$$

Critical values: $\boxed{}$, $\dfrac{1}{2}$, and $\boxed{}$

Interval	Test Value	Sign of $f(x)$
$(-\infty,\ -4)$	$f(-5) = 7.7$	$+$
$\left(-4,\ \dfrac{1}{2}\right)$	$f(-2) = -2.5$	$\boxed{}$
$\left(\dfrac{1}{2},\ 5\right)$	$f(3) = 2.5$	$\boxed{}$
$(5,\ \infty)$	$f(6) = -7.7$	$-$

The critical values -4 and 5 are not in the domain. The critical value $\dfrac{1}{2}$ is.

The solution set is $\left(-\infty, \boxed{}\right) \cup \left[\dfrac{1}{2}, 5\right)$, or $\left\{ x \,\middle|\, x < \boxed{} \ or \ \dfrac{1}{2} \leq x < \boxed{} \right\}$.

Section 9.1 The Composition of Functions

Composition of Functions

The **composite function** $f \circ g$, the **composition** of f and g, is defined as

$$(f \circ g)(x) = f(g(x)),$$

where x is in the domain of g and $g(x)$ is in the domain of f.

Example 1   Given that $f(x) = 2x - 5$ and $g(x) = x^2 - 3x + 8$, find each of the following.

a) $(f \circ g)(x)$ and $(g \circ f)(x)$

$$(f \circ g)(x) = f(g(x)) = f\left(x^2 - 3x + \boxed{}\right)$$

$$= 2\left(x^2 - 3x + 8\right) - 5 = 2x^2 - 6x + \boxed{} - 5$$

$$= 2x^2 - 6x + 11$$

$$(g \circ f)(x) = g(f(x)) = g\left(2x - \boxed{}\right)$$

$$= (2x - 5)^2 - 3(2x - 5) + 8$$

$$= 4x^2 - \boxed{}x + 25 - 6x + 15 + 8$$

$$= 4x^2 - 26x + \boxed{}$$

b) $(f \circ g)(7)$ and $(g \circ f)(7)$

$$(f \circ g)(7) = f\left(7^2 - 3 \cdot \boxed{} + 8\right) = f(36)$$

$$= 2 \cdot \boxed{} - 5 = 67$$

$$(g \circ f)(7) = g(2 \cdot 7 - 5) = g\left(\boxed{}\right)$$

$$= 9^2 - 3 \cdot 9 + \boxed{} = 62$$

We could also use the results from part (a) to find $(f \circ g)(7)$ and $(g \circ f)(7)$.

$$(f \circ g)(x) = 2x^2 - 6x + 11$$

$$(f \circ g)(7) = 2 \cdot 7^2 - 6 \cdot \boxed{} + 11 = 67$$

$$(g \circ f)(x) = 4x^2 - 26x + 48$$

$$(g \circ f)(7) = 4 \cdot 7^2 - 26 \cdot 7 + \boxed{} = 62$$

c) $(g \circ g)(1)$

$$(g \circ g)(1) = g(g(1)) = g\left(1^2 - 3 \cdot \boxed{} + 8\right)$$

$$= g(6)$$

$$= \boxed{}^2 - 3 \cdot 6 + 8 = 26$$

d) $(f \circ f)(x)$

$$(f \circ f)(x) = f(f(x)) = f(2x - 5)$$

$$= 2\left(2x - \boxed{}\right) - 5$$

$$= \boxed{}x - 10 - 5$$

$$= 4x - 15$$

Example 2

a) Given that $f(x) = \sqrt{x}$ and $g(x) = x - 3$, find $f \circ g$ and $g \circ f$.

$$(f \circ g)(x) = f(g(x)) = \sqrt{x - 3}$$

$$(g \circ f)(x) = g(f(x)) = \boxed{} - 3$$

b) Given that $f(x) = \sqrt{x}$ and $g(x) = x - 3$, find the domain of $f \circ g$ and the domain of $g \circ f$.

$$(f \circ g)(x) = \sqrt{x - 3}$$

Domain of $f(x) = \left[\boxed{}, \infty\right)$

Domain of $g(x) = (-\infty, \infty)$

Domain of $f \circ g = \{x | x \geq 3\}$, or $\left[\boxed{}, \infty\right)$ because $g(x) \geq 0$, or $x - 3 \geq 0$, when $x \geq 3$.

We could also get this by considering the domain of $(f \circ g)(x) = \sqrt{x - 3}$.

$$(g \circ f)(x) = \sqrt{x} - 3$$

Domain of $g \circ f =$ Domain of $f(x) = [0, \infty)$

We could also find this by considering the domain of $(g \circ f)(x) = \sqrt{x} - 3$.

Example 3 Given that $f(x) = \dfrac{1}{x-2}$ and $g(x) = \dfrac{5}{x}$, find $f \circ g$ and $g \circ f$ and the domain of each.

$$(f \circ g)(x) = f(g(x)) = \dfrac{1}{\dfrac{5}{x} - 2}$$

$$= \dfrac{1}{\dfrac{5-2x}{x}} = 1 \cdot \dfrac{x}{5-2x}$$

$$= \dfrac{x}{5-2x}$$

$$(g \circ f)(x) = g(f(x)) = \dfrac{5}{\dfrac{1}{x-2}}$$

$$= 5 \cdot \dfrac{x-2}{1}$$

$$= 5(x-2)$$

Domain of g: $\{x \mid x \neq 0\}$

Domain of f: $\{x \mid x \neq 2\}$

In addition to 0, values of x for which $g(x) = 2$ must be excluded from the domain of $f \circ g$. We have

$$\dfrac{5}{x} = 2$$

$$5 = 2x$$

$$\dfrac{5}{2} = x.$$

Both $\dfrac{5}{2}$ and $\boxed{}$ must be excluded from the domain of $f \circ g$.

Domain of $f \circ g$: $\left(-\infty, \boxed{}\right) \cup \left(0, \dfrac{5}{2}\right) \cup \left(\boxed{}, \infty\right)$

In addition to 2, values of x for which $f(x) = 0$ must be excluded from the domain of $g \circ f$. We have

$$\frac{1}{x-2} = 0$$

$$\boxed{} = 0 \qquad \text{False}$$

Thus, only 2 must be excluded.

Domain of $g \circ f$: $\left(-\infty, \boxed{}\right) \cup (2, \infty)$

Example 4   If $h(x) = (2x-3)^5$, find $f(x)$ and $g(x)$ such that $h(x) = (f \circ g)(x)$.

For $f(x) = x^5$ and $g(x) = 2x - \boxed{}$,

$$h(x) = (f \circ g)(x) = f(g(x))$$
$$= f(2x-3)$$
$$= \left(\boxed{}\right)^5.$$

These functions give us $h(x)$, but there are other choices for $f(x)$ and $g(x)$.

For $f(x) = (x+7)^5$ and $g(x) = 2x - 10$,

$$h(x) = (f \circ g)(x) = f(g(x))$$
$$= f\left(2x - \boxed{}\right)$$
$$= (2x - 10 + 7)^5$$
$$= \left(2x - \boxed{}\right)^5.$$

This is also a correct choice. There are others as well.

Example 5   If $h(x) = \dfrac{1}{(x+3)^3}$, find $f(x)$ and $g(x)$ such that $h(x) = (f \circ g)(x)$.

For $f(x) = \dfrac{1}{x^3}$ and $g(x) = x + \boxed{}$,

$$h(x) = (f \circ g)(x) = f(g(x))$$
$$= f\left(\boxed{}\right)$$
$$= \dfrac{1}{(x+3)^3}.$$

These functions give us $h(x)$, but there are other correct choices for $f(x)$ and $g(x)$.

For $f(x) = \dfrac{1}{x}$ and $g(x) = (x+3)^3$,

$$h(x) = (f \circ g)(x) = f(g(x))$$
$$= f\left((x+3)^3\right)$$
$$= \dfrac{1}{\boxed{}}.$$

This is also a correct choice. There are others as well.

Section 9.2   Inverse Functions

Inverse Relation

Interchanging the first and second coordinates of each ordered pair in a relation produces the **inverse relation**.

Example 1   Consider the relation g given by

$$g = \{(2, 4), (-1, 3), (-2, 0)\}.$$

Graph the relation in blue. Find the inverse relation and graph it in red.

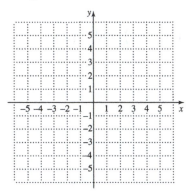

Inverse of $g = \left\{ \boxed{} \right\}$

Inverse Relation

If a relation is defined by an equation, interchanging the variables produces an equation of the inverse relation.

Example 2   Find an equation for the inverse of the relation: $y = x^2 - 5x.$

The equation of the inverse relation is $\boxed{}$.

One-to-One Functions

A function f is **one-to-one** if different inputs have different outputs – that is,

if $a \neq b$, then $f(a) \neq f(b)$.

Or a function is **one-to-one** if when the outputs are the same, the inputs are the same – that is,

if $f(a) = f(b)$, then $a = b$.

If the inverse of a function f is also a function, it is named f^{-1} (read "f-inverse").

One-to-One Functions and Inverses
- If a function f is one-to-one, then its inverse f^{-1} is a function.
- The domain of a one-to-one function f is the range of the inverse f^{-1}.
- The range of a one-to-one function f is the domain of the inverse f^{-1}.
- A function that is increasing over its entire domain or is decreasing over its entire domain is a one-to-one function.

Example 3   Given the function f described by $f(x) = 2x - 3$, prove that f is one-to-one (that is, it has an inverse that is a function).

Given $f(x) = 2x - 3$.

If $f(a) = f(b)$, then $\boxed{}$.

$\qquad f(a) = \boxed{}$ $\qquad$ $f(b) = \boxed{}$

$\qquad 2a - 3 = 2b - 3$

$\qquad \boxed{} = \boxed{}$

$\qquad \boxed{} = \boxed{}$

Thus, if $f(a) = f(b)$, then $a = b$.

This tells us that $f(x)$ is one-to-one.

Example 4   Given the function g described by $g(x) = x^2$, prove that g is not one-to-one.

To show that g is not a one-to-one function, we need to find two numbers such that $a \neq b$ and $g(a) = g(b)$.

Try the values $x = 3$ and $x = -3$.

$g\left(\boxed{}\right) = 3^2 = \boxed{}$

$g\left(\boxed{}\right) = (-3)^2 = \boxed{}$

Thus, g _____ one-to-one.
$\qquad\qquad$ is / is not

Horizontal-Line Test

If it is possible for a horizontal line to intersect the graph of a function more than once, then the function is *not* one-to-one and its inverse is *not* a function.

Example 5   From the graphs shown, determine whether each function is one-to-one and thus has an inverse that is a function.

a)

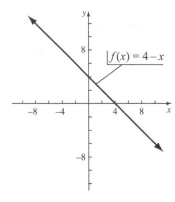

one-to-one / not one-to-one

b)

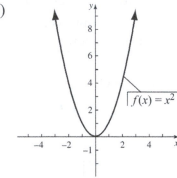

one-to-one / not one-to-one

c)

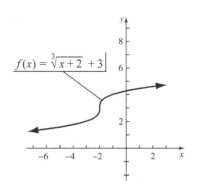

one-to-one / not one-to-one

d)

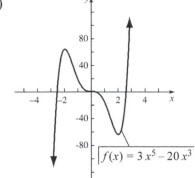

one-to-one / not one-to-one

Obtaining a Formula for an Inverse

If a function *f* is one-to-one, a formula for its inverse can generally be found as follows.

1. Replace $f(x)$ with y.

2. Interchange x and y.

3. Solve for y.

4. Replace y with $f^{-1}(x)$.

Example 6   Determine whether the function $f(x) = 2x - 3$ is one-to-one, and if it is, find a formula for $f^{-1}(x)$.

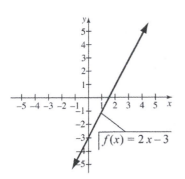

$f(x) = 2x - 3$

The graph passes the horizontal-line test, so it is one-to-one. Find the inverse.

1.   Replace $f(x)$ with y:

   ⬚

2.   Interchange x and y.

   ⬚

3.   Solve for y:
$$x + 3 = 2y$$

   ⬚ $= y$

4.   Replace y with $f^{-1}(x)$:

$$f^{-1}(x) = \boxed{}$$

Example 7   Graph $f(x) = 2x - 3$ and $f^{-1}(x) = \dfrac{x+3}{2}$ using the same set of axes. Then compare the two graphs.

x	$f(x) = 2x - 3$
-1	-5
0	⬚
2	⬚
3	⬚

x	$f^{-1}(x) = \dfrac{x+3}{2}$
-5	-1
-3	⬚
1	⬚
3	⬚

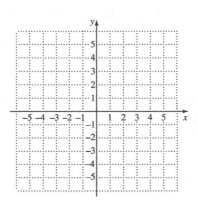

The graphs are reflections of each other across the line $y = x$.

The graph of f^{-1} is a reflection of the graph of f across the line $y = x$.

Example 8   Consider $g(x) = x^3 + 2$.

    a) Determine whether the function is one-to-one.

    b) If it is one-to-one, find a formula for its inverse.

    c) Graph the function and its inverse.

a) $g(x)$ passes the horizontal-line test, so it is one-to-one.

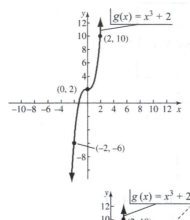

b) **1.** Replace $g(x)$ with y.

$$y = x^3 + 2$$

    2. Interchange x and y.

    3. Solve for y.

$$x = y^3 + 2$$

$$\boxed{} = y^3$$

$$\boxed{} = y$$

    4. Replace y with $g^{-1}(x)$

$$g^{-1}(x) = \boxed{}$$

c)

If a function f is one-to-one, then f^{-1} is the unique function such that each of the following holds:

$$\left(f^{-1} \circ f\right)(x) = f^{-1}\left(f(x)\right) = x, \quad \text{for each } x \text{ in the domain of } f, \text{ and}$$

$$\left(f \circ f^{-1}\right)(x) = f\left(f^{-1}(x)\right) = x, \quad \text{for each } x \text{ in the domain of } f^{-1}.$$

Example 9   Given that $f(x) = 5x + 8$, use composition of functions to show that

$$f^{-1}(x) = \frac{x-8}{5}.$$

$$\left(f^{-1} \circ f\right)(x) = f^{-1}\left(f(x)\right) = f^{-1}\left(\boxed{}\right) = \frac{\left(\boxed{} - 8\right)}{5} = \frac{5x}{5} = \boxed{}$$

$$\left(f \circ f^{-1}\right)(x) = f\left(f^{-1}(x)\right) = f\left(\boxed{}\right) = 5\left(\boxed{}\right) + 8 = x - 8 + 8 = \boxed{}$$

These two compositions show us that $f^{-1}(x) = \dfrac{x-8}{5}$ is the inverse of $f(x) = 5x + 8$.

Section 9.3   Exponential Functions and Graphs

Exponential Function

The function $f(x) = a^x$, where x is a real number, $a > 0$ and $a \neq 1$, is called the **exponential function, base a.**

Example 1   Graph the exponential function $y = f(x) = 2^x$.

x	$y = f(x) = 2^x$	(x, y)
0	1	$(0, 1)$
1	$\square$	$(1, 2)$
2	4	$\left(2, \square\right)$
3	8	$(3, 8)$
−1	$\dfrac{1}{2}$	$\left(-1, \dfrac{1}{2}\right)$
−2	$\square$	$\left(-2, \dfrac{1}{4}\right)$
−3	$\dfrac{1}{8}$	$\left(\square, \dfrac{1}{8}\right)$

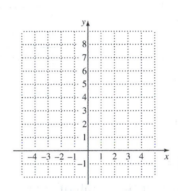

Example 2   Graph the exponential function $y = f(x) = \left(\dfrac{1}{2}\right)^x$.

Points of $g(x) = 2^x$	Points of $f(x) = \left(\dfrac{1}{2}\right)^x = 2^{-x}$
$(0, 1)$	$(0, 1)$
$(1, 2)$	$(-1, 2)$
$(2, 4)$	$\left(\square, 4\right)$
$(3, 8)$	$(-3, 8)$
$\left(-1, \dfrac{1}{2}\right)$	$\left(\square, \dfrac{1}{2}\right)$
$\left(-2, \dfrac{1}{4}\right)$	$\left(2, \dfrac{1}{4}\right)$
$\left(-3, \dfrac{1}{8}\right)$	$\left(3, \dfrac{1}{8}\right)$

$$y = f(x) = \left(\frac{1}{2}\right)^x = 2^{-x}$$

$$y = g(x) = 2^x$$

Example 3   Graph each of the following. Before doing so, describe how each graph can be obtained from the graph of $f(x) = 2^x$.

a) $f(x) = 2^{x-2}$

b) $f(x) = 2^x - 4$

c) $f(x) = 5 - 0.5^x$

a) Graph $f(x) = 2^{x-2}$.

The graph of $f(x) = 2^{x-2}$ is the graph of $f(x) = 2^x$ shifted _____ 2 units.
left / right

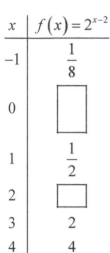

x	$f(x) = 2^{x-2}$
-1	$\dfrac{1}{8}$
0	
1	$\dfrac{1}{2}$
2	
3	2
4	4

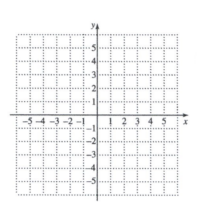

b) Graph $f(x) = 2^x - 4$.

Next, we consider the graph of $f(x) = 2^x - 4$, which is the graph of $f(x) = 2^x$ shifted

_____ 4 units.
up / down

x	$f(x) = 2^x - 4$
-2	$-3\dfrac{3}{4}$
-1	
0	-3
1	
2	0
3	4

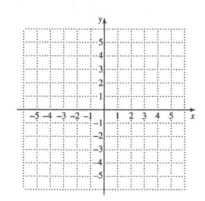

c) Graph $f(x) = 5 - 0.5^x$.

$$f(x) = 5 - 0.5^x = 5 - \left(\frac{1}{2}\right)^x$$

$$= 5 - \left(2^{-1}\right)^x$$

$$= 5 - 2^{-x}$$

The graph of $f(x) = 5 - 0.5^x$ is the graph of $f(x) = 2^x$ reflected across the

_____, then reflected across the _____, and finally shifted
  x-axis / y-axis   x-axis / y-axis

_____ 5 units.
  up / down

x	$f(x) = 5 - 0.5^x$
−3	−3
−2	$\boxed{}$
−1	3
0	$\boxed{}$
1	$4\dfrac{1}{2}$
2	$4\dfrac{3}{4}$

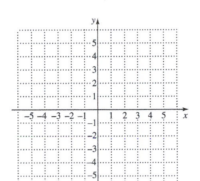

Example 4 *Compound Interest.* The amount of money A to which a principal P will grow after t years at interest rate r (in decimal form), compounded n times per year, is given by the formula

$$A = P\left(1 + \frac{r}{n}\right)^{nt}.$$

Suppose that $100,000 is invested at 6.5% interest, compounded semiannually.

a) Find a function for the amount to which the investment grows after t years.

$$A = P\left(1 + \frac{r}{n}\right)^{nt}$$

$P = 100{,}000 \qquad r = \boxed{} \qquad n = 2$

$$A(t) = 100{,}000\left(1 + \frac{0.065}{2}\right)^{2t}$$

$$A(t) = 100{,}000\left(\boxed{}\right)^{2t}$$

b) Find the amount of money in the account at $t = 0, 4, 8,$ and 10 years.

$$A(0) = 100,000(1.0325)^{2 \cdot \square} = \boxed{}$$

$$A(4) = 100,000(1.0325)^{2 \cdot 4} \approx \$129,157.75$$

$$A(8) = 100,000(1.0325)^{2 \cdot \square} \approx \boxed{}$$

$$A(10) = 100,000(1.0325)^{2 \cdot 10} \approx \$189,583.79$$

c) Graph the function.

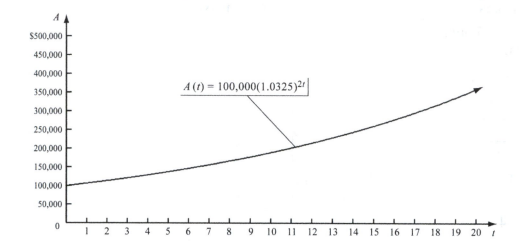

$$A(t) = 100,000(1.0325)^{2t}$$

$e = 2.7182818284...$

Example 5 Find each value of e^x, to four decimal places, using the e^x key on a calculator.

a) $e^3 = e\wedge(3) \approx$ [   ]

b) $e^{-0.23} = e\wedge(-0.23) \approx$ [   ]

c) $e^0 = e\wedge(0) =$ [   ]

Example 6 Graph $f(x) = e^x$ and $g(x) = e^{-x}$.

x	$f(x) = e^x$	$g(x) = e^{-x}$
-2	0.135	7.389
-1	0.368	[ ]
0	[ ]	1
1	2.718	0.368
2	7.389	0.135

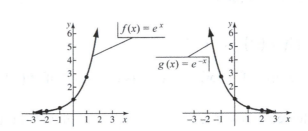

Example 7 Graph each of the following. Before doing so, describe how each graph can be obtained from the graph of $y = e^x$.

a) $f(x) = e^{x+3}$ b) $f(x) = e^{-0.5x}$ c) $f(x) = 1 - e^{-2x}$

a) Graph $f(x) = e^{x+3}$.

The graph of $f(x) = e^{x+3}$ is the graph of $f(x) = e^x$ shifted _____ 3 units.
                                                                left / right

x	$f(x) = e^{x+3}$
-5	0.135
-4	0.368
-3	[ ]
-2	[ ]
-1	7.389
0	20.086
1	54.598

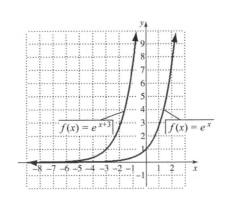

b) Graph $f(x) = e^{-0.5x}$.

The graph of $f(x) = e^{-0.5x}$ is the graph of $f(x) = e^x$ stretched

_____ and reflected across the _____.
    horizontally / vertically   x-axis / y-axis

x	$f(x) = e^{-0.5x}$
−3	4.482
−2	[ ]
−1	1.649
0	[ ]
1	0.607
2	0.368

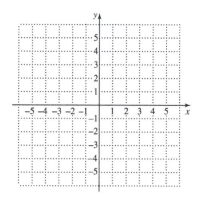

c) Graph $f(x) = 1 - e^{-2x}$.

The graph of $f(x) = 1 - e^{-2x}$ is the graph of $f(x) = e^x$ shrunk _____,
                                                                        horizontally / vertically

reflected across the _____ and also reflected across the _____,
                        x-axis / y-axis   x-axis / y-axis

and finally shifted _____ 1 unit.
                       up / down

x	$f(x) = 1 - e^{-2x}$
−1	−6.389
0	[ ]
1	0.865
2	0.982
3	[ ]

Section 9.4 Logarithmic Functions and Graphs

Example 1   Graph $x = 2^y$.

$x \ \left(x = 2^y\right)$	y	(x, y)
1	0	$(1, 0)$
☐	1	☐
4	2	$(4, 2)$
☐	-1	☐
☐	-2	☐
$\dfrac{1}{8}$	-3	$\left(\dfrac{1}{8}, -3\right)$

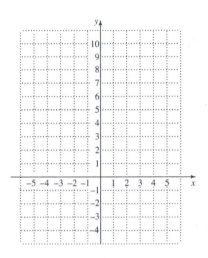

Logarithmic Function, Base a

We define $y = \log_a x$ as that number y such that $x = a^y$, where $x > 0$ and a is a positive constant other than 1.

Example 2

a) Find $\log_{10} 10{,}000$.

$$10^{?} = 10{,}000$$

$$\log_{10} 10{,}000 = \boxed{}$$

b) Find $\log_{10} 0.01$.

$$10^{?} = \frac{1}{100} = 10^{\square}$$

$$\log_{10} 0.01 = \boxed{}$$

c) Find $\log_2 8$.

$$2^{?} = 8$$

$$\log_2 8 = \boxed{}$$

d) Find $\log_9 3$.

$$9^{?} = 3 = 9^{1/2}$$

$$\log_9 3 = \boxed{}$$

e) Find $\log_6 1$.

$$6^{?} = 1$$

$$\log_6 1 = \boxed{}$$

f) Find $\log_8 8$.

$$8^{?} = 8$$

$$\log_8 8 = \boxed{}$$

$\log_a 1 = 0$ and $\log_a a = 1$, for any logarithmic base a.

$\log_a x = y \longleftrightarrow x = a^y$ A logarithm is an exponent!

Example 3

Convert each of the following to a logarithmic equation.

a) $16 = 2^x$ $\log_{\square} 16 = \boxed{}$

b) $10^{-3} = 0.001$ $\log_{\square} 0.001 = \boxed{}$

c) $e^t = 70$ $\log_{\square} 70 = \boxed{}$

Example 4

Convert each of the following to an exponential equation.

a) $\log_2 32 = 5$ $2^{\square} = \boxed{}$

b) $\log_a Q = 8$ $a^{\square} = \boxed{}$

c) $x = \log_t M$ $t^{\square} = \boxed{}$

Common Logarithms (Base 10): $\log M = \log_{10} M$

Natural Logarithms (Base e): $\ln M = \log_e M$

Example 5   Find each of the following common logarithms on a calculator. If you are using a graphing calculator, set the calculator in REAL mode. Round to four decimal places.

a) $\log(645{,}778) \approx \boxed{}$

b) $\log(0.0000239) \approx \boxed{}$

c) $\log(-3)$

The domain of the log function is $(0, \infty)$; -3 is not in this domain.

$\log(-3)$ _____ exist.
             does / does not

Example 6 Find each of the following natural logarithms on a calculator. If you are using a graphing calculator, set the calculator in REAL mode. Round to four decimal places.

a) $\ln(645{,}778) \approx \boxed{}$

b) $\ln(0.0000239) \approx \boxed{}$

c) $\ln(-5)$

The domain of the ln function is $(0, \infty)$; -5 is not in this domain.

$\ln(-5)$ $\underline{}$ exist.
$\quad\quad\quad$ does / does not

d) $\ln(e) = 1$ $\qquad\qquad$ $e^? = e$

$\qquad\qquad\qquad\qquad$ $e^{\boxed{}} = e$

e) $\ln(1) = 0$ $\qquad\qquad$ $e^? = 1$

$\qquad\qquad\qquad\qquad$ $e^{\boxed{}} = 1$

$\boxed{\ln 1 = 0 \text{ and } \ln e = 1, \text{ for the logarithmic base } e.}$

The Change-of-Base Formula

For any logarithmic bases a and b, and any positive number M,
$$\log_b M = \frac{\log_a M}{\log_a b}.$$

Example 7 Find $\log_5 8$ using common logarithms.

Change-of-base formula:
$$\log_b M = \frac{\log_a M}{\log_a b}$$

Let $a = \boxed{}$, $b = \boxed{}$, and $M = 8$.

$$\log_5 8 = \frac{\log_{10} \boxed{}}{\log_{10} 5} \approx \boxed{}$$

Example 8   Find $\log_5 8$ using natural logarithms.

Change-of-base formula:

$$\log_b M = \frac{\log_a M}{\log_a b}$$

Let $a = \boxed{}$, $b = 5$, and $M = \boxed{}$.

$$\log_5 8 = \frac{\log_\square 8}{\log_\square 5} = \frac{\ln 8}{\ln 5} \approx \boxed{}$$

Example 9   Graph $y = f(x) = \log_5 x$.

Think: $x = \boxed{}$

x, or 5^y	y
1	0
$\boxed{}$	1
$\boxed{}$	-1
$\dfrac{1}{25}$	-2

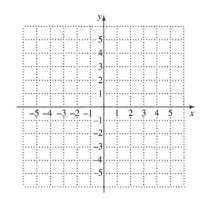

Example 10   Graph $g(x) = \ln x$.

x	$g(x) = \ln x$
0.5	-0.7
1	$\boxed{}$
2	$\boxed{}$
3	$\boxed{}$
4	1.4
5	1.6

Example 11   Graph each of the following. Before doing so, describe how each graph can be obtained from the graph of $y = \ln x$. Give the domain and the vertical asymptote of each function.

a) $f(x) = \ln(x+3)$ b) $f(x) = 3 - \dfrac{1}{2}\ln x$ c) $f(x) = \left|\ln(x-1)\right|$

a) Graph $f(x) = \ln(x+3)$.

The graph of $f(x) = \ln(x+3)$ is a shift of the graph of $y = \ln x$ _____ 3 units.
$\underset{\text{left / right}}{}$

x	$f(x)$
−2.9	−2.303
−2	0
0	1.099
2	1.609
4	1.946

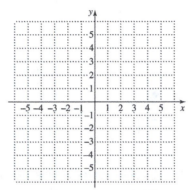

Domain: $\left(\boxed{}, \infty\right)$

Vertical asymptote: $\boxed{}$

b) Graph $f(x) = 3 - \dfrac{1}{2}\ln x$.

The graph of $f(x) = 3 - \dfrac{1}{2}\ln x$ is a _____ shrinking of the graph of
$\underset{\text{vertical / horizontal}}{}$
$y = \ln x$, followed by a reflection across the _____ , and then a translation
$\underset{x\text{-axis} / y\text{-axis}}{}$
_____ 3 units.
$\underset{\text{up / down}}{}$

x	$f(x)$
0.1	4.151
1	3
3	2.451
6	2.104
9	1.901

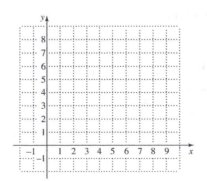

Domain: $\left(\boxed{}, \infty\right)$

Vertical asymptote: $\boxed{}$

c) Graph $f(x) = |\ln(x-1)|$.

The graph of $f(x) = |\ln(x-1)|$ is a shift of the graph of $y = \ln x$ _____ 1 unit
$$\underset{\text{left / right}}{}$$

followed by a reflection of negative outputs across the _____ .
$$\underset{\text{x-axis / y-axis}}{}$$

x	$f(x)$
1.1	2.303
2	0
4	1.099
6	1.609
8	1.946

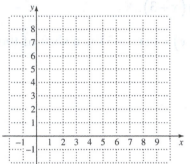

Domain: $\left(\boxed{}, \infty\right)$

Vertical asymptote: $\boxed{}$

Example 12   In a study by psychologists Bornstein and Bornstein, it was found that the average walking speed w, in feet per second, of a person living in a city of population P, in thousands, is given by the function

$$w(P) = 0.37\ln P + 0.05.$$

a) The population of Billings, Montana, is 106,954. Find the average walking speed of people living in Billings.

$106,954 = 106.954$ thousands

$w(106.954) = 0.37\ln\left(\boxed{}\right) + 0.05$

$\approx \boxed{}$ ft/sec

b) The population of Chicago, Illinois, is 2,714,856. Find the average walking speed of people living in Chicago.

$2,714,856 = 2714.856$ thousands

$w(2714.856) = 0.37\ln\left(\boxed{}\right) + 0.05$

$\approx \boxed{}$ ft/sec

Example 13 Measured on the Richter scale, the magnitude R of an earthquake of intensity I is defined as

$$R = \log \frac{I}{I_0},$$

where I_0 is a minimum intensity used for comparison. We can think of I_0 as a threshold intensity that is the weakest earthquake that can be recorded on a seismograph. If one earthquake is 10 times as intense as another, its magnitude on the Richter scale is 1 greater than that of the other. If one earthquake is 100 times as intense as another, its magnitude on the Richter scale is 2 higher, and so on. Thus an earthquake whose magnitude is 7 on the Richter scale is 10 times as intense as an earthquake whose magnitude is 6. Earthquake intensities can be interpreted as multiples of the minimum intensity I_0.

The undersea Tohoku earthquake and tsunami, near the northeast coast of Honshu, Japan, on March 1, 2011, had an intensity of $10^{9.0} \cdot I_0$ (*Source*: earthquake.usgs.gov). They caused extensive loss of life and severe structural damage to buildings, railways, and roads. What was the magnitude on the Richter scale?

$$R = \log \frac{I}{I_0} \qquad\qquad I = 10^{9.0} \cdot I_0$$

$$R = \log \frac{\boxed{}}{I_0}$$

$$R = \log \boxed{}$$

$$R = 9.0$$

The magnitude of the earthquake was $\boxed{}$ on the Richter scale.

Section 9.5 Properties of Logarithmic Functions

Properties of Logarithms

For any positive numbers M and N, any logarithmic base a, and any real number p:

    Product Rule: $\log_a MN = \log_a M + \log_a N$.

    Power Rule: $\log_a M^p = p \log_a M$.

    Quotient Rule: $\log_a \dfrac{M}{N} = \log_a M - \log_a N$.

Example 1   Express as a sum of logarithms.

$\log_3 (9 \cdot 27) = \boxed{} + \boxed{}$ Using the product rule

Example 2   Express as a single logarithm.

$\log_2 p^3 + \log_2 q = \log_2 \left(\boxed{} \right)$ Using the product rule

Example 3

a) Express $\log_a 11^{-3}$ as a product.

    Power Rule: $\log_a M^p = p \log_a M$

    $M = 11$ and $p = -3$

    $\log_a 11^{-3} = \boxed{} \log_a \boxed{}$

b) Express $\log_a \sqrt[4]{7}$ as a product.

    First rewrite $\sqrt[4]{7}$ with an exponent using $\sqrt[n]{x} = x^{\frac{1}{n}}$.

    $\log_a \sqrt[4]{7} = \log_a 7^{\boxed{}}$

    $= \boxed{} \log_a 7$ Using the power rule

c) Express $\ln x^6$ as a product.

    Using the power rule:

    $\ln x^6 = \boxed{} \ln \boxed{}$

Example 4

Express as a difference of logarithms.

$$\log_t \frac{8}{w} = \log_t \boxed{} - \log_t \boxed{} \qquad \text{Using the quotient rule}$$

Example 5

Express as a single logarithm.

$$\log_b 64 - \log_b 16 = \boxed{} \qquad \text{Using the quotient rule}$$

$$= \boxed{}$$

Example 6

a) Express $\log_a \dfrac{x^2 y^5}{z^4}$ in terms of sums and differences of logarithms.

$$\log_a \frac{x^2 y^5}{z^4} = \log_a \left(\boxed{} \right) - \log_a \boxed{}$$

$$= \log_{\boxed{}} x^2 + \log_{\boxed{}} y^5 - \log_a z^4$$

$$= \boxed{} \log_a x + \boxed{} \log_a y - \boxed{} \log_a z$$

b) Express $\log_a \sqrt[3]{\dfrac{a^2 b}{c^5}}$ in terms of sums and differences of logarithms.

$$\log_a \sqrt[3]{\frac{a^2 b}{c^5}} = \log_a \left(\frac{a^2 b}{c^5} \right)^{\boxed{}}$$

$$= \boxed{} \log_a \left(\frac{a^2 b}{c^5} \right)$$

$$= \frac{1}{3} \left(\log_a \left(\boxed{} \right) - \log_a \boxed{} \right)$$

$$= \frac{1}{3} \left(\log_a \boxed{} + \log_a \boxed{} - \log_a c^5 \right)$$

$$= \frac{1}{3} \left(\boxed{} \log_a a + \log_a b - \boxed{} \log_a c \right)$$

$$= \boxed{} + \boxed{} \log_a b - \boxed{} \log_a c$$

c) Express $\log_b \dfrac{ay^5}{m^3n^4}$ in terms of sums and differences of logarithms.

$$\log_b \frac{ay^5}{m^3n^4} = \log_b \left(ay^5\right) - \log_\square \left(m^3n^4\right)$$

$$= \log_b \boxed{} + \log_b \boxed{} - \left(\log_\square m^3 + \log_\square n^4\right)$$

$$= \log_b a + \log_b y^5 \boxed{} \log_b m^3 \boxed{} \log_b n^4$$

$$= \log_b a + \boxed{} \log_b y - \boxed{} \log_b m - \boxed{} \log_b n$$

Example 7   Express as a single logarithm.

$$5\log_b x - \log_b y + \frac{1}{4}\log_b z$$

$$= \log_b x^{\boxed{}} - \log_b y + \log_b z^{\boxed{}}$$

$$= \log_b \frac{\boxed{}}{\boxed{}} + \log_b z^{\frac{1}{4}}$$

$$= \log_b \frac{\boxed{}}{\boxed{}}$$

Example 8   Express as a single logarithm.

$$\ln\left(3x+1\right) - \ln\left(3x^2 - 5x - 2\right)$$

$$= \ln \frac{\boxed{}}{3x^2 - 5x - 2} = \ln \frac{3x+1}{(3x+1)\left(\boxed{}\right)}$$

$$= \ln \frac{1}{\boxed{}}$$

Example 9

a) Given that $\log_a 2 \approx 0.301$ and $\log_a 3 \approx 0.477$, find $\log_a 6$, if possible.

$$\log_a 6 = \log_a \left(2 \cdot 3\right)$$

$$= \log_a \boxed{} + \log_a \boxed{}$$

$$\approx 0.301 + \boxed{}$$

$$\approx \boxed{}$$

b) Given that $\log_a 2 \approx 0.301$ and $\log_a 3 \approx 0.477$, find $\log_a \frac{2}{3}$, if possible.

$$\log_a \frac{2}{3} = \log_a 2 - \log_a \boxed{}$$
$$\approx \boxed{} - 0.477$$
$$\approx \boxed{}$$

c) Given that $\log_a 2 \approx 0.301$ and $\log_a 3 \approx 0.477$, find $\log_a 81$, if possible.

$$\log_a 81 = \log_a \boxed{}$$
$$= \boxed{} \log_a 3$$
$$\approx 4 \left(\boxed{} \right)$$
$$\approx 1.908$$

d) Given that $\log_a 2 \approx 0.301$ and $\log_a 3 \approx 0.477$, find $\log_a \frac{1}{4}$, if possible.

$$\log_a \frac{1}{4} = \log_a 1 - \log_a \boxed{}$$
$$= \boxed{} - \log_a 2^2$$
$$= -2 \log_a \boxed{}$$
$$\approx -2 \left(\boxed{} \right)$$
$$\approx -0.602$$

e) Given that $\log_a 2 \approx 0.301$ and $\log_a 3 \approx 0.477$, find $\log_a 5$, if possible.

$$\log_a 5 = \log_a (2+3) \neq \log_a 2 + \log_a 3$$

Finding $\log_a 5$ using the given logarithms _____ possible.

is / is not

f) Given that $\log_a 2 \approx 0.301$ and $\log_a 3 \approx 0.477$, find $\frac{\log_a 3}{\log_a 2}$, if possible.

$$\frac{\log_a 3}{\log_a 2} \approx \frac{\boxed{}}{\boxed{}} \approx 1.585$$

> ### The Logarithm of a Base to a Power
>
> For any base a and any positive real number x,
>
> $\log_a a^x = x.$

Example 10

a) Simplify.

$\log_a a^8 = \boxed{}$

b) Simplify.

$\ln e^{-t} = \log_{\boxed{}} e^{-t}$

$= \boxed{}$

c) Simplify.

$\log 10^{3k} = \log_{\boxed{}} 10^{3k}$

$= \boxed{}$

> ### A Base to a Logarithmic Power
>
> For any base a and any positive real number x,
>
> $a^{\log_a x} = x.$

Example 11

a) Simplify.

$4^{\log_4 k} = \boxed{}$

b) Simplify.

$e^{\ln 5} = e^{\boxed{}} = \boxed{}$

c) Simplify.

$10^{\log 7t} = 10^{\boxed{}} = \boxed{}$

Section 9.6 Solving Exponential Equations and Logarithmic Equations

---
Base-Exponent Property

For any $a > 0$, $a \neq 1$,

$$a^x = a^y \longleftrightarrow x = y.$$

Example 1   Solve $2^{3x-7} = 32$.

$2^{3x-7} = 32$

$2^{3x-7} = \boxed{}$

$3x - 7 = 5$

$3x = \boxed{}$

$x = \boxed{}$

Check:

$$\frac{2^{3x-7} = 32}{2^{3 \cdot \boxed{} - 7} \ ? \ 32}$$

$2^{\boxed{}} \quad \Big|$

$32 \ \Big| \ 32 \quad$ TRUE

The solution is $\boxed{}$.

Property of Logarithmic Equality

For any $M > 0$, $N > 0$, $a > 0$, and $a \neq 1$,

$$\log_a M = \log_a N \longleftrightarrow M = N.$$

Example 2   Solve $3^x = 20$.

$3^x = 20$

$\log 3^x = \log 20$

$\boxed{} \log 3 = \log 20$

$x = \dfrac{\log 20}{\boxed{}}$

$x \approx 2.7268$

Check: $3^{2.7268} \approx \boxed{}$

The solution is about $\boxed{}$.

Example 3   Solve $100e^{0.08t} = 2500$.

$$100e^{0.08t} = 2500$$

$$e^{0.08t} = \boxed{}$$

$$\ln e^{0.08t} = \ln 25$$

$$\boxed{} = \ln 25$$

$$t = \frac{\ln 25}{\boxed{}}$$

$$t \approx 40.2$$

The solution is about $\boxed{}$.

Example 4   Solve $4^{x+3} = 3^{-x}$.

$$4^{x+3} = 3^{-x}$$

$$\log 4^{x+3} = \log 3^{-x}$$

$$\left(\boxed{}\right)\log 4 = \boxed{}\log 3$$

$$x\log 4 + \boxed{}\log 4 = -x\log 3$$

$$x\log 4 + \boxed{} = -3\log 4$$

$$\boxed{}\left(\log 4 + \log 3\right) = -3\log 4$$

$$x = \frac{-3\log 4}{\boxed{}}$$

$$x \approx -1.6737$$

The solution is about $\boxed{}$.

Example 5   Solve $e^x + e^{-x} - 6 = 0$.

$$e^x + e^{-x} - 6 = 0$$

$$e^x + \frac{1}{\boxed{}} - 6 = 0$$

$$\boxed{}\left(e^x + \frac{1}{e^x} - 6\right) = \boxed{} \cdot 0$$

$$e^{2x} + 1 - 6e^x = 0$$

$$e^{2x} - 6e^x + 1 = 0 \qquad\qquad e^{2x} = \left(e^x\right)^2$$

$$\left(\boxed{}\right)^2 - 6e^x + 1 = 0$$

$$u^2 - 6 \cdot \boxed{} + 1 = 0 \qquad\qquad \text{Let } u = e^x.$$

$$a = 1 \qquad b = -6 \qquad c = 1$$

$$u = \frac{-b \pm \sqrt{b^2 - 4ac}}{2a}$$

$$= \frac{-\left(\boxed{}\right) \pm \sqrt{(-6)^2 - 4 \cdot 1 \cdot 1}}{2 \cdot 1}$$

$$= \frac{6 \pm \sqrt{36 - 4}}{2}$$

$$= \frac{6 \pm \sqrt{\boxed{}}}{2} = \frac{6 \pm \boxed{} \cdot \sqrt{2}}{2} = 3 \pm \boxed{}$$

$$e^x = 3 \pm 2\sqrt{2}$$

$$\ln e^x = \boxed{}\left(3 \pm 2\sqrt{2}\right)$$

$$\boxed{} \cdot \ln e = \ln\left(3 \pm 2\sqrt{2}\right)$$

$$x = \ln\left(3 \pm 2\sqrt{2}\right)$$

The solutions are approximately $\boxed{}$ and $\boxed{}$.

Example 6   Solve $\log_3 x = -2$.

$\log_3 x = -2$

$3^{\square} = x$

$\dfrac{1}{3^2} = x$

$\boxed{} = x$

Check:

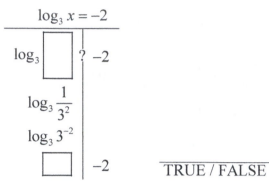

$$\begin{array}{c|c} \log_3 x = -2 \\ \hline \log_3 \boxed{} & ?\ -2 \\ \log_3 \dfrac{1}{3^2} \\ \log_3 3^{-2} \\ \boxed{} & -2 \end{array}$$

$\overline{\text{TRUE / FALSE}}$

The solution is $\dfrac{1}{9}$.

Example 7   Solve $\log x + \log(x+3) = 1$.

$\log x + \log(x+3) = 1$

$\log_{10}\left[x\left(\boxed{} \right) \right] = 1$

$x(x+3) = \boxed{}^1$

$x^2 + \boxed{} = 10$

$x^2 + 3x - 10 = 0$

$(x-2)\left(\boxed{} \right) = 0$

$x - 2 = 0 \quad or \quad x + 5 = 0$

$x = \boxed{} \quad or \quad x = \boxed{}$

Check: For 2:

$$\begin{array}{c|c} \log x + \log(x+3) = 1 \\ \hline \log\boxed{} + \log\left(\boxed{} + 3 \right) & ?\ 1 \\ \log 2 + \log 5 \\ \log(2 \cdot 5) \\ \log 10 \\ 1 & 1 \quad \text{TRUE} \end{array}$$

The solution is $\boxed{}$.

Check: For -5:

$$\begin{array}{c|c} \log x + \log(x+3) = 1 \\ \hline \log\left(\boxed{} \right) + \log(-5+3) & ?\ 1 \end{array}$$

We see that -5 $\underline{\hspace{2cm}}$ be a
 can / cannot

solution because the logarithmic function is not defined for negative values of x.

Example 8   Solve $\log_3(2x-1)-\log_3(x-4)=2$.

$$\log_3(2x-1)-\log_3(x-4)=2$$

$$\log_3\frac{2x-1}{\boxed{}}=2$$

$$\frac{2x-1}{x-4}=\boxed{}^2$$

$$\frac{2x-1}{x-4}=9$$

$$(x-4)\cdot\frac{2x-1}{x-4}=(x-4)\cdot 9$$

$$2x-1=9x-36$$

$$35=7x$$

$$5=x$$

Check:

$$\log_3(2x-1)-\log_3(x-4)=2$$

$$\overline{\log_3(2\cdot 5-1)-\log_3(5-4)\;?\;2}$$

$$\log_3\boxed{}-\log_3\boxed{}$$

$$2-\boxed{}$$

$$2\;|\;2\qquad\text{TRUE}$$

The solution is $\boxed{}$.

Example 9   Solve $\ln(4x+6)-\ln(x+5)=\ln x$.

$$\ln(4x+6)-\ln(x+5)=\ln x$$

$$\ln\frac{\boxed{}}{x+5}=\ln x$$

$$\frac{4x+6}{x+5}=\boxed{}$$

$$(x+5)\cdot\frac{4x+6}{x+5}=(x+5)\cdot x$$

$$4x+6=\boxed{}+5x$$

$$0=x^2+x-6$$

$$0=(x+3)\left(\boxed{}\right)$$

$$x+3=0\qquad or\quad x-2=0$$

$$x=\boxed{}\qquad or\qquad x=\boxed{}$$

The number -3 _____ a solution because $4(-3)+6=-6$ and $\ln(-6)$ is not a real
is / is not

number. The value $\boxed{}$ checks and is the solution.

Section 9.7 Applications and Models: Growth and Decay; Compound Interest

The **population growth function** is

$$P(t) = P_0 e^{kt}, \; k > 0,$$

where P_0 is the population at time 0, $P(t)$ is the population after time t, and k is the **exponential growth rate**.

Example 1   In 2013, the population of Ghana, located on the west coast of Africa, was about 25.2 million, and the exponential growth rate was 2.19% per year (*Source: CIA World Factbook, 2014*).

a) Find the exponential growth function.

$P(t) = 25.2e^{\boxed{} \cdot t}$, where t = number of years after 2013

b) Estimate the population in 2018.

$P\left(\boxed{}\right) = 25.2e^{0.0219\left(\boxed{}\right)} \approx 28.1$ million

c) After how long will the population be double what it was in 2013?

$$P(T) = 25.2e^{0.0219T}$$

$$\boxed{} = 25.2e^{0.0219T}$$

$$\boxed{} = e^{0.0219T}$$

$$\ln 2 = \ln e^{0.0219T}$$

$$\ln 2 = \boxed{}$$

$$\frac{\ln 2}{\boxed{}} = T$$

$$31.7 \approx T$$

The population of Ghana will be double what it was in 2013 about $\boxed{}$ years after 2013.

d) At this growth rate, when will the population be 40 million?

$$\boxed{} = 25.2e^{0.0219t}$$

$$\frac{40}{25.2} = e^{0.0219t}$$

$$\ln\left(\frac{40}{25.2}\right) = \boxed{}$$

$$\frac{\ln\left(\dfrac{40}{25.2}\right)}{\boxed{}} = t$$

$$21 \approx t$$

The population of Ghana will be 40 million about $\boxed{}$ years after 2013.

Interest Compounded Continuously

When an amount P_0 is invested in a savings account at interest rate r compounded continuously, the amount $P(t)$ in the account after t years is given by the function

$$P(t) = P_0 e^{kt}, \ k > 0.$$

Example 2 Suppose that $2000 is invested at interest rate k, compounded continuously, and grows to $2504.65 in 5 years.

a) What is the interest rate?

$$P(t) = 2000e^{kt}$$

$$\boxed{} = 2000e^{k \cdot \boxed{}}$$

$$\frac{2504.65}{2000} = e^{5k}$$

$$\ln\frac{2504.65}{2000} = \ln e^{5k}$$

$$\ln\frac{2504.65}{2000} = \boxed{}$$

$$\frac{\ln\dfrac{2504.65}{2000}}{\boxed{}} = k$$

$$0.045 \approx k$$

The interest rate is about $\boxed{}$.

b) Find the exponential growth function.

$$P(t) = P_0 e^{kt}$$

$$P(t) = \boxed{} e^{\boxed{} \cdot t}$$

c) What will the balance be after 10 years?

$$P(t) = 2000 e^{0.045t}$$

$$P(10) = 2000 e^{0.045(\boxed{})}$$

$$= 2000 e^{\boxed{}}$$

$$\approx 3136.62$$

The balance after 10 years is $\boxed{}$.

d) After how long will the $2000 have doubled?

$$\boxed{} = 2000 e^{0.045T}$$

$$\boxed{} = e^{0.045T}$$

$$\ln 2 = \ln e^{0.045T}$$

$$\ln 2 = \boxed{}$$

$$\frac{\ln 2}{\boxed{}} = T$$

$$15.4 \approx T$$

The investment of $2000 will double in about $\boxed{}$ years.

Growth Rate and Doubling Time

The **growth rate k** and the **doubling time T** are related by

$$kT = \ln 2, \text{ or } k = \frac{\ln 2}{T}, \text{ or } T = \frac{\ln 2}{k}.$$

Example 3   The population of the Philippines is now doubling every 37.7 years. What is the exponential growth rate?

$$k = \frac{\ln 2}{T}$$

$$k = \frac{\ln 2}{\boxed{}} \approx \boxed{} \approx 1.84\%$$

The growth rate of the population of the Philippines is about $\boxed{}$ per year.

Example 4   A lake is stocked with 400 fish of a new variety. The size of the lake, the availability of food, and the number of other fish restrict the growth of that type of fish in the lake to a limiting value of 2500. The population gets closer and closer to this limiting value, but never reaches it. The population of fish in the lake after time *t*, in months, is given by the function

$$P(t) = \frac{2500}{1 + 5.25e^{-0.32t}}.$$

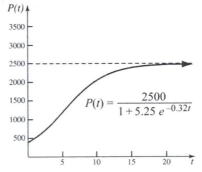

Note that this function increases toward a limiting value of 2500. The graph has $y = 2500$ as a horizontal asymptote. Find the population after 0, 1, 5, 10, 15, and 20 months.

$$P(0) = \frac{2500}{1 + 5.25e^{-0.32(\boxed{})}} = \boxed{}$$

$$P(1) \approx \boxed{}$$

$$P(5) \approx \boxed{}$$

$$P(10) \approx \boxed{}$$

$$P(15) \approx \boxed{}$$

$$P(20) \approx \boxed{}$$

The **exponential decay function** is

$$P(t) = P_0 e^{-kt}, \ k > 0,$$

where P_0 is the population at time t, and k is the **decay rate**.

Decay Rate and Half-Life

The decay rate k and the half-life T are related by

$$kT = \ln 2, \ \text{or} \ k = \frac{\ln 2}{T}, \ \text{or} \ T = \frac{\ln 2}{k}.$$

Example 5   The radioactive element carbon-14 has a half-life of 5750 years. The percentage of carbon-14 present in the remains of organic matter can be used to determine the age of that organic matter. Archaeologists discovered that the linen wrapping from one of the Dead Sea Scrolls had lost 22.3% of its carbon-14 at the time it was found. How old was the linen wrapping?

$$P(t) = P_0 e^{-kt}, \ k > 0$$

Find k:

$$k = \frac{\ln 2}{T} = \frac{\ln 2}{\boxed{}} \approx 0.00012$$

$$P(t) = P_0 e^{\boxed{} \cdot t}$$

$$\boxed{} = P_0 e^{-0.00012t} \qquad \text{(77.7\% of the carbon-14 remains.)}$$

$$0.777 P_0 = P_0 e^{-0.00012t}$$

$$0.777 = \boxed{}$$

$$\ln 0.777 = \ln e^{-0.00012t}$$

$$\ln 0.777 = \boxed{}$$

$$\frac{\ln 0.777}{\boxed{}} = t$$

$$2103 \approx t$$

The linen wrapping on the Dead Sea Scrolls was about $\boxed{}$ years old.

Section 10.1 Matrices and Systems of Equations

We can use matrices and Gaussian elimination to solve systems of equations. A **matrix** is a rectangular array of numbers. We can use the following operations to transform a matrix like

$$\begin{bmatrix} 2 & -1 & 4 & | & -3 \\ 1 & -2 & -10 & | & -6 \\ 3 & 0 & 4 & | & 7 \end{bmatrix}$$

to a row-equivalent matrix of the form

$$\begin{bmatrix} 1 & a & b & | & c \\ 0 & 1 & d & | & e \\ 0 & 0 & 1 & | & f \end{bmatrix}.$$

Row-Equivalent Operations

1. Interchange any two rows.
2. Multiply each entry in a row by the same nonzero constant.
3. Add a nonzero multiple of one row to another row.

Example 1   Solve the following system:

$$2x - y + 4z = -3,$$
$$x - 2y - 10z = -6,$$
$$3x + 4z = 7.$$

First we write the augmented matrix for this system of equations. This matrix consists of the coefficients of the x-, y-, and z-terms, along with the constant terms. We write a 0 for the missing y-term in the third equation.

$$\begin{bmatrix} 2 & -1 & 4 & | & \square \\ 1 & -2 & \square & | & -6 \\ 3 & \square & 4 & | & 7 \end{bmatrix}$$

Interchanging rows 1 and 2 puts a 1 in the upper left-hand corner and will help to simplify the process of putting the first column in the proper form.

$$\begin{bmatrix} 1 & \square & -10 & | & -6 \\ \square & -1 & 4 & | & -3 \\ 3 & 0 & 4 & | & \square \end{bmatrix} \begin{array}{l} \text{New row 1 = row 2} \\ \text{New row 2 = row 1} \\ \end{array}$$

Now we multiply the first row by −2 and add it to row 2, and then multiply the first row by −3 and add it to row 3.

$$\left[\begin{array}{ccc|c} 1 & -2 & -10 & -6 \\ \Box & 3 & 24 & \Box \\ 0 & \Box & 34 & \Box \end{array}\right]$$

New row 2 = −2(row 1) + row 2
New row 3 = −3(row 1) + row 3

To get a 1 in the second row, second column, multiply row 2 by $\frac{1}{3}$.

$$\left[\begin{array}{ccc|c} 1 & -2 & -10 & -6 \\ 0 & \Box & 8 & 3 \\ \Box & 6 & 34 & \Box \end{array}\right]$$

New row 2 = $\frac{1}{3}$(row 2)

We can get a 0 in the third row, second column by multiplying the second row by −6 and adding it to row 3.

$$\left[\begin{array}{ccc|c} \Box & -2 & -10 & \Box \\ 0 & \Box & 8 & 3 \\ 0 & \Box & -14 & 7 \end{array}\right]$$

New row 3 = −6(row 2) + row 3

Finally, we multiply the last row of the matrix by $-\frac{1}{14}$ to get a 1 in the third row, third column.

$$\left[\begin{array}{ccc|c} 1 & -2 & -10 & \Box \\ 0 & \Box & 8 & 3 \\ \Box & 0 & \Box & -\frac{1}{2} \end{array}\right]$$

New row 3 = $-\frac{1}{14}$(row 3)

Now we reinsert the variables and write the system of equations that corresponds to the last matrix.

$$x - 2y - 10z = -6, \quad (1)$$
$$\Box + 8z = 3, \quad (2)$$
$$z = \Box \quad (3)$$

Substitute $-\dfrac{1}{2}$ for z in equation (2) and solve for y.

$$y + 8\left(-\dfrac{1}{2}\right) = 3$$

$$y - 4 = 3$$

$$y = \boxed{}$$

Now substitute for y and z in equation (1) and solve for x.

$$x - 2 \cdot \boxed{} - 10\left(-\dfrac{1}{2}\right) = -6$$

$$\boxed{} - 14 + 5 = -6$$

$$x = \boxed{}$$

The solution of the system of equations is $\left(3, 7, -\dfrac{1}{2}\right)$.

The goal of Gaussian elimination is to find a matrix in the following form:

$$\left[\begin{array}{ccc|c} 1 & a & b & c \\ 0 & 1 & d & e \\ 0 & 0 & 1 & f \end{array}\right].$$

This is called **row-echelon form**.

Row-Echelon Form

1. If a row does not consist entirely of 0's, then the first nonzero element in the row is a 1 (called a **leading 1**).

2. For any two successive nonzero rows, the leading 1 in the lower row is farther to the right than the leading 1 in the higher row.

3. All the rows consisting entirely of 0's are at the bottom of the matrix.

If a fourth property is also satisfied, a matrix is said to be in **reduced row-echelon form**.

4. Each column that contains a leading 1 has 0's everywhere else.

Example 2   Which of the following matrices are in row-echelon form? Which, if any, are in reduced row-echelon form?

a) $\begin{bmatrix} 1 & -3 & 5 & | & -2 \\ 0 & 1 & -4 & | & 3 \\ 0 & 0 & 1 & | & 10 \end{bmatrix}$ b) $\begin{bmatrix} 0 & -1 & | & 2 \\ 0 & 1 & | & 5 \end{bmatrix}$ c) $\begin{bmatrix} 1 & -2 & -6 & 4 & | & 7 \\ 0 & 3 & 5 & -8 & | & -1 \\ 0 & 0 & 1 & 9 & | & 2 \end{bmatrix}$

d) $\begin{bmatrix} 1 & 0 & 0 & | & -2.4 \\ 0 & 1 & 0 & | & 0.8 \\ 0 & 0 & 1 & | & 5.6 \end{bmatrix}$ e) $\begin{bmatrix} 1 & 0 & 0 & 0 & | & 2/3 \\ 0 & 1 & 0 & 0 & | & -1/4 \\ 0 & 0 & 1 & 0 & | & 6/7 \\ 0 & 0 & 0 & 1 & | & 0 \end{bmatrix}$ f) $\begin{bmatrix} 1 & -4 & 2 & | & 5 \\ 0 & 0 & 0 & | & 0 \\ 0 & 1 & -3 & | & 8 \end{bmatrix}$

a) This matrix _____ in row-echelon form. It _____ in reduced
         is / is not   is / is not
    row-echelon form.

b) This matrix _____ in row-echelon form.
         is / is not

c) This matrix _____ in row-echelon form.
         is / is not

d) This matrix _____ in row-echelon form. It _____ in reduced
         is / is not   is / is not
    row-echelon form.

e) This matrix _____ in row-echelon form. It _____ in reduced
         is / is not   is / is not
    row-echelon form.

f) This matrix _____ in row-echelon form.
         is / is not

The goal of **Gauss-Jordan elimination** is to find a *reduced* row-echelon matrix that is equivalent to a given system of equations.

Example 3   Use Gauss-Jordan elimination to solve the system of equations

$$2x - y + 4z = -3, \quad (1)$$
$$x - 2y - 10z = -6 \quad (2)$$
$$3x \quad\;\; + 4z = 7. \quad (3)$$

Earlier we solved this system using Gaussian elimination and obtain the matrix

$$\begin{bmatrix} 1 & -2 & -10 & | & -6 \\ 0 & 1 & 8 & | & 3 \\ 0 & 0 & 1 & | & -1/2 \end{bmatrix}.$$

We continue to perform row-equivalent operations until we have a matrix in reduced row-echelon form.

$$\left[\begin{array}{ccc|c} 1 & -2 & \Box & -11 \\ 0 & 1 & \Box & 7 \\ 0 & 0 & \Box & -1/2 \end{array}\right] \quad \begin{array}{l} \text{New row } 1 = 10(\text{row } 3) + \text{row } 1 \\ \text{New row } 2 = -8(\text{row } 3) + \text{row } 2 \end{array}$$

$$\left[\begin{array}{ccc|c} 1 & \Box & 0 & \Box \\ 0 & \Box & 0 & \Box \\ \Box & 0 & 1 & \Box \end{array}\right] \quad \text{New row } 1 = 2(\text{row } 2) + \text{row } 1$$

The matrix is now in reduced row-echelon form. The system of equations that corresponds to the matrix is

$$\begin{aligned} x \quad &= 3, \\ y \quad &= \Box, \\ z &= -\frac{1}{2}. \end{aligned}$$

The solution is $\left(3, 7, -\dfrac{1}{2}\right)$.

Example 4 Solve the following system:

$$\begin{aligned} 3x - 4y - z &= 6 && (1) \\ 2x - y + z &= -1, && (2) \\ 4x - 7y - 3z &= 13. && (3) \end{aligned}$$

First we write the augmented matrix. We will use Gauss-Jordan elimination.

$$\left[\begin{array}{ccc|c} 3 & -4 & \Box & 6 \\ 2 & -1 & \Box & -1 \\ \Box & -7 & -3 & \Box \end{array}\right]$$

We then multiply the second and third rows by 3 so that each number in the first column below the first number, 3, is a multiple of that number.

$$\left[\begin{array}{ccc|c} 3 & -4 & -1 & 6 \\ 6 & \Box & 3 & \Box \\ 12 & \Box & \Box & 39 \end{array}\right] \quad \begin{array}{l} \text{New row } 2 = 3(\text{row } 2) \\ \text{New row } 3 = 3(\text{row } 3) \end{array}$$

Now to get zeroes below the 3 in the first column, we can multiply the first row by -2 and add it to the second row and multiply the first row by -4 and add it to the third row.

$$\left[\begin{array}{ccc|c} 3 & -4 & \Box & 6 \\ \Box & 5 & 5 & -15 \\ \Box & -5 & \Box & 15 \end{array}\right] \quad \begin{array}{l} \text{New row } 2 = -2(\text{row } 1) + \text{row } 2 \\ \text{New row } 3 = -4(\text{row } 1) + \text{row } 3 \end{array}$$

To get a zero in the second column of the third row, we can add the second row to the third row.

$$\left[\begin{array}{ccc|c} 3 & \Box & -1 & 6 \\ 0 & 5 & \Box & -15 \\ 0 & \Box & 0 & \Box \end{array}\right] \quad \text{New row } 3 = \text{row } 2 + \text{row } 3$$

We have a row of all zeros, so the equations are dependent, and the system is equivalent to the following system.

$$3x - 4y - z = 6,$$
$$5y + 5z = -15$$

This particular system has infinitely many solutions. (A system containing dependent equations could be inconsistent.) We solve the second equation for y.

$$5y + 5z = -15$$
$$5y = \boxed{} - 15$$
$$y = -z - \boxed{}$$

Now we substitute $-z - 3$ for y in the first equation and solve for x.

$$3x - 4(-z - 3) - z = 6$$
$$3x + 4z + \boxed{} - z = 6$$
$$3x + 3z + 12 = 6$$
$$3x = -3z - \boxed{}$$
$$x = -z - \boxed{}$$

The solutions are of the form $\left(-z - 2, \ -z - 3, \ \boxed{}\right)$.

Section 10.2 Matrix Operations

We generally use a capital letter to represent a matrix and a lower-case letter with double subscripts to represent the entries. For example, a_{34} represents the entry which is in the third row and fourth column. The general term of a matrix is represented by a_{ij}, and we can write a matrix as

$$\mathbf{A} = \begin{bmatrix} a_{ij} \end{bmatrix} = \begin{bmatrix} a_{11} & a_{12} & a_{13} & \cdots & a_{1n} \\ a_{21} & a_{22} & a_{23} & \cdots & a_{2n} \\ a_{31} & a_{32} & a_{33} & \cdots & a_{3n} \\ \vdots & \vdots & \vdots & & \vdots \\ a_{m1} & a_{m2} & a_{m3} & \cdots & a_{mn} \end{bmatrix}.$$

A matrix with m rows and n columns (as above) has order $m \times n$.

Addition and Subtraction of Matrices

Given two $m \times n$ matrices $\mathbf{A} = \begin{bmatrix} a_{ij} \end{bmatrix}$ and $\mathbf{B} = \begin{bmatrix} b_{ij} \end{bmatrix}$, their sum is

$$\mathbf{A} + \mathbf{B} = \begin{bmatrix} a_{ij} + b_{ij} \end{bmatrix}$$

and their difference is

$$\mathbf{A} - \mathbf{B} = \begin{bmatrix} a_{ij} - b_{ij} \end{bmatrix}.$$

Example 1 Find $\mathbf{A} + \mathbf{B}$ for each of the following.

a) $\mathbf{A} = \begin{bmatrix} -5 & 0 \\ 4 & 1/2 \end{bmatrix}$, $\mathbf{B} = \begin{bmatrix} 6 & -3 \\ 2 & 3 \end{bmatrix}$

b) $\mathbf{A} = \begin{bmatrix} 1 & 3 \\ -1 & 5 \\ 6 & 0 \end{bmatrix}$, $\mathbf{B} = \begin{bmatrix} -1 & -2 \\ 1 & -2 \\ -3 & 1 \end{bmatrix}$

a) **A** and **B** have the same order, 2×2, so we can add them.

$$\mathbf{A} + \mathbf{B} = \begin{bmatrix} -5+6 & 0+(-3) \\ 4+\square & 1/2+\square \end{bmatrix} = \begin{bmatrix} 1 & \square \\ 6 & \square \end{bmatrix}$$

b) **A** and **B** have the same order, 3×2, so we can add them.

$$\mathbf{A} + \mathbf{B} = \begin{bmatrix} 1+(-1) & \square+(-2) \\ -1+\square & 5+(-2) \\ \square+(-3) & 0+\square \end{bmatrix} = \begin{bmatrix} \square & 1 \\ 0 & \square \\ 3 & 1 \end{bmatrix}$$

Example 2 Find $C - D$ for each of the following.

a) $C = \begin{bmatrix} 1 & 2 \\ -2 & 0 \\ -3 & -1 \end{bmatrix}$, $D = \begin{bmatrix} 1 & -1 \\ 1 & 3 \\ 2 & 3 \end{bmatrix}$

b) $C = \begin{bmatrix} 5 & -6 \\ -3 & 4 \end{bmatrix}$, $D = \begin{bmatrix} -4 \\ 1 \end{bmatrix}$

a) C and D have the same order, 3×2, so we can subtract them.

$$C - D = \begin{bmatrix} 1 - \square & 2 - (-1) \\ -2 - \square & 0 - 3 \\ \square - 2 & -1 - \square \end{bmatrix} = \begin{bmatrix} \square & 3 \\ -3 & \square \\ -5 & \square \end{bmatrix}$$

b) C is a 2×2 matrix, and D is a 2×1 matrix. Since the order of C _____ the order

$$= / \neq$$

of D, we _____ subtract.

can / cannot

Example 3 Find $-A$ and $A + (-A)$ for

$$A = \begin{bmatrix} 1 & 0 & 2 \\ 3 & -1 & 5 \end{bmatrix}.$$

To find $-A$, we replace each entry of A by its opposite.

$$-A = \begin{bmatrix} -1 & \square & -2 \\ \square & 1 & \square \end{bmatrix}$$

$$A + (-A) = \begin{bmatrix} 1 & 0 & 2 \\ 3 & -1 & 5 \end{bmatrix} + \begin{bmatrix} -1 & 0 & -2 \\ -3 & 1 & -5 \end{bmatrix}$$

$$= \begin{bmatrix} 0 & \square & 0 \\ \square & 0 & \square \end{bmatrix}$$

Scalar Product

The **scalar product** of a number k and a matrix A is the matrix denoted by kA, obtained by multiplying each entry of A by the number k. The number k is called a **scalar**.

Example 4   Find $3\mathbf{A}$ and $(-1)\mathbf{A}$ for

$$\mathbf{A} = \begin{bmatrix} -3 & 0 \\ 4 & 5 \end{bmatrix}.$$

$$3\mathbf{A} = 3\begin{bmatrix} -3 & 0 \\ 4 & 5 \end{bmatrix} = \begin{bmatrix} \boxed{}\cdot(-3) & 3\cdot\boxed{} \\ 3\cdot\boxed{} & \boxed{}\cdot5 \end{bmatrix} = \begin{bmatrix} -9 & \boxed{} \\ 12 & \boxed{} \end{bmatrix}$$

$$(-1)\mathbf{A} = (-1)\begin{bmatrix} -3 & 0 \\ 4 & 5 \end{bmatrix} = \begin{bmatrix} 3 & \boxed{} \\ \boxed{} & -5 \end{bmatrix}$$

Example 5   Waterworks, Inc., manufactures three types of kayaks in its two plants. The table below lists the number of each style produced at each plant in April.

	Whitewater Kayak	Ocean Kayak	Crossover Kayak
Madison Plant	150	120	100
Greensburg Plant	180	90	130

a) Write a 2×3 matrix $\mathbf{A}$ that represents the information in the table.

b) The manufacturer increased production by 20% in May. Find a matrix $\mathbf{M}$ that represents the increased production figures.

c) Find the matrix $\mathbf{A}+\mathbf{M}$ and tell what it represents.

a) We record the information from the table in the matrix $\mathbf{A}$ below.

$$\mathbf{A} = \begin{bmatrix} 150 & \boxed{} & 100 \\ \boxed{} & 90 & \boxed{} \end{bmatrix}.$$

b) We write a matrix $\mathbf{M}$ that represents an increase of 20% in production.

$$\mathbf{M} = \mathbf{A} + 20\% \cdot \mathbf{A} = \mathbf{A} + 0.2\mathbf{A} = \boxed{}\mathbf{A}$$

$$= 1.2 \cdot \begin{bmatrix} 150 & 120 & 100 \\ 180 & 90 & 130 \end{bmatrix} = \begin{bmatrix} 180 & \boxed{} & 120 \\ \boxed{} & 108 & \boxed{} \end{bmatrix}.$$

c) We find the sum $\mathbf{A}+\mathbf{M}$.

$$\mathbf{A}+\mathbf{M} = \begin{bmatrix} 330 & \boxed{} & 220 \\ \boxed{} & 198 & \boxed{} \end{bmatrix}.$$

The sum $\mathbf{A}+\mathbf{M}$ represents the number of each type of kayak produced in each plant during $\boxed{}$ and $\boxed{}$.

Matrix Multiplication

For an $m \times n$ matrix $A = \begin{bmatrix} a_{ij} \end{bmatrix}$ and an $n \times p$ matrix $B = \begin{bmatrix} b_{ij} \end{bmatrix}$, the **product** $AB = \begin{bmatrix} c_{ij} \end{bmatrix}$ is an $m \times p$ matrix, where

$$c_{ij} = a_{i1} \cdot b_{1j} + a_{i2} \cdot b_{2j} + a_{i3} \cdot b_{3j} + \cdots + a_{in} \cdot b_{nj}$$

Example 6   For

$$A = \begin{bmatrix} 3 & 1 & -1 \\ 2 & 0 & 3 \end{bmatrix}, \quad B = \begin{bmatrix} 1 & 6 \\ 3 & -5 \\ -2 & 4 \end{bmatrix}, \quad C = \begin{bmatrix} 4 & -6 \\ 1 & 2 \end{bmatrix},$$

find each of the following.

a) **AB**   b) **BA**   c) **BC**   d) **AC**

a) We first note that **A** has $\boxed{}$ columns, and **B** has $\boxed{}$ rows. Therefore, we

$\overline{\text{can perform this multiplication / cannot perform this multiplication}}$.

$$AB = \begin{bmatrix} 3 & 1 & -1 \\ 2 & 0 & 3 \end{bmatrix} \cdot \begin{bmatrix} 1 & 6 \\ 3 & -5 \\ -2 & 4 \end{bmatrix}$$

$$= \begin{bmatrix} 3 \cdot 1 + \boxed{} \cdot \boxed{} + (-1) \cdot (-2) & 3 \cdot 6 + \boxed{} \cdot (-5) + (-1) \cdot \boxed{} \\ 2 \cdot \boxed{} + 0 \cdot 3 + \boxed{} \cdot (-2) & \boxed{} \cdot 6 + 0 \cdot (-5) + \boxed{} \cdot 4 \end{bmatrix}$$

$$= \begin{bmatrix} 8 & \boxed{} \\ \boxed{} & 24 \end{bmatrix}$$

b) We first note that **B** has $\boxed{}$ columns, and **A** has $\boxed{}$ rows. Therefore, we

$\overline{\text{can perform this multiplication / cannot perform this multiplication}}$.

$$BA = \begin{bmatrix} 1 & 6 \\ 3 & -5 \\ -2 & 4 \end{bmatrix} \cdot \begin{bmatrix} 3 & 1 & -1 \\ 2 & 0 & 3 \end{bmatrix}$$

$$= \begin{bmatrix} 1 \cdot 3 + \boxed{} \cdot 2 & 1 \cdot 1 + 6 \cdot \boxed{} & 1 \cdot (-1) + \boxed{} \cdot \boxed{} \\ 3 \cdot \boxed{} + (-5) \cdot \boxed{} & 3 \cdot \boxed{} + (-5) \cdot 0 & \boxed{} \cdot (-1) + (-5) \cdot \boxed{} \\ -2 \cdot 3 + \boxed{} \cdot 2 & \boxed{} \cdot 1 + 4 \cdot \boxed{} & -2 \cdot (-1) + \boxed{} \cdot 3 \end{bmatrix}$$

$$= \begin{bmatrix} 15 & \boxed{} & 17 \\ -1 & 3 & \boxed{} \\ \boxed{} & -2 & \boxed{} \end{bmatrix}$$

c) We first note that **B** has [   ] columns, and **C** has [   ] rows. Therefore, we

can perform this multiplication / cannot perform this multiplication .

$$\mathbf{BC} = \begin{bmatrix} 1 & 6 \\ 3 & -5 \\ -2 & 4 \end{bmatrix} \cdot \begin{bmatrix} 4 & -6 \\ 1 & 2 \end{bmatrix}$$

$$= \begin{bmatrix} 1\cdot 4 + \boxed{}\cdot 1 & \boxed{}\cdot(-6) + 6\cdot\boxed{} \\ 3\cdot\boxed{} + (-5)\cdot\boxed{} & 3\cdot(-6) + (-5)\cdot\boxed{} \\ -2\cdot\boxed{} + \boxed{}\cdot 1 & \boxed{}\cdot(-6) + \boxed{}\cdot 2 \end{bmatrix}$$

$$= \begin{bmatrix} 10 & \boxed{} \\ 7 & -28 \\ \boxed{} & \boxed{} \end{bmatrix}$$

d) We first note that **A** has [   ] columns, and **C** has [   ] rows. Therefore, we

can perform this multiplication / cannot perform this multiplication .

Thus, **AC** _____ defined.
    is / is not

Example 7   Two of the items sold at Sweet Treats Bakery are gluten-free bagels and gluten-free doughnuts. The table below lists the number of dozens of each product that are sold at the bakery's three stores one week.

	Main Street Store	Avon Road Store	Dalton Avenue Store
Bagels (in dozens)	25	30	20
Doughnuts (in dozens)	40	35	15

The bakery's profit on one dozen bagels is $5, and its profit on one dozen doughnuts is $6. Use matrices to find the total profit on these items at each store for the given week.

We can write the information from the table above in a matrix **S**.

$$\mathbf{S} = \begin{bmatrix} 25 & \boxed{} & 20 \\ 40 & \boxed{} & \boxed{} \end{bmatrix}$$

We can also write the information about the profit in matrix form.

$$\mathbf{P} = \begin{bmatrix} 5 & \boxed{} \end{bmatrix}$$

To find the total profit in each of the three stores, we find the product of the two matrices, **PS**.

$$\mathbf{PS} = \begin{bmatrix} 5 & 6 \end{bmatrix} \cdot \begin{bmatrix} 25 & 30 & \square \\ \square & 35 & \square \end{bmatrix}$$

$$= \begin{bmatrix} 5 \cdot \square + 6 \cdot \square & 5 \cdot 30 + \square \cdot 35 & 5 \cdot \square + 6 \cdot \square \end{bmatrix}$$

$$= \begin{bmatrix} 125 + \square & \square + 210 & 100 + \square \end{bmatrix}$$

$$= \begin{bmatrix} 365 & \square & 190 \end{bmatrix}$$

At the Main Street store the total profit was $365 for the given week, at the Avon Road store the total profit was $\square$, and at the Dalton Avenue store the total profit was $\square$.

Example 8   Write a matrix equation equivalent to the following system of equations:

$$4x + 2y - z = 3,$$
$$9x \quad\quad + z = 5,$$
$$4x + 5y - 2z = 1.$$

First, we can form a matrix of the coefficients from the equations above.

$$\mathbf{A} = \begin{bmatrix} 4 & \square & -1 \\ 9 & 0 & \square \\ \square & 5 & -2 \end{bmatrix}$$

Next we write a variable matrix.

$$\mathbf{X} = \begin{bmatrix} x \\ y \\ z \end{bmatrix}$$

Finally we write a constant matrix.

$$\mathbf{B} = \begin{bmatrix} 3 \\ \square \\ 1 \end{bmatrix}$$

Matrix equation: $\mathbf{AX} = \square$

Section 10.3 Inverses of Matrices

Identity Matrix

For any positive integer n, the $n \times n$ **identity matrix** is an $n \times n$ matrix with 1's on the main diagonal and 0's elsewhere and is denoted by

$$I = \begin{bmatrix} 1 & 0 & 0 & \cdots & 0 \\ 0 & 1 & 0 & \cdots & 0 \\ 0 & 0 & 1 & \cdots & 0 \\ \vdots & \vdots & \vdots & & \vdots \\ 0 & 0 & 0 & \cdots & 1 \end{bmatrix}.$$

Then $AI = IA = A$, for any $n \times n$ matrix A.

Example 1 For

$$A = \begin{bmatrix} 4 & -7 \\ -3 & 2 \end{bmatrix} \text{ and } I = \begin{bmatrix} 1 & 0 \\ 0 & 1 \end{bmatrix},$$

find each of the following.

a) **AI**   b) **IA**

a) $AI = \begin{bmatrix} 4 & -7 \\ -3 & 2 \end{bmatrix} \cdot \begin{bmatrix} 1 & 0 \\ 0 & 1 \end{bmatrix} = \begin{bmatrix} 4 \cdot \square + (-7) \cdot \square & 4 \cdot \square + (-7) \cdot \square \\ \square \cdot 1 + \square \cdot 0 & -3 \cdot \square + \square \cdot 1 \end{bmatrix}$

$ = \begin{bmatrix} 4 & \square \\ \square & 2 \end{bmatrix}$

b) $IA = \begin{bmatrix} 1 & 0 \\ 0 & 1 \end{bmatrix} \cdot \begin{bmatrix} 4 & -7 \\ -3 & 2 \end{bmatrix} = \begin{bmatrix} \square & -7 \\ -3 & \square \end{bmatrix}$

Therefore, $AI = IA = \boxed{}$.

Inverse of a Matrix

For an $n \times n$ matrix A, if there is a matrix A^{-1} for which $A^{-1} \cdot A = I = A \cdot A^{-1}$, then A^{-1} is the **inverse** of A.

Example 2   Verify that

$$\mathbf{B} = \begin{bmatrix} 4 & -3 \\ 3 & -2 \end{bmatrix} \text{ is the inverse of } \mathbf{A} = \begin{bmatrix} -2 & 3 \\ -3 & 4 \end{bmatrix}.$$

Show that $\mathbf{BA} = \mathbf{I} = \mathbf{AB}$.

$$\mathbf{BA} = \begin{bmatrix} 4 & -3 \\ 3 & -2 \end{bmatrix} \cdot \begin{bmatrix} -2 & 3 \\ -3 & 4 \end{bmatrix} = \begin{bmatrix} 4(-2)+(-3)(-3) & 4(\boxed{})+(-3)(4) \\ 3(-2)+(-2)(\boxed{}) & 3(3)+(-2)(\boxed{}) \end{bmatrix}$$

$$= \begin{bmatrix} 1 & \boxed{} \\ \boxed{} & 1 \end{bmatrix} = \mathbf{I}$$

$$\mathbf{AB} = \begin{bmatrix} -2 & 3 \\ -3 & 4 \end{bmatrix} \cdot \begin{bmatrix} 4 & -3 \\ 3 & -2 \end{bmatrix} = \begin{bmatrix} -2(\boxed{})+3(3) & -2(-3)+\boxed{}(-2) \\ -3(4)+4(\boxed{}) & -3(-3)+4(\boxed{}) \end{bmatrix}$$

$$= \begin{bmatrix} \boxed{} & 0 \\ 0 & \boxed{} \end{bmatrix} = \mathbf{I}$$

To find the inverse of a square matrix $\mathbf{A}$ we first form the **augmented matrix** consisting of $\mathbf{A}$ on the left side and the identity matrix with the same dimensions on the right side. The goal is to use row-equivalent operations in order to transform this matrix to an augmented matrix with the identity matrix on the left side. Then, the inverse of $\mathbf{A}$ is the matrix on the right side. For example,

$$\mathbf{A} = \begin{bmatrix} -2 & 3 \\ -3 & 4 \end{bmatrix} \xrightarrow[\substack{\text{augment} \\ \text{matrix}}]{} \left[\begin{array}{cc|cc} -2 & 3 & 1 & 0 \\ -3 & 4 & 0 & 1 \end{array}\right] \xrightarrow[\substack{\text{row} \\ \text{operations}}]{} \left[\begin{array}{cc|cc} 1 & 0 & a & b \\ 0 & 1 & c & d \end{array}\right] \longrightarrow \mathbf{A}^{-1} = \begin{bmatrix} a & b \\ c & d \end{bmatrix}.$$

Example 3   Find $\mathbf{A}^{-1}$, where

$$\mathbf{A} = \begin{bmatrix} -2 & 3 \\ -3 & 4 \end{bmatrix}.$$

We start by forming the augmented matrix.

$$\left[\begin{array}{cc|cc} -2 & \boxed{} & 1 & \boxed{} \\ -3 & \boxed{} & \boxed{} & 1 \end{array}\right]$$

Our goal is to transform this matrix to one of the form

$$\left[\begin{array}{cc|cc} 1 & 0 & a & b \\ 0 & 1 & c & d \end{array}\right].$$

First, we will get a 1 in the upper left corner of the matrix.

$$\begin{bmatrix} 1 & -3/2 & | & -1/2 & \Box \\ -3 & 4 & | & \Box & \Box \end{bmatrix} \qquad \text{New row } 1 = -\frac{1}{2}(\text{row } 1)$$

Next, we want a 0 in the entry in the lower left corner.

$$\begin{bmatrix} 1 & -3/2 & | & -1/2 & 0 \\ 0 & -1/2 & | & \Box & \Box \end{bmatrix} \qquad \text{New row } 2 = 3(\text{row } 1) + \text{row } 2$$

Now we want a 1 in row 2, column 2.

$$\begin{bmatrix} 1 & \Box & | & -1/2 & 0 \\ 0 & 1 & | & \Box & \Box \end{bmatrix} \qquad \text{New row } 2 = -2(\text{row } 2)$$

Finally, we want a 0 in row 1, column 2.

$$\begin{bmatrix} 1 & 0 & | & \Box & -3 \\ 0 & 1 & | & 3 & \Box \end{bmatrix} \qquad \text{New row } 1 = \frac{3}{2}(\text{row } 2) + \text{row } 1$$

Thus,

$$\mathbf{A}^{-1} = \begin{bmatrix} 4 & \Box \\ \Box & -2 \end{bmatrix}.$$

The $\boxed{x^{-1}}$ key on a graphing calculator can also be used to find the inverse of a matrix.

Example 4 Find $\mathbf{A}^{-1}$, where

$$\mathbf{A} = \begin{bmatrix} 1 & 2 & -1 \\ 3 & 5 & 3 \\ 2 & 4 & 3 \end{bmatrix}.$$

We start by forming the augmented matrix.

$$\begin{bmatrix} 1 & 2 & -1 & | & 1 & \Box & \Box \\ 3 & 5 & 3 & | & \Box & 1 & \Box \\ 2 & 4 & 3 & | & \Box & \Box & 1 \end{bmatrix}$$

Our goal is to transform this matrix to one of the form

$$\begin{bmatrix} 1 & 0 & 0 & | & a & b & c \\ 0 & 1 & 0 & | & d & e & f \\ 0 & 0 & 1 & | & g & h & i \end{bmatrix}.$$

$$-3R_1 + R_2 \rightarrow \quad -2R_1 + R_3 \rightarrow \begin{bmatrix} 1 & 2 & -1 & 1 & 0 & 0 \\ \square & -1 & 6 & \square & 1 & 0 \\ 0 & \square & 5 & -2 & \square & 1 \end{bmatrix}$$

$$\frac{1}{5}R_3 \rightarrow \begin{bmatrix} 1 & 2 & -1 & 1 & 0 & 0 \\ 0 & -1 & 6 & -3 & 1 & 0 \\ 0 & 0 & \square & -2/5 & 0 & \square \end{bmatrix}$$

$$R_3 + R_1 \rightarrow \quad -6R_3 + R_2 \rightarrow \begin{bmatrix} 1 & 2 & \square & 3/5 & \square & 1/5 \\ 0 & -1 & \square & -3/5 & 1 & -6/5 \\ 0 & 0 & 1 & -2/5 & 0 & 1/5 \end{bmatrix}$$

$$2R_2 + R_1 \rightarrow \begin{bmatrix} 1 & 0 & \square & -3/5 & \square & -11/5 \\ 0 & -1 & 0 & -3/5 & 1 & -6/5 \\ 0 & 0 & 1 & -2/5 & 0 & 1/5 \end{bmatrix}$$

$$-1 \cdot R_2 \rightarrow \begin{bmatrix} 1 & 0 & 0 & -3/5 & 2 & -11/5 \\ 0 & \square & 0 & 3/5 & -1 & 6/5 \\ 0 & 0 & 1 & -2/5 & 0 & 1/5 \end{bmatrix}$$

We have

$$\mathbf{A}^{-1} = \begin{bmatrix} -3/5 & 2 & -11/5 \\ 3/5 & -1 & 6/5 \\ -2/5 & 0 & 1/5 \end{bmatrix}.$$

We can write a system of n linear equations in n variables as a matrix equation $\mathbf{AX} = \mathbf{B}$. If $\mathbf{A}$ has an inverse, then the system of equations has a unique solution that can be found by solving for $\mathbf{X}$, as follows:

$$\mathbf{AX} = \mathbf{B}$$
$$\mathbf{A}^{-1}(\mathbf{AX}) = \mathbf{A}^{-1}\mathbf{B}$$
$$(\mathbf{A}^{-1}\mathbf{A})\mathbf{X} = \mathbf{A}^{-1}\mathbf{B}$$
$$\mathbf{IX} = \mathbf{A}^{-1}\mathbf{B}$$
$$\mathbf{X} = \mathbf{A}^{-1}\mathbf{B}.$$

Example 5   Use an inverse matrix to solve the following system of equations:

$$-2x + 3y = 4,$$
$$-3x + 4y = 5.$$

First, we can write this as a matrix equation $\mathbf{AX} = \mathbf{B}$.

$$\begin{bmatrix} -2 & 3 \\ \square & \square \end{bmatrix} \cdot \begin{bmatrix} x \\ y \end{bmatrix} = \begin{bmatrix} 4 \\ \square \end{bmatrix}.$$

Then $\mathbf{X} = \mathbf{A}^{-1}\mathbf{B}$. Earlier, we found $\mathbf{A}^{-1}$. We have

$$\mathbf{A}^{-1} = \begin{bmatrix} 4 & -3 \\ 3 & -2 \end{bmatrix}.$$

Then

$$\begin{bmatrix} x \\ y \end{bmatrix} = \begin{bmatrix} 4 & -3 \\ 3 & -2 \end{bmatrix} \cdot \begin{bmatrix} 4 \\ \square \end{bmatrix} = \begin{bmatrix} 1 \\ \square \end{bmatrix}.$$

Thus, the solution of the system is $\left(\square, \square \right)$.

Section 10.4   Determinants and Cramer's Rule

With every square matrix, we associate a number called its determinant.

Determinant of a 2×2 Matrix

The **determinant** of the matrix $\begin{bmatrix} a & c \\ b & d \end{bmatrix}$ is denoted $\begin{vmatrix} a & c \\ b & d \end{vmatrix}$ and is defined as

$$\begin{vmatrix} a & c \\ b & d \end{vmatrix} = ad - bc.$$

Example 1   Evaluate $\begin{vmatrix} \sqrt{2} & -3 \\ -4 & -\sqrt{2} \end{vmatrix}$.

$$\begin{vmatrix} \sqrt{2} & -3 \\ -4 & -\sqrt{2} \end{vmatrix} = \left(\sqrt{2}\right)\left(-\sqrt{2}\right) - \left(\boxed{}\right)(-3)$$

$$= \boxed{} - (12) = \boxed{}$$

Often we first find minors and cofactors of matrices in order to evaluate determinants.

Minor

For a square matrix $\mathbf{A} = \left[a_{ij} \right]$, the **minor** M_{ij} of an entry a_{ij} is the determinant of the matrix formed by deleting the ith row and the jth column of $\mathbf{A}$.

Example 2   For the matrix

$$\mathbf{A} = \left[a_{ij} \right] = \begin{bmatrix} -8 & 0 & 6 \\ 4 & -6 & 7 \\ -1 & -3 & 5 \end{bmatrix},$$

find each of the following.

a) M_{11} b) M_{23}

a)   The minor M_{11} is the determinant of the matrix formed by eliminating the first row and the first column of $\mathbf{A}$.

$$M_{11} = \begin{vmatrix} -6 & 7 \\ -3 & 5 \end{vmatrix} = -6 \cdot \boxed{} - (-3) \cdot \boxed{}$$

$$= \boxed{} - (-21) = -30 + 21 = \boxed{}$$

b) The minor M_{23} is the determinant of the matrix formed by eliminating the second row and third column of **A**.

$$M_{23} = \begin{vmatrix} -8 & \boxed{} \\ -1 & \boxed{} \end{vmatrix} = \boxed{} \cdot (-3) - (-1) \cdot \boxed{}$$

$$= 24 - \boxed{} = \boxed{}$$

Cofactor

For a square matrix $\mathbf{A} = \left[a_{ij} \right]$, the **cofactor** A_{ij} of an entry a_{ij} is given by

$$A_{ij} = (-1)^{i+j} M_{ij},$$

where M_{ij} is the minor of a_{ij}.

Example 3   For the matrix

$$\mathbf{A} = \left[a_{ij} \right] = \begin{bmatrix} -8 & 0 & 6 \\ 4 & -6 & 7 \\ -1 & -3 & 5 \end{bmatrix}$$

find each of the following.

a) A_{11} b) A_{23}

a) $A_{11} = (-1)^{1+1} \cdot \begin{vmatrix} -6 & 7 \\ -3 & 5 \end{vmatrix} = 1 \cdot \left[(-6) \cdot \boxed{} - (-3) \cdot \boxed{} \right]$

$= \boxed{} + 21 = \boxed{}$

b) $A_{23} = (-1)^{2+3} \cdot M_{23} = (-1)^5 \begin{vmatrix} -8 & \boxed{} \\ -1 & \boxed{} \end{vmatrix} = -1 \cdot \left[\boxed{} \cdot (-3) - (-1) \cdot \boxed{} \right]$

$= -1 \cdot (24) = \boxed{}$

Determinant of Any Square Matrix

For any square matrix $\mathbf{A}$ of order $n \times n$ $(n > 1)$, we define the **determinant** of $\mathbf{A}$, denoted $|\mathbf{A}|$, as follows. Choose any row or column. Multiply each element in that row or column by its cofactor and add the results. The determinant of a 1×1 matrix is simply the element of the matrix. The value of a determinant will be the same no matter which row or column is chosen.

Example 4   Evaluate $|\mathbf{A}|$ by expanding across the third row.

$$\mathbf{A} = \begin{bmatrix} -8 & 0 & 6 \\ 4 & -6 & 7 \\ -1 & -3 & 5 \end{bmatrix}$$

The entries in the third row are -1, -3, and 5, so

$|\mathbf{A}| = (-1)A_{31} + (-3)A_{32} + 5A_{33}$

$= (-1)(-1)^{3+1}\begin{vmatrix} 0 & 6 \\ -6 & 7 \end{vmatrix} + (-3)(-1)^{3+2}\begin{vmatrix} -8 & \square \\ 4 & \square \end{vmatrix} + 5(-1)^{3+3}\begin{vmatrix} -8 & \square \\ 4 & \square \end{vmatrix}$

$= -1 \cdot \left[0 \cdot 7 - (-6) \cdot \square\right] + 3 \cdot \left[-8 \cdot \square - 4 \cdot \square\right] + 5 \cdot \left[\square \cdot (-6) - 4 \cdot \square\right]$

$= -36 - \square + 240 = \square$

Cramer's Rule for 2×2 Systems

The solution of the system of equations

$$a_1 x + b_1 y = c_1,$$
$$a_2 x + b_2 y = c_2$$

is given by

$$x = \frac{D_x}{D}, \quad y = \frac{D_y}{D},$$

where

$$D = \begin{vmatrix} a_1 & b_1 \\ a_2 & b_2 \end{vmatrix}, \quad D_x = \begin{vmatrix} c_1 & b_1 \\ c_2 & b_2 \end{vmatrix}, \quad D_y = \begin{vmatrix} a_1 & c_1 \\ a_2 & c_2 \end{vmatrix}, \text{ and } D \neq 0.$$

Example 5 Solve using Cramer's rule:

$$2x + 5y = 7,$$
$$5x - 2y = -3.$$

First we need to calculate D, D_x, and D_y.

$$D = \begin{vmatrix} 2 & \boxed{} \\ 5 & \boxed{} \end{vmatrix} = -4 - 25 = \boxed{}$$

$$D_x = \begin{vmatrix} 7 & 5 \\ \boxed{} & -2 \end{vmatrix} = -14 + \boxed{} = 1$$

$$D_y = \begin{vmatrix} 2 & \boxed{} \\ 5 & \boxed{} \end{vmatrix} = -6 - 35 = \boxed{}$$

Cramer's rule tells us that

$$x = \frac{D_x}{D} = \frac{1}{-29} = -\frac{\boxed{}}{\boxed{}}, \qquad y = \frac{D_y}{D} = \frac{-41}{-29} = \frac{\boxed{}}{\boxed{}}.$$

The solution of the system of equations is $\left(-\dfrac{1}{29}, \boxed{}\right)$.

We can also use a graphing calculator to solve this system of equations using Cramer's rule.

Cramer's Rule for 3×3 Systems

The solution of the system of equations

$$a_1 x + b_1 y + c_1 z = d_1,$$
$$a_2 x + b_2 y + c_2 z = d_2,$$
$$a_3 x + b_3 y + c_3 z = d_3$$

is given by

$$x = \frac{D_x}{D}, \qquad y = \frac{D_y}{D}, \qquad z = \frac{D_z}{D},$$

where

$$D = \begin{vmatrix} a_1 & b_1 & c_1 \\ a_2 & b_2 & c_2 \\ a_3 & b_3 & c_3 \end{vmatrix}, \quad D_x = \begin{vmatrix} d_1 & b_1 & c_1 \\ d_2 & b_2 & c_2 \\ d_3 & b_3 & c_3 \end{vmatrix}, \quad D_y = \begin{vmatrix} a_1 & d_1 & c_1 \\ a_2 & d_2 & c_2 \\ a_3 & d_3 & c_3 \end{vmatrix}, \quad D_z = \begin{vmatrix} a_1 & b_1 & d_1 \\ a_2 & b_2 & d_2 \\ a_3 & b_3 & d_3 \end{vmatrix}, \text{ and } D \neq 0.$$

Example 6 Solve using Cramer's rule:

$$x - 3y + 7z = 13,$$
$$x + y + z = 1,$$
$$x - 2y + 3z = 4.$$

$$D = \begin{vmatrix} 1 & \boxed{} & 7 \\ \boxed{} & 1 & 1 \\ 1 & -2 & 3 \end{vmatrix} = -10 \qquad\qquad D_y = \begin{vmatrix} 1 & \boxed{} & 7 \\ 1 & 1 & 1 \\ 1 & \boxed{} & 3 \end{vmatrix} = -6$$

$$D_x = \begin{vmatrix} 13 & -3 & \boxed{} \\ 1 & \boxed{} & 1 \\ \boxed{} & -2 & 3 \end{vmatrix} = 20 \qquad\qquad D_z = \begin{vmatrix} 1 & -3 & 13 \\ 1 & 1 & 1 \\ \boxed{} & -2 & \boxed{} \end{vmatrix} = -24$$

$$x = \frac{D_x}{D} = \frac{20}{\boxed{}} = \boxed{}, \qquad y = \frac{D_y}{D} = \frac{\boxed{}}{-10} = \frac{3}{5}, \qquad z = \frac{D_z}{D} = \frac{\boxed{}}{-10} = \frac{12}{\boxed{}}$$

Section 11.1   The Parabola

Parabola

A **parabola** is the set of all points in a plane equidistant from a fixed line (the **directrix**) and a fixed point not on the line (the **focus**).

The graph of the quadratic function $f(x) = ax^2 + bx + c,\ a \neq 0$, is a parabola. The line that is perpendicular to the directrix and contains the focus is the **axis of symmetry**. The **vertex** is the midpoint of the segment between the focus and the directrix.

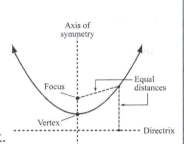

Standard Equation of a Parabola With Vertex $(0,0)$ and Vertical Axis of Symmetry

The standard equation of a parabola with vertex $(0,0)$ and directrix $y = -p$ is

$$x^2 = 4py.$$

The focus is $(0, p)$, and the y-axis is the axis of symmetry. When $p > 0$, the parabola opens up; when $p < 0$, the parabola opens down.

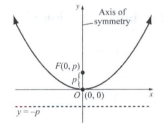

Standard Equation of a Parabola With Vertex $(0,0)$ and Horizontal Axis of Symmetry

The standard equation of a parabola with vertex $(0,0)$ and directrix $x = -p$ is

$$y^2 = 4px.$$

The focus is $(p, 0)$, and the x-axis is the axis of symmetry.

When $p > 0$, the parabola opens to the right; when $p < 0$, the parabola opens to the left.

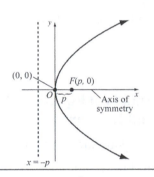

Example 1   Find the vertex, the focus, and the directrix of the parabola $y = -\dfrac{1}{12}x^2$. Then graph the parabola.

$y = -\dfrac{1}{12}x^2$

We want to write this in the form $x^2 = 4py$.

$x^2 = \boxed{} \cdot y$

$x^2 = 4\left(\boxed{}\right) y$

Vertex: $\left(0, \boxed{}\right)$

Focus $(0, p)$: $\left(0, \boxed{}\right)$

Directrix $y = -p$: $y = \boxed{}$

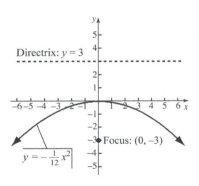

Example 2   Find an equation of the parabola with vertex $(0, 0)$ and focus $(5, 0)$. Then graph the parabola.

$y^2 = 4px$

Focus: $(5, 0)$ $p = \boxed{}$

$y^2 = 4\left(\boxed{}\right)x$

$y^2 = \boxed{} \cdot x$

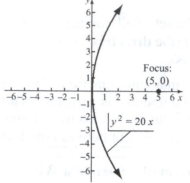

Focus: (5, 0)

$y^2 = 20x$

Standard Equation of a Parabola with Vertex (h, k) and Vertical Axis of Symmetry

The standard equation of a parabola with vertex (h, k) and vertical axis of symmetry is

$$(x - h)^2 = 4p(y - k),$$

where the vertex is (h, k), the focus is $(h, k + p)$, and the directrix is $y = k - p$.

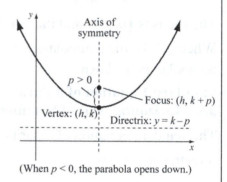

Axis of symmetry

$p > 0$

Focus: $(h, k + p)$

Vertex: (h, k) Directrix: $y = k - p$

(When $p < 0$, the parabola opens down.)

Standard Equation of a Parabola with Vertex (h, k) and Horizontal Axis of Symmetry

The standard equation of a parabola with vertex (h, k) and horizontal axis of symmetry is

$$(y - k)^2 = 4p(x - h),$$

where the vertex is (h, k), the focus is $(h + p, k)$, and the directrix is $x = h - p.$

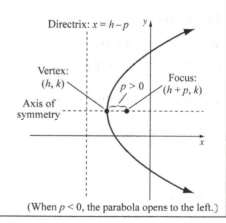

Directrix: $x = h - p$

Vertex: (h, k)

Focus: $(h + p, k)$

Axis of symmetry $p > 0$

(When $p < 0$, the parabola opens to the left.)

Example 3   For the parabola

$$x^2 + 6x + 4y + 5 = 0,$$

find the vertex, the focus, and the directrix. Then draw the graph.

$$(x - h)^2 = 4p(y - k)$$

$$x^2 + 6x \qquad = -4y - 5$$

$$x^2 + 6x + \boxed{} = -4y - 5 + \boxed{}$$

$$\left(\boxed{}\right)^2 = -4y + \boxed{}$$

$$(x + 3)^2 = -4(y - 1)$$

$$\left[x - \left(\boxed{}\right)\right]^2 = 4\left(\boxed{}\right)(y - 1)$$

$$h = -3, \qquad k = \boxed{}, \qquad p = \boxed{}$$

Vertex (h, k): $(-3, 1)$

Focus $(h, k + p)$: $\left(-3, \boxed{}\right)$

Directrix $y = k - p$:

$$y = 1 - (-1)$$

$$y = \boxed{}$$

Some points on the graph:

$\left(-4, \dfrac{3}{4}\right)$

$\left(-2, \dfrac{3}{4}\right)$

$(-5, 0)$

$(-1, 0)$

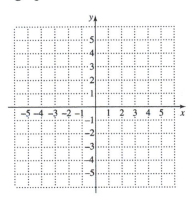

Example 4   For the parabola

$$y^2 - 2y - 8x - 31 = 0,$$

find the vertex, the focus, and the directrix. Then draw the graph.

$$(y-k)^2 = 4p(x-h)$$

$$y^2 - 2y \qquad = 8x + 31$$

$$y^2 - 2y + \boxed{} = 8x + 31 + \boxed{}$$

$$\left(\boxed{}\right)^2 = 8x + \boxed{}$$

$$(y-1)^2 = 8(x+4)$$

$$(y-1)^2 = 4(2)\left[x - \left(\boxed{}\right)\right]$$

$$h = \boxed{}, \qquad k = 1, \qquad p = \boxed{}$$

Vertex (h, k): $(-4, 1)$

Focus $(h + p, k)$: $(-2, 1)$

Directrix $x = h - p$:

$$x = -4 - 2$$

$$x = \boxed{}$$

Some points on the graph:

$$\left(-\frac{31}{8}, 2\right)$$

$$\left(-\frac{31}{8}, 0\right)$$

$$\left(-\frac{7}{2}, 3\right)$$

$$\left(-\frac{7}{2}, -1\right)$$

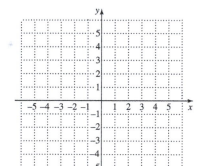

Section 11.2   The Circle and the Ellipse

Standard Equation of a Circle

A **circle** is the set of all points in a plane that are at a fixed distance from a fixed point (the **center**) in the plane.

The standard equation of a circle with center (h, k) and radius r is

$$(x-h)^2 + (y-k)^2 = r^2.$$

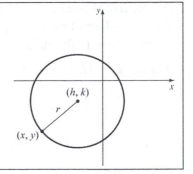

Example 1   For the circle

$$x^2 + y^2 - 16x + 14y + 32 = 0,$$

find the center and the radius. Then graph the circle.

$$(x-h)^2 + (y-k)^2 = r^2$$

$$x^2 + y^2 - 16x + 14y + 32 = 0$$

$$x^2 - 16x \qquad + y^2 + 14y \qquad = -32$$

$$x^2 - 16x + \boxed{} + y^2 + 14y + \boxed{} = -32 + \boxed{} + \boxed{}$$

$$(x-8)^2 + \left(\boxed{}\right)^2 = \boxed{}$$

$$(x-8)^2 + \left[y-(-7)\right]^2 = 9^2$$

Center: $\left(8, \boxed{}\right)$

Radius: $\boxed{}$

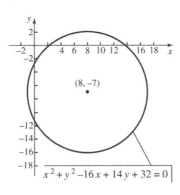

(8, –7)

$$x^2 + y^2 - 16x + 14y + 32 = 0$$

Standard Equation of an Ellipse with Center at the Origin

An **ellipse** is the set of all points in a plane, the sum of whose distances from two fixed points (the **foci**) is constant. The **center** of an ellipse is the midpoint of the segment between the foci.

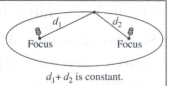

$d_1 + d_2$ is constant.

Major Axis Horizontal

$$\frac{x^2}{a^2} + \frac{y^2}{b^2} = 1, \quad a > b > 0$$

Vertices: $(-a, 0)$, $(a, 0)$

y-intercepts: $(0, -b)$, $(0, b)$

Foci: $(-c, 0)$, $(c, 0)$, where $c^2 = a^2 - b^2$

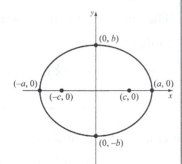

Major Axis Vertical

$$\frac{x^2}{b^2} + \frac{y^2}{a^2} = 1, \quad a > b > 0$$

Vertices: $(0, -a)$, $(0, a)$

x-intercepts: $(-b, 0)$, $(b, 0)$

Foci: $(0, -c)$, $(0, c)$, where $c^2 = a^2 - b^2$

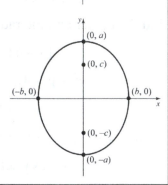

Example 2   Find the standard equation of the ellipse with vertices $(-5, 0)$ and $(5, 0)$ and foci $(-3, 0)$ and $(3, 0)$. Then graph the ellipse.

$$\frac{x^2}{a^2} + \frac{y^2}{b^2} = 1$$

Vertices: $(-5, 0)$, $(5, 0)$ $a = \boxed{}$

Foci: $(-3, 0)$, $(3, 0)$ $c = \boxed{}$

$$c^2 = a^2 - b^2$$

$$3^2 = \boxed{}^2 - b^2$$

$$9 = 25 - b^2$$

$$b^2 = 16$$

$$b = \boxed{}$$

$$\frac{x^2}{\boxed{}^2} + \frac{y^2}{\boxed{}^2} = 1$$

$$\frac{x^2}{25} + \frac{y^2}{16} = 1$$

y-intercepts: $\left(0, \boxed{}\right)$, $(0, 4)$

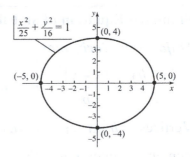

Example 3 For the ellipse

$$9x^2 + 4y^2 = 36,$$

find the vertices and the foci. Then draw the graph.

$$\frac{9x^2}{\boxed{}} + \frac{4y^2}{36} = \frac{36}{36}$$

$$\frac{x^2}{4} + \frac{y^2}{\boxed{}} = 1$$

$$\frac{x^2}{\boxed{}^2} + \frac{y^2}{3^2} = 1$$

$a = \boxed{}$, $\qquad b = 2$

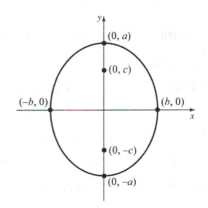

Vertices: $(0, a)$, $(0, -a)$

$\qquad\qquad (0, 3)$, $\left(0, \boxed{}\right)$

Foci: $(0, c)$, $(0, -c)$

$\qquad \left(0, \boxed{}\right)$, $\left(0, -\sqrt{5}\right)$

$\qquad\qquad c^2 = a^2 - b^2$

$\qquad\qquad c^2 = \boxed{} - 4$

$\qquad\qquad c^2 = 5$

$\qquad\qquad c = \boxed{}$

x-intercepts: $(-b, 0)$, $(b, 0)$

$\qquad\qquad \left(\boxed{}, 0\right)$, $(2, 0)$

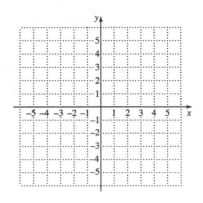

Standard Equation of an Ellipse with Center at (h, k)

Major Axis Horizontal

$$\frac{(x-h)^2}{a^2} + \frac{(y-k)^2}{b^2} = 1, \quad a > b > 0$$

Vertices: $(h-a, k)$, $(h+a, k)$

Length of minor axis: $2b$

Foci: $(h-c, k)$, $(h+c, k)$, where $c^2 = a^2 - b^2$

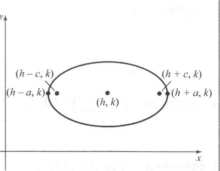

Major Axis Vertical

$$\frac{(x-h)^2}{b^2} + \frac{(y-k)^2}{a^2} = 1, \quad a > b > 0$$

Vertices: $(h, k-a)$, $(h, k+a)$

Length of minor axis: $2b$

Foci: $(h, k-c)$, $(h, k+c)$, where $c^2 = a^2 - b^2$

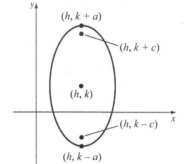

Example 4 For the ellipse

$$4x^2 + y^2 + 24x - 2y + 21 = 0,$$

find the center, the vertices, and the foci. Then draw the graph.

$$4\left(x^2 + 6x \quad\right) + \left(y^2 - 2y \quad\right) = -21$$

$$4\left(x^2 + 6x + \boxed{}\right) + \left(y^2 - 2y + 1\right) = -21 + \boxed{} + 1$$

$$4\left(\boxed{}\right)^2 + \left(\boxed{}\right)^2 = 16$$

$$\frac{1}{16}\left[4(x+3)^2 + (y-1)^2\right] = \frac{1}{16} \cdot 16$$

$$\frac{(x+3)^2}{\boxed{}} + \frac{(y-1)^2}{\boxed{}} = \boxed{}$$

$$\frac{\left[x-(-3)\right]^2}{2^2} + \frac{(y-1)^2}{4^2} = 1$$

General form: $\dfrac{(x-h)^2}{b^2}+\dfrac{(y-k)^2}{a^2}=1$

Center: $\left(\boxed{},1\right)$

Vertices: $(-3,1+4)$, $(-3,1-4)$, or

$\left(-3,\boxed{}\right)$, $(-3,-3)$

Foci: $\left(-3,\boxed{}+2\sqrt{3}\right)$, $\left(-3,1-2\sqrt{3}\right)$

$c^2=a^2-b^2$

$c^2=16-4=12$

$c=\sqrt{12}=\boxed{}$

Endpoints of minor axis: $(-3+2,1)$, $(-3-2,1)$, or

$\left(\boxed{},1\right)$, $\left(\boxed{},1\right)$

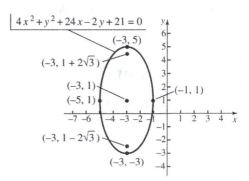

$4x^2+y^2+24x-2y+21=0$

(−3, 5)

(−3, 1 + 2√3)

(−3, 1)

(−5, 1)

(−1, 1)

(−3, 1 − 2√3)

(−3, −3)

Section 11.3   The Hyperbola

Hyperbola

A **hyperbola** is the set of all points in a plane for which the absolute value of the difference of the distances from two fixed points (the **foci**) is constant. The midpoint of the segment between the foci is the **center** of the hyperbola.

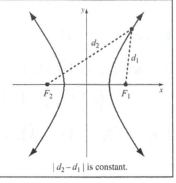

$|d_2 - d_1|$ is constant.

Standard Equation of a Hyperbola with Center at the Origin

Transverse Axis Horizontal

$$\frac{x^2}{a^2} - \frac{y^2}{b^2} = 1$$

Vertices: $(-a, 0)$, $(a, 0)$

Foci: $(-c, 0)$, $(c, 0)$, where $c^2 = a^2 + b^2$

Asymptotes: $y = \frac{b}{a}x$,

$$y = -\frac{b}{a}x$$

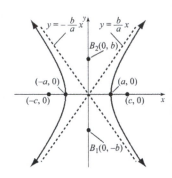

Transverse Axis Vertical

$$\frac{y^2}{a^2} - \frac{x^2}{b^2} = 1$$

Vertices: $(0, -a)$, $(0, a)$

Foci: $(0, -c)$, $(0, c)$, where $c^2 = a^2 + b^2$

Asymptotes: $y = \frac{a}{b}x$,

$$y = -\frac{a}{b}x$$

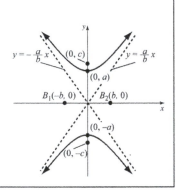

Example 1   Find an equation of the hyperbola with vertices $(0, -4)$ and $(0, 4)$ and foci $(0, -6)$ and $(0, 6)$.

$a = 4, \quad c = 6$

$c^2 = a^2 + b^2$

$6^2 = \boxed{}^2 + b^2$

$36 = 16 + b^2$

$\boxed{} = b^2$

$\dfrac{y^2}{a^2} - \dfrac{x^2}{b^2} = 1$

$\dfrac{y^2}{\boxed{}} - \dfrac{x^2}{\boxed{}} = 1$

Example 2 For the hyperbola given by

$$9x^2 - 16y^2 = 144,$$

find the vertices, the foci, and the asymptotes. Then graph the hyperbola.

$$\boxed{} \cdot \left(9x^2 - 16y^2\right) = \frac{1}{144} \cdot 144$$

$$\frac{x^2}{16} - \frac{y^2}{\boxed{}} = 1$$

$$\frac{x^2}{\boxed{}^2} - \frac{y^2}{3^2} = 1$$

$$a = 4, \quad b = 3$$

Vertices: $(-4, 0), \left(\boxed{}, 0\right)$

$$c^2 = a^2 + b^2 = 16 + 9 = 25$$

$$c = \boxed{}$$

Foci: $\left(\boxed{}, 0\right), (5, 0)$

Asymptotes: $\quad y = -\dfrac{b}{a}x \qquad\qquad y = \dfrac{b}{a}x$

$$y = \boxed{}\,x \qquad\qquad y = \frac{3}{4}x$$

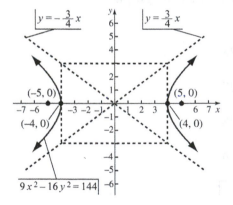

Standard Equation of a Hyperbola with Center at (h, k)

Transverse Axis Horizontal

$$\frac{(x-h)^2}{a^2} - \frac{(y-k)^2}{b^2} = 1$$

Vertices: $(h-a, k), (h+a, k)$

Asymptotes: $y-k = \frac{b}{a}(x-h), y-k = -\frac{b}{a}(x-h)$

Foci: $(h-c, k), (h+c, k)$, where $c^2 = a^2 + b^2$

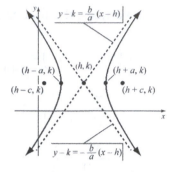

Transverse Axis Vertical

$$\frac{(y-k)^2}{a^2} - \frac{(x-h)^2}{b^2} = 1$$

Vertices: $(h, k-a), (h, k+a)$

Asymptotes: $y-k = \frac{a}{b}(x-h), y-k = -\frac{a}{b}(x-h)$

Foci: $(h, k-c), (h, k+c)$, where $c^2 = a^2 + b^2$

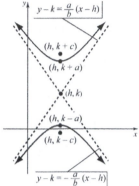

Example 3   For the hyperbola given by

$$4y^2 - x^2 + 24y + 4x + 28 = 0,$$

find the center, the vertices, the foci, and the asymptotes. Then draw the graph.

$$4y^2 - x^2 + 24y + 4x + 28 = 0$$

$$4\left(y^2 + 6y \qquad\right) - \left(x^2 - 4x \qquad\right) = -28$$

$$4\left(y^2 + 6y + \boxed{}\right) - \left(x^2 - 4x + \boxed{}\right) = -28$$

$$4\left(y^2 + 6y + 9\right) + 4\left(\boxed{}\right) - \left(x^2 - 4x + 4\right) + (-1)\left(\boxed{}\right) = -28$$

$$4\left(y^2 + 6y + 9\right) - \boxed{} - \left(x^2 - 4x + 4\right) + 4 = -28$$

$$4\left(\boxed{}\right)^2 - (x-2)^2 = \boxed{}$$

$$\frac{(y+3)^2}{1} - \frac{(x-2)^2}{\boxed{}} = 1$$

$$\frac{\left[y-(-3)\right]^2}{1^2} - \frac{(x-2)^2}{2^2} = 1 \qquad\qquad \text{Standard form: } \frac{(y-k)^2}{a^2} - \frac{(x-h)^2}{b^2} = 1$$

Center: $\left(\boxed{}, -3\right)$

Vertices: $(2, -3-1),\ (2, -3+1),$ or

$\qquad (2, -4),\ \left(2, \boxed{}\right)$

Foci: $\left(2, -3-\sqrt{5}\right),\ \left(2, -3+\sqrt{5}\right)$ $\qquad\qquad a^2 = 1, \quad b^2 = 4$

$$c^2 = a^2 + b^2$$

$$c^2 = 1 + 4$$

$$= 5$$

$$c = \boxed{}$$

Asymptotes:

$$y - k = \frac{a}{b}(x - h), \qquad\qquad y - k = -\frac{a}{b}(x - h)$$

$$y - (-3) = \boxed{}(x - 2), \qquad y - (-3) = \boxed{}(x - 2)$$

$$y + 3 = \frac{1}{2}(x - 2), \qquad\qquad y + 3 = -\frac{1}{2}(x - 2)$$

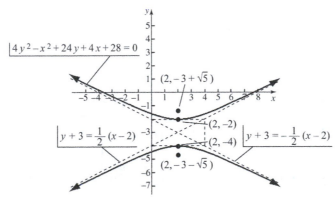

Section 11.4   Nonlinear Systems of Equations and Inequalities

Example 1   Solve the following system of equations:

$x^2 + y^2 = 25,$ (1)   The graph is a circle.

$3x - 4y = 0.$ (2)   The graph is a line.

$3x - 4y = 0$

$\qquad 3x = 4y$

$\qquad x = \boxed{}$

$x^2 + y^2 = 25$ (1)

$\left(\boxed{}\right)^2 + y^2 = 25$

$\dfrac{16}{9}y^2 + y^2 = 25$

$\boxed{} \cdot y^2 = 25$

$\qquad y^2 = 9$

$\qquad y = \boxed{}$

$x = \dfrac{4}{3}y$

$x = \dfrac{4}{3} \cdot \boxed{} = 4 \qquad \left(\boxed{}, 3\right)$

$x = \dfrac{4}{3} \cdot \left(\boxed{}\right) = -4 \qquad \left(\boxed{}, -3\right)$

Check for $(4, 3)$:

$$\begin{array}{c|c}
\multicolumn{2}{c}{x^2 + y^2 = 25} \\
\hline
(4)^2 + \left(\boxed{}\right)^2 \ ?\ 25 & \\
16 + 9 & \\
25 & 25 \quad \text{TRUE}
\end{array} \qquad
\begin{array}{c|c}
\multicolumn{2}{c}{3x - 4y = 0} \\
\hline
3\left(\boxed{}\right) - 4(3) \ ?\ 0 & \\
12 - 12 & \\
0 & 0 \quad \text{TRUE}
\end{array}$$

Check for $(-4, -3)$:

$$\begin{array}{c|c}
\multicolumn{2}{c}{x^2 + y^2 = 25} \\
\hline
\left(\boxed{}\right)^2 + (-3)^2 \ ?\ 25 & \\
16 + 9 & \\
25 & 25 \quad \text{TRUE}
\end{array} \qquad
\begin{array}{c|c}
\multicolumn{2}{c}{3x - 4y = 0} \\
\hline
3(-4) - 4\left(\boxed{}\right) \ ?\ 0 & \\
-12 + 12 & \\
0 & 0 \quad \text{TRUE}
\end{array}$$

The solutions are $(4, 3)$ and $(-4, -3)$.

Example 2   Solve the following system of equations:

$x + y = 5$,   (1)   The graph is a line.

$y = 3 - x^2$.   (2)   The graph is a parabola.

$x + \boxed{} = 5$

$x - 2 - x^2 = 0$

$x^2 - x + 2 = 0$ $a = 1$, $b = \boxed{}$, $c = 2$

$$x = \frac{-b \pm \sqrt{b^2 - 4ac}}{2a} = \frac{-(-1) \pm \sqrt{\left(\boxed{}\right)^2 - 4 \cdot 1 \cdot \boxed{}}}{2 \cdot 1}$$

$$= \frac{1 \pm \sqrt{1 - 8}}{2} = \frac{1 \pm \sqrt{\boxed{}}}{2}$$

$$= \frac{1 \pm \boxed{} \cdot \sqrt{7}}{2} = \frac{1}{2} \pm \frac{\sqrt{7}}{2} i$$

$\boxed{} + y = 5$

$$y = \frac{9}{2} - \frac{\sqrt{7}}{2} i$$

$\boxed{} + y = 5$

$$y = \frac{9}{2} + \frac{\sqrt{7}}{2} i$$

The solutions are

$$\left(\frac{1}{2} + \frac{\sqrt{7}}{2} i, \frac{9}{2} - \frac{\sqrt{7}}{2} i \right) \text{ and } \left(\frac{1}{2} - \frac{\sqrt{7}}{2} i, \frac{9}{2} + \frac{\sqrt{7}}{2} i \right).$$

Example 3   Solve the following system of equations:

$2x^2 + 5y^2 = 39$,   (1)   The graph is an ellipse.

$3x^2 - y^2 = -1$,   (2)   The graph is a hyperbola.

$$2x^2 + 5y^2 = 39$$
$$\underline{15x^2 - 5y^2 = \boxed{}}$$
$$17x^2 \qquad = 34$$
$$x^2 = 2$$
$$x = \boxed{}$$
$$3x^2 - y^2 = -1 \qquad\qquad (2)$$
$$3\left(\boxed{}\right)^2 - y^2 = -1$$
$$3 \cdot 2 - y^2 = -1$$
$$6 - y^2 = -1$$
$$-y^2 = -7$$
$$y^2 = 7$$
$$y = \boxed{}$$

The solutions are

$\left(\sqrt{2}, \boxed{}\right)$,

$\left(\sqrt{2}, \sqrt{7}\right)$,

$\left(\boxed{}, \sqrt{7}\right)$,

$\left(-\sqrt{2}, -\sqrt{7}\right)$.

Example 4   Solve the following system of equations:

$x^2 - 3y^2 = 6$,   (1)

$xy = 3$.   (2)

$$xy = 3$$
$$y = \frac{3}{\boxed{}} \qquad x^2 - 3y^2 = 6$$

$$x^2 - 3\left(\boxed{}\right)^2 = 6$$

$$x^2 - 3 \cdot \frac{9}{x^2} = 6$$

$$x^2 - \frac{27}{x^2} = 6$$

$$x^4 - 27 = \boxed{}$$

$$x^4 - 6x^2 - 27 = 0 \qquad u = \boxed{}$$

$$u^2 - 6 \cdot \boxed{} - 27 = 0$$

$$\left(\boxed{}\right)(u+3) = 0$$

$$u - 9 = 0 \qquad or \quad u + 3 = 0$$

$$u = 9 \qquad or \qquad u = -3$$

$$x^2 = 9 \qquad or \qquad x^2 = -3$$

$$x = \boxed{} \quad or \qquad x = \boxed{}$$

$$y = \frac{3}{x}$$

The solutions are

$$x = 3;\ y = \frac{3}{3} = \boxed{}$$

$(3,1),$

$$x = -3;\ y = \frac{3}{-3} = -1$$

$\left(\boxed{}, -1\right),$

$$x = i\sqrt{3};\ y = \frac{3}{i\sqrt{3}}$$

$$y = \frac{3}{i\sqrt{3}} \cdot \frac{-i\sqrt{3}}{-i\sqrt{3}} = \boxed{}$$

$\left(i\sqrt{3}, -i\sqrt{3}\right),$ and

$$x = -i\sqrt{3};\ y = \frac{3}{-i\sqrt{3}}$$

$$y = \frac{3}{-i\sqrt{3}} \cdot \frac{i\sqrt{3}}{i\sqrt{3}} = i\sqrt{3}$$

$\left(-i\sqrt{3}, \boxed{}\right).$

Example 5   *Dimensions of a Piece of Land.* For a student recreation building at Southport Community College, an architect wants to lay out a rectangular piece of land that has a perimeter of 204 m and an area of 2565 m². Find the dimensions of the piece of land.

1. **Familiarize.**

l = length in meters

w = width in meters

$P = 204$ m, $A = 2565$ m²

2. **Translate.**

$2w + 2l = \boxed{}$

$lw = \boxed{}$

3. **Carry Out.**

$$2w + 2l = 204$$

$$lw = 2565$$

$$l = \frac{2565}{\boxed{}}$$

$$2w + 2\left(\boxed{}\right) = 204$$

$$2w^2 + 2(2565) = 204w$$

$$2w^2 - 204w + 2(2565) = 0$$

$$w^2 - \boxed{}w + 2565 = 0$$

$$(w - 57)\left(\boxed{}\right) = 0$$

$$w - 57 = 0 \qquad or \quad w - 45 = 0$$

$$w = \boxed{} \quad or \qquad w = 45$$

Substituting for w in $l = \dfrac{2565}{w}$, we have

$$w = 57: \quad l = \frac{2565}{\boxed{}} = 45,$$

$$w = 45: \quad l = \frac{2565}{\boxed{}} = 57.$$

Considering l longer than w, we have $l = \boxed{}$ and $w = 45$.

4. **Check.**

   Perimeter is $2 \cdot \boxed{} + 2 \cdot 57 = \boxed{}$.

   Area is $57 \cdot 45 = \boxed{}$.

   The answer checks.

5. **State.**

   The length of the piece of land is 57 m, and the width is 45 m.

Example 6   Graph the solution set of the system

$$x^2 + y^2 \le 25,$$
$$3x - 4y > 0.$$

Related equations:

$$x^2 + y^2 \boxed{} 25 \qquad\qquad 3x - 4y \boxed{} 0$$
$$y = \frac{3x}{4}$$

Determine which region to shade.

Test $(0, 0)$: $\dfrac{x^2 + y^2 \le 25}{0^2 + \boxed{}^2 \ ?\ 25}$ Test $(0, 2)$: $\dfrac{3x - 4y > 0}{3 \cdot 0 - 4 \cdot \boxed{}\ ?\ 0}$

$\qquad\qquad\qquad 0 \mid 25$ TRUE $\qquad\qquad\qquad -8 \mid 0$ FALSE

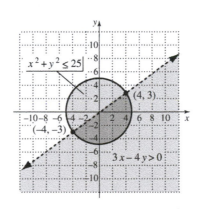

Example 7 Graph the solution set of the system

$$y \le 4 - x^2,$$
$$x + y \ge 2.$$

$y \boxed{} 4 - x^2$

Vertex: $\left(0, \boxed{}\right)$

x-intercepts: $\left(\boxed{}, 0\right), (2, 0)$

$x + y \boxed{} 2$

x-intercept: $\left(\boxed{}, 0\right)$

y-intercept: $\left(0, \boxed{}\right)$

Test $(0, 0)$:

$$\frac{y \le 4 - x^2}{0 \stackrel{?}{} 4 - 0^2}$$
$$0 \mid 4 \qquad \text{TRUE}$$

Test $(0, 0)$:

$$\frac{x + y \ge 2}{0 + 0 \stackrel{?}{} 2}$$
$$0 \mid 2 \qquad \text{FALSE}$$

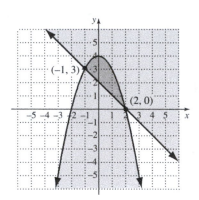

Section 12.1   Sequences and Series

> **Sequences**
>
> - An **infinite sequence** is a function having for its domain the set of positive integers, $\{1, 2, 3, 4, 5, \ldots\}$.
>
> - A **finite sequence** is a function having for its domain a set of positive integers, $\{1, 2, 3, 4, 5, \ldots, n\}$, for some positive integer n.
>
> - The first term of a sequence is denoted as a_1, the fifth term as a_5, and the nth term, or **general term**, as a_n.

Example 1   Find the first 4 terms and the 23rd term of the sequence whose general term is given by $a_n = (-1)^n n^2$.

$$a_1 = (-1)^1 \cdot 1^2 = -1 \cdot 1 = \boxed{}$$

$$a_2 = (-1)^2 \cdot 2^2 = 1 \cdot 4 = \boxed{}$$

$$a_3 = (-1)^{\boxed{}} \cdot \boxed{}^2 = \boxed{} \cdot 9 = -9$$

$$a_4 = (-1)^{\boxed{}} \cdot \boxed{}^2 = \boxed{} \cdot 16 = 16$$

$$a_{23} = (-1)^{23} \cdot 23^2 = -1 \cdot 529 = \boxed{}$$

Example 2   For each of the following sequences, predict the general term.

a) $1, \sqrt{2}, \sqrt{3}, 2, \ldots$

$\sqrt{1}, \sqrt{2}, \sqrt{3}, \boxed{}, \ldots$

The general term might be $\sqrt{n}$.

b) $-1, 3, -9, 27, -81, \ldots$

First look at the numbers disregarding their signs.

$1, 3, 9, \boxed{}, 81, \ldots$

Disregarding signs we have 3^{n-1}.
Now look at the signs. They alternate between negative and positive.

The general term might be $(-1)^n 3^{\boxed{}}$.

c) 2, 4, 8, ...

   We could look at this sequence as powers of 2.

   2, 4, 8, 16, 32, ...

   Then one general term might be $2^{\square}$.

   We might also observe that 4 is $2+2$, 8 is $4+4$, and so on so we would have

   $$2, 4, \boxed{}, 14, \boxed{}, \ldots.$$

   If we work with these numbers, we see that another general term might be $n^2 - n + 2$.

Series

Given the infinite sequence

$$a_1, a_2, a_3, a_4, \ldots, a_n, \ldots,$$

the sum of the terms

$$a_1 + a_2 + a_3 + \cdots + a_n + \cdots$$

is called an **infinite series**. A **partial sum** is the sum of the first n terms:

$$a_1 + a_2 + a_3 + \cdots + a_n.$$

A partial sum is also called a **finite series**, or **nth partial sum**, and is denoted S_n.

Example 3   For the sequence $-2, 4, -6, 8, -10, 12, -14, \ldots$, find each of the following sums.

a) $S_1 = \boxed{}$

b) $S_4 = -2 + 4 + \left(\boxed{}\right) + 8 = \boxed{}$

c) $S_5 = -2 + \boxed{} + (-6) + 8 + \left(\boxed{}\right) = \boxed{}$

Example 4   Find and evaluate each of the following sums.

a) $\displaystyle\sum_{k=1}^{5} k^3 = 1^3 + \boxed{}^3 + 3^3 + 4^3 + \boxed{}^3$

   $\phantom{\sum_{k=1}^{5} k^3} = 1 + 8 + \boxed{} + \boxed{} + 125$

   $\phantom{\sum_{k=1}^{5} k^3} = \boxed{}$

b) $\displaystyle\sum_{k=0}^{4} (-1)^k 5^k = (-1)^0 \cdot 5^0 + (-1)^1 \cdot 5^1 + (-1)^{\square} \cdot 5^{\square} + (-1)^3 \cdot 5^3 + (-1)^4 \cdot 5^4$

   $\phantom{\sum_{k=0}^{4}} = 1 \cdot \boxed{} + (-1) \cdot 5 + (1) \cdot \boxed{} + \left(\boxed{}\right) \cdot 125 + (1) \cdot 625$

   $\phantom{\sum_{k=0}^{4}} = 1 - 5 + 25 - \boxed{} + 625$

   $\phantom{\sum_{k=0}^{4}} = \boxed{}$

c) $\displaystyle\sum_{i=8}^{11}\left(2+\frac{1}{i}\right)=\left(2+\frac{1}{\boxed{}}\right)+\left(2+\frac{1}{\boxed{}}\right)+\left(2+\frac{1}{10}\right)+\left(2+\frac{1}{11}\right)$

$\qquad\qquad =\boxed{}+\dfrac{1}{8}+\dfrac{1}{9}+\dfrac{1}{10}+\dfrac{1}{11}$

$\qquad\qquad =8\dfrac{\boxed{}}{3960}$

Example 5 Write sigma notation for each sum.

a) $1+2+4+8+16+32+64$

$\qquad =2^{\boxed{}}+2^{1}+\boxed{}^{2}+2^{3}+2^{\boxed{}}+2^{5}+2^{6}$

$\qquad =\displaystyle\sum_{k=0}^{\boxed{}}2^{\boxed{}}$

b) $-2+4-6+8-10$

$\qquad =(-1)(2)+(1)(4)+\left(\boxed{}\right)(6)+(1)(8)+(-1)(10)$

$\qquad =(-1)\left(2\cdot\boxed{}\right)+(1)\left(2\cdot\boxed{}\right)+(-1)(2\cdot3)+(1)(2\cdot4)+(-1)(2\cdot5)$

$\qquad =(-1)^{1}(2\cdot1)+\left(\boxed{}\right)^{2}(2\cdot2)+(-1)^{\boxed{}}(2\cdot3)+(-1)^{4}(2\cdot4)+(-1)^{5}(2\cdot5)$

$\qquad =\displaystyle\sum_{k=1}^{5}\left(\boxed{}\right)^{k}\left(2\cdot\boxed{}\right)$

c) $x+\dfrac{x^{2}}{2}+\dfrac{x^{3}}{3}+\dfrac{x^{4}}{4}+\cdots$

$\qquad =\dfrac{x^{1}}{\boxed{}}+\dfrac{x^{2}}{2}+\dfrac{x^{3}}{3}+\dfrac{x^{4}}{4}+\cdots$

$\qquad =\displaystyle\sum_{k=1}^{\boxed{}}\dfrac{x^{k}}{\boxed{}}$

Example 6 Find the first 5 terms of the sequence defined by

$a_{1}=5,\qquad a_{n+1}=2a_{n}-3,\qquad$ for $n\geq1.$

$a_{1}=5$

$a_{2}=2\cdot\boxed{}-3=7$

$a_{3}=2\cdot7-3=\boxed{}$

$a_{4}=2\cdot\boxed{}-3=19$

$a_{5}=2\cdot19-3=\boxed{}$

Section 12.2 Arithmetic Sequences and Series

Arithmetic Sequence

A sequence is **arithmetic** if there exists a number d, called the **common difference**, such that $a_{n+1} = a_n + d$ for any integer $n \geq 1$.

Example 1 For each of the following arithmetic sequences, identify the first term, a_1, and the common difference, d.

a) $4, 9, 14, 19, 24, \ldots$

First term: $\boxed{}$

Common difference $d = 9 - 4 = \boxed{}$

$$14 - 9 = 5$$
$$19 - 14 = 5$$

b) $34, 27, 20, 13, 6, -1, -8, \ldots$

First term: $\boxed{}$

$d = \boxed{} - 34 = -7$

$$20 - 27 = -7$$
$$13 - 20 = -7$$

c) $2, 2\frac{1}{2}, 3, 3\frac{1}{2}, 4, 4\frac{1}{2}, \ldots$

First term: $\boxed{}$

$d = 2\frac{1}{2} - 2 = \boxed{}$

$$3 - 2\frac{1}{2} = \frac{1}{2}$$
$$3\frac{1}{2} - 3 = \frac{1}{2}$$

---

nth Term of an Arithmetic Sequence

The **nth term** of an arithmetic sequence is given by

$$a_n = a_1 + (n-1)d, \text{ for any integer } n \geq 1.$$

Example 2 Find the 14th term of the arithmetic sequence $4, 7, 10, 13, \ldots$.

$$a_n = a_1 + (n-1)d \qquad\qquad a_1 = 4$$
$$a_{14} = \boxed{} + (14-1) \cdot \boxed{} \qquad d = 7 - 4 = 3$$
$$= 4 + 13 \cdot 3 \qquad\qquad n = 14$$
$$= 4 + 39$$
$$= 43$$

The 14th term is $\boxed{}$.

Example 3   In the sequence of $4, 7, 10, 13, \ldots$, which term is 301? That is, find n if $a_n = 301$.

$$a_n = a_1 + (n-1)d \qquad\qquad a_1 = 4$$
$$\boxed{} = 4 + (n-1) \cdot 3 \qquad\qquad d = 7 - 4 = 3$$
$$301 = 4 + 3n - \boxed{} \qquad\qquad a_n = 301$$
$$301 = 3n + 1 \qquad\qquad n = ?$$
$$300 = 3n$$
$$100 = n$$

The term 301 is the $\boxed{}$ th term.

Example 4   The 3rd term of an arithmetic sequence is 8, and the 16th term is 47. Find a_1 and d and construct the sequence.

$$8 + \boxed{} = 47$$
$$13d = 39$$
$$d = 3$$
$$a_1 = \boxed{} - 2 \cdot 3 = \boxed{}$$
$$2, \boxed{}, 8, 11, \ldots$$

Sum of the First n Terms

The sum of the first n terms of an arithmetic sequence is given by

$$S_n = \frac{n}{2}(a_1 + a_n).$$

Example 5   Find the sum of the first 100 natural numbers.

$$1 + 2 + 3 + \cdots + 99 + 100$$

$$S_n = \frac{n}{2}(a_1 + a_n) \qquad\qquad a_1 = 1$$
$$\qquad\qquad a_n = \boxed{}$$
$$S_{100} = \frac{\boxed{}}{2}(1 + 100) = \boxed{} \qquad\qquad n = 100$$

Example 6   Find the sum of the first 15 terms of the arithmetic sequence $4, 7, 10, 13, \ldots$.

$$a_1 = 4, \quad d = 3, \quad n = \boxed{}$$

$$S_n = \frac{n}{2}(a_1 + a_n) \qquad\qquad\qquad a_n = a_1 + (n-1)d$$

$$S_{15} = \frac{\boxed{}}{2}(4 + 46) \qquad\qquad a_{15} = \boxed{} + (15-1)3$$

$$= \frac{15}{2}(50) \qquad\qquad\qquad\quad = 4 + 14 \cdot 3 = \boxed{}$$

$$= \boxed{}$$

Example 7   Find the sum: $\displaystyle\sum_{k=1}^{130}(4k + 5)$.

We write the first few terms.

$$9 + \boxed{} + 17 + \cdots$$

$$a_1 = 9, \quad d = \boxed{}, \quad n = 130$$

$$a_{130} = 4 \cdot \boxed{} + 5$$

$$a_{130} = 525$$

$$S_{130} = \frac{130}{2}\left(\boxed{} + 525\right)$$

$$S_{130} = \boxed{}$$

Example 8   Rachel accepts a job, starting with an hourly wage of $14.25, and is promised a raise of 15¢ per hour every 2 months for 5 years. At the end of 5 years, what will Rachel's hourly wage be?

$14.25 $14.25, 14.40, 14.55, \ldots$ $a_1 = 14.25$

$14.40 $a_n = a_1 + (n-1)d$ $d = 0.15$

$14.55   1 year $\dfrac{12}{2} = \boxed{}$ $n = \boxed{}$

$\cdot$

$\cdot$

$\cdot$ 5 years $5 \cdot 6 = \boxed{}$

We have 31 terms, the original wage plus 30 raises.

$$a_{31} = \boxed{} + (31-1)0.15$$

$$a_{31} = 18.75$$

At the end of 5 years, Rachel's hourly wage will be $\boxed{}$.

Example 9   A stack of electric poles has 30 poles in the bottom row. There are 29 poles in the second row, 28 in the next row, and so on. How many poles are in the stack if there are 5 poles in the top row?

$$S_n = \frac{n}{2}(a_1 + a_n)$$

$$S_{26} = \frac{\boxed{}}{2}(30 + 5)$$

$$= 13(35)$$

$$= 455$$

$n = \boxed{}$

$a_1 = 30$

$a_{26} = \boxed{}$

There are $\boxed{}$ poles in the stack.

Section 12.3 Geometric Sequences and Series

Geometric Sequence

A sequence is **geometric** if there is a number r, called the **common ratio**, such that

$$\frac{a_{n+1}}{a_n} = r, \text{ or } a_{n+1} = a_n r, \text{ for any integer } n \geq 1.$$

Example 1   For each of the following geometric sequences, identify the common ratio.

a) $3, 6, 12, 24, 48, \ldots$

$$r = \frac{6}{\boxed{}} = 2$$

b) $1, -\frac{1}{2}, \frac{1}{4}, -\frac{1}{8}, \ldots$

$$r = \frac{-\frac{1}{2}}{1} = \boxed{}$$

c) $\$5200, \$3900, \$2925, \$2193.75, \ldots$

$$r = \frac{\boxed{}}{5200} = 0.75$$

d) $\$1000, \$1060, \$1123.60, \ldots$

$$r = \frac{1060}{1000} = \boxed{}$$

---

nth Term of a Geometric Sequence

The **nth term** of a geometric sequence is given by

$$a_n = a_1 r^{n-1}, \text{ for any integer } n \geq 1.$$

Example 2   Find the 7th term of the geometric sequence $4, 20, 100, \ldots$.

$$a_n = a_1 r^{n-1}, \; n \geq 1$$

$$a_1 = 4 \quad n = \boxed{}, \quad r = \frac{20}{4} = \boxed{}$$

$$a_7 = \boxed{} \cdot 5^{7-1} = 4 \cdot 5^6$$

$$= 4 \cdot 15,625$$

$$= \boxed{}$$

The 7th term is 62,500.

Example 3 Find the 10th term of the geometric sequence $64, -32, 16, -8, \ldots$.

$$a_n = a_1 r^{n-1}, \ n \geq 1$$

$$a_1 = \boxed{}, \qquad n = 10, \qquad r = \frac{\boxed{}}{64} = -\frac{1}{2}$$

$$a_{10} = 64 \cdot \left(\boxed{}\right)^{10-1} = 64 \cdot \left(-\frac{1}{2}\right)^9$$

$$= 64 \cdot \left(-\frac{1}{2^9}\right)$$

$$= \boxed{} \cdot \left(-\frac{1}{2^9}\right)$$

$$= -\frac{1}{2^3} = \boxed{}$$

The 10th term is $-\dfrac{1}{8}$.

Sum of the First n Terms

The sum of the first n terms of a geometric sequence is given by

$$S_n = \frac{a_1\left(1 - r^n\right)}{1 - r}, \text{ for any } r \neq 1.$$

Example 4 Find the sum of the first 7 terms of the geometric sequence $3, 15, 75, 375, \ldots$.

$$a_1 = 3, \qquad n = 7, \qquad r = \frac{15}{\boxed{}} = 5$$

$$S_n = \frac{a_1\left(1 - r^n\right)}{1 - r}, \ r \neq 1$$

$$S_7 = \frac{3\left(1 - 5^{\boxed{}}\right)}{1 - \boxed{}} = \frac{3(1 - 78,125)}{-4}$$

$$= \boxed{}$$

The sum of the first 7 terms is 58,593.

Example 5   Find the sum: $\displaystyle\sum_{k=1}^{11}(0.3)^k$.

$$S_n = \frac{a_1\left(1-r^n\right)}{1-r} \qquad\qquad a_1 = 0.3$$

$$S_{11} = \frac{0.3\left(1-0.3^{\boxed{}}\right)}{1-0.3} \qquad\qquad r = \boxed{}$$

$$\qquad\qquad\qquad\qquad n = 11$$

$$S_{11} \approx \boxed{}$$

Limit or Sum of an Infinite Geometric Series

When $|r| < 1$, the limit or sum of an infinite geometric series is given by

$$S_\infty = \frac{a_1}{1-r}.$$

Example 6   Determine whether each of the following infinite geometric series has a limit. If a limit exists, find it.

a) $1 + 3 + 9 + 27 + \cdots$ $\qquad\qquad\qquad r = \dfrac{3}{1} = 3$

An infinite geometric series has a limit if $|r| < 1$. $\qquad\qquad |r| = |3| = \boxed{}$

Since 3 is not less than $\boxed{}$, this series $\underset{\text{does / does not}}{\underline{}}$ have a limit.

b) $-2 + 1 - \dfrac{1}{2} + \dfrac{1}{4} - \dfrac{1}{8} + \cdots$ $\qquad\qquad r = \dfrac{1}{-2} = \boxed{}$

An infinite geometric series has a limit if $|r| < 1$. $\qquad\qquad |r| = \left|-\dfrac{1}{2}\right| = \dfrac{1}{2}$

Since $\boxed{} < 1$, this series has a limit.

$$S_\infty = \frac{a_1}{1-r} = \frac{-2}{1-\left(\boxed{}\right)} = \frac{-2}{\dfrac{3}{2}} = -\frac{4}{3}$$

The limit of this series is $\boxed{}$.

Example 7   Find fraction notation for $0.78787878\ldots$, or $0.\overline{78}$.

$0.78 + 0.0078 + 0.000078 + \cdots$

$a_1 = 0.78 \qquad r = \boxed{}$

$|r| = |0.01| = 0.01 < \boxed{}$

This infinite geometric series has a limit.

$$S_\infty = \frac{a_1}{1-r} = \frac{0.78}{1-\boxed{}} = \frac{0.78}{0.99} = \frac{78}{99} = \frac{\boxed{}}{33}$$

Example 8   Suppose someone offered you a job for the month of September (30 days) under the following conditions. You will be paid $0.01 for the first day, $0.02 for the second, $0.04 for the third, and so on, doubling your previous day's salary each day. How much would you earn altogether for the month? (Would you take the job? Make a conjecture before reading further.)

$0.01 \qquad 0.02 \qquad \boxed{} \ldots$

$0.01 \qquad 0.01(2) \qquad 0.01(2)(2)\ldots$

$0.01 + 0.01 \cdot 2^1 + 0.01 \cdot 2^{\boxed{}} + 0.01 \cdot 2^3 + \cdots$

$a_1 = 0.01 \qquad r = \boxed{} \qquad n = \boxed{}$

$$S_n = \frac{a_1\left(1 - r^n\right)}{1-r}$$

$$S_{30} = \frac{0.01\left(1 - 2^{30}\right)}{1 - \boxed{}} = \$10{,}737{,}418.23$$

Your pay exceeds $\boxed{}$ million for the month.

Example 9   An **annuity** is a sequence of equal payments, made at equal time intervals, that earns interest. Fixed deposits in a savings account are an example of an annuity. Suppose that to save money to buy a car, Jacob deposits $2000 at the *end* of each of 5 years in an account that pays 3% interest, compounded annually. The total amount in the account at the end of 5 years is called the **amount of the annuity**. Find that amount.

End of Year	Amount of Deposit		Amount Grows To	
0	0		0	
1	2000	$\rightarrow$	$2000\left(\boxed{}\right)^4$	$\left(\text{at 3\% per year}\right)$
2	2000	$\rightarrow$	$2000(1.03)^{\boxed{}}$	
3	2000	$\rightarrow$	$2000(1.03)^2$	
4	2000	$\rightarrow$	$2000(1.03)^1$	
5	2000	$\rightarrow$	2000	

$$2000 + 2000(1.03)^1 + 2000(1.03)^2 + 2000(1.03)^3 + 2000(1.03)^4$$

$$a_1 = 2000 \qquad\qquad S_n = \frac{a_1\left(1-r^n\right)}{1-r}$$

$$r = \boxed{}$$

$$n = 5 \qquad\qquad S_5 = \frac{\boxed{}\left(1-1.03^5\right)}{1-1.03}$$

$$\approx \boxed{}$$

Jacob has been able to save $10,618.27.

Example 10   Large sporting events have a significant impact on the economy of the host city. Super Bowl XLVII, hosted by New Orleans, generated a $480-million net impact for the region. Assume that 60% of that amount is spent again in the area, and then 60% of that amount is spent again, and so on. This is known as the *economic multiplier effect*. Find the total effect on the economy.

$$480,000,000 + 480,000,000(0.60) + 480,000,000(0.60)^2 + \cdots$$

$$|r| < 1 \ \text{ because } r = \boxed{}$$

$$S_\infty = \frac{a_1}{1-r} = \frac{480,000,000}{1-0.6} = \frac{480,000,000}{\boxed{}}$$

$$= 1,200,000,000$$

$$= \boxed{} \text{ billion}$$

Section 12.4 Mathematical Induction

The Principle of Mathematical Induction

We can prove an infinite sequence of statements S_n by showing the following.

(1) *Basis step.* S_1 is true.

(2) *Induction step.* For all natural numbers k, $S_k \to S_{k+1}$.

Example 1   Prove. For every natural number n,

$$1+3+5+\cdots+(2n-1)=n^2.$$

$$S_n : 1+3+5+\cdots+(2n-1)=n^2$$

For $n=1$

$$S_1 : 1 = \boxed{}^2$$

For $n=k$

$$S_k : 1+3+5+\cdots+\left(\boxed{}\right)=\boxed{}^2$$

For $n=k+1$

$$S_{k+1} : 1+3+5+\cdots+(2k-1)+\left[2\left(\boxed{}\right)-1\right]=\left(\boxed{}\right)^2$$

1. Basis step: S_1 is true.

2. Induction step: Assume $\boxed{}$ is true.

$$\underbrace{1+3+5+\cdots+(2k-1)}+\left[2(k+1)-1\right]$$

$$=\boxed{}+\left[2(k+1)-1\right]$$
$$=k^2+2k+2-1$$
$$=k^2+2k+1$$
$$=\left(\boxed{}\right)^2$$

We have shown that for all natural numbers k, $S_k \to S_{k+1}$. This completes the induction step. It and the Basis Step tell us that the proof is complete.

Example 2 Prove: For every natural number n,

$$\frac{1}{2} + \frac{1}{4} + \frac{1}{8} + \cdots + \frac{1}{2^n} = \frac{2^n - 1}{2^n}.$$

$S_n:$ $\frac{1}{2} + \frac{1}{4} + \frac{1}{8} + \cdots + \frac{1}{2^n} = \frac{2^n - 1}{2^n}$

$S_1:$ $\frac{1}{2^\square} = \frac{2^\square - 1}{2^\square}$

$S_k:$ $\frac{1}{2} + \frac{1}{4} + \frac{1}{8} + \cdots + \frac{1}{2^\square} = \frac{2^\square - 1}{2^\square}$

$S_{k+1}:$ $\frac{1}{2} + \frac{1}{4} + \frac{1}{8} + \cdots + \frac{1}{2^k} + \frac{1}{2^\square} = \frac{2^\square - 1}{2^\square}$

1. Basis step. S_1 is true.

2. Induction step: Assume S_k is true.

$$\underbrace{\frac{1}{2} + \frac{1}{4} + \frac{1}{8} + \cdots + \frac{1}{2^k}} + \frac{1}{2^{k+1}}$$

$$= \boxed{} + \frac{1}{2^{k+1}} = \frac{2^k - 1}{2^k} \cdot \boxed{} + \frac{1}{2^{k+1}}$$

$$= \frac{2(2^k - 1)}{2^{k+1}} + \frac{1}{2^{k+1}} = \frac{2^{k+1} - 2 + 1}{2^{k+1}}$$

$$= \frac{2^{k+1} - 1}{2^{k+1}}$$

We have shown that for all natural numbers k, $S_k \to S_{k+1}$. This completes the Induction Step. It and the Basis Step tell us that the proof is complete.

Example 3   Prove: For every natural number n, $n < 2^n$.

$S_n : n < 2^n$

$S_1 : 1 < 2^{\boxed{}}$

$S_k : \boxed{} < 2^k$

$S_{k+1} : k+1 < 2^{\boxed{}}$

1. Basis step: S_1 is true.

2. Induction step: Assume S_k is true.

    $k < 2^k$ This is S_k.

    $2k < 2 \cdot 2^k$ Multiplying by 2 on both sides

    $2k < 2^{\boxed{}}$ Adding exponents on the right

    $k+k < 2^{k+1}$ Rewriting $2k$ as $k+k$

    Since k is any natural number, we know that $1 \le \boxed{}$.

    $k+1 \le k+k$ Adding k on both sides of $1 \le k$.

    Putting $k+k < 2^{k+1}$ and $k+1 \le k+k$ together, we have

    $k+1 \le \boxed{} < 2^{k+1}$

    $k+1 < 2^{\boxed{}}$

    We have shown for all natural numbers k, $S_k \rightarrow S_{k+1}$. This completes the Induction Step. It and the Basis Step tell us that proof is complete.

Section 12.5 Combinatorics: Permutations

In order to study probability, it is first necessary that we learn about **combinatorics**, the theory of counting. The study of permutations involves *order* and *arrangements*.

Example 1   How many 3-letter code symbols can be formed with the letters A, B, C *without* repetition (that is, using each letter only once)?

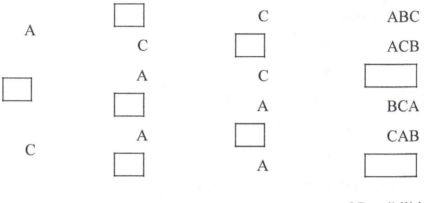

6 Possibilities

The set of all permutations is $\{$ABC, $\boxed{}$, BAC, BCA, CAB, CBA$\}$.

The Fundamental Counting Principle

Given a combined action, or *event*, in which the first action can be performed in n_1 ways, the second action can be performed in n_2 ways, and so on, the total number of ways in which the combined action can be performed is the product

$$n_1 \cdot n_2 \cdot n_3 \cdot \ \cdots \ \cdot n_k.$$

Example 2   How many 3-letter code symbols can be formed with the letters A, B, C, D, E *with* repetition (that is, allowing letters to be repeated)?

$5 \cdot 5 \cdot \boxed{} = \boxed{}$

Thus, there are $\boxed{}$ code symbols.

Permutations

A **permutation** of a set of n objects is an ordered arrangement of all n objects.

The total number of permutations of n objects, denoted $_nP_n$ is given by

$$_nP_n = n(n-1)(n-2)\cdots 3\cdot 2\cdot 1.$$

Example 3 Find each of the following.

a) $_4P_4 = 4\cdot\boxed{}\cdot 2\cdot 1 = \boxed{}$

b) $_7P_7 = \boxed{}\cdot 6\cdot\boxed{}\cdot 4\cdot 3\cdot 2\cdot 1 = \boxed{}$

Example 4 In how many ways can 9 packages be placed in 9 mailboxes, one package in a box?

$$_nP_n = n(n-1)(n-2)\cdots 3\cdot 2\cdot 1$$

$$_9P_9 = 9\cdot\boxed{}\cdot 7\cdot 6\cdot\boxed{}\cdot 4\cdot\boxed{}\cdot 2\cdot 1$$

$$= \boxed{}$$

Factorial Notation

- For any natural number n, $n! = n(n-1)(n-2)\cdots 3\cdot 2\cdot 1.$

- For any natural number n, $n! = n(n-1)!.$

- For the number 0, $0! = 1.$

- $_nP_n = n!$

Example 5 Rewrite 7! with a factor of 5!.

$$7! = 7\cdot\boxed{} = 7\cdot 6\cdot\boxed{}$$

The Number of Permutations of n Objects Taken k at a Time

The number of permutations of a set of n objects taken k at a time, denoted $_nP_k$, is given by

$$_nP_k = \underbrace{n(n-1)(n-2)\cdots\left[n-(k-1)\right]}_{k \text{ factors}} \quad (1)$$

$$= \frac{n!}{(n-k)!}. \quad (2)$$

Example 6   Compute $_8P_4$ using both forms of the formula.

$$_8P_4 = 8 \cdot \boxed{} \cdot 6 \cdot 5 = \boxed{}$$

$$_8P_4 = \frac{8!}{(8-4)!} = \frac{8!}{\boxed{}}$$

$$= \frac{8 \cdot 7 \cdot 6 \cdot 5 \cdot \boxed{}}{4!}$$

$$= 8 \cdot 7 \cdot 6 \cdot 5 = \boxed{}$$

Example 7   *Flags of Nations.* The flags of many nations consist of three vertical stripes. For example, the flag of Ireland has its first stripe green, its second white, and its third orange. Suppose that the following 9 colors are available:

{black, yellow, red, blue, white, gold, orange, pink, purple}.

How many different flags of three horizontal stripes can be made without repetition of colors in a flag? (This assumes that the order in which the stripes appear is considered.)

$$_9P_3 = 9 \cdot 8 \cdot \boxed{} = \boxed{}$$

Example 8   *Batting Orders.* A baseball manager arranges the batting order as follows: The 4 infielders will bat first. Then the 3 outfielders, the catcher, and the pitcher will follow, not necessarily in that order. How many different batting orders are possible?

4 Infielders $\boxed{}$ other players

$$_4P_4 \cdot {}_5P_5 = 4! \boxed{} = \boxed{}$$

The number of distinct arrangements of n objects taken k at a time, allowing repetition, is n^k.

Example 9   How many 5-letter code symbols can be formed with the letters A, B, C, and D if we allow a letter to occur more than once?

$$4 \cdot \boxed{} \cdot 4 \cdot 4 \cdot 4 = 4^{\boxed{}}$$

$$= \boxed{}$$

For a set of n objects in which n_1 are of one kind, n_2 are of another kind, …, and n_k are of a kth kind, the number of distinguishable permutations is $$\frac{n!}{n_1! \cdot n_2! \cdot \cdots \cdot n_k!}.$$

Example 10   In how many distinguishable ways can the letters of the word CINCINNATI be arranged?

   2 C's   3 I's   3 N's   1 A   1 T

$$N = \frac{\boxed{}}{2! \cdot \boxed{} \cdot 3! \cdot 1! \cdot 1!}$$

$$= 50,400$$

The letters of the word CINCINNATI can be arranged in 50,400 distinguishable ways.

Section 12.6 Combinatorics: Combinations

We sometimes make a selection from a set *without regard to order*. Such a selection is called a *combination*.

Example 1   Find all the combinations of 3 letters taken from the set of 5 letters $\{A, B, C, D, E\}$.

$\{A, B, C\}$ $\{A, B, \boxed{}\}$ $\{A, B, \boxed{}\}$

$\{A, \boxed{}, D\}$ $\{A, C, E\}$ $\{A, \boxed{}, E\}$

$\{\boxed{}, C, D\}$ $\{B, C, \boxed{}\}$ $\{B, D, E\}$

$\{C, \boxed{}, E\}$

There are 10 combinations of the 5 letters taken 3 at a time.

Combinations of *n* Objects Taken *k* at a Time

The total number of combinations of *n* objects taken *k* at a time, denoted $_nC_k$, is given by

$$_nC_k = \binom{n}{k} = \frac{n!}{k!(n-k)!}, \qquad (1)$$

or

$$_nC_k = \binom{n}{k} = \frac{_nP_k}{k!} = \frac{n(n-1)(n-2)\cdots[n-(k-1)]}{k!}. \qquad (2)$$

Example 2   Evaluate $\binom{7}{5}$, using forms (1) and (2).

$$\binom{7}{5} = \frac{7!}{5!(\boxed{})!} = \frac{7!}{5!\,2!} = \frac{7 \cdot 6 \cdot \boxed{}}{5!\,2!}$$

$$= \frac{7 \cdot 6}{\boxed{} \cdot 1} = \boxed{}$$

$$\binom{7}{5} = \frac{7 \cdot 6 \cdot 5 \cdot 4 \cdot \boxed{}}{5 \cdot 4 \cdot 3 \cdot \boxed{} \cdot 1}$$

$$= \frac{7 \cdot 6}{2 \cdot 1} = \boxed{}$$

Example 3 Evaluate $\binom{n}{0}$ and $\binom{n}{2}$.

$$\binom{n}{0} = \frac{n!}{\boxed{}(n-0)!} = \frac{n!}{1\cdot\boxed{}} = \boxed{} \qquad \text{Using form 1}$$

$$\binom{n}{2} = \frac{n\left(\boxed{}\right)}{2!} = \frac{n(n-1)}{\boxed{}}, \text{ or } \frac{n^2-n}{2} \qquad \text{Using form 2}$$

Subsets of Size k and of Size $n-k$

$$\binom{n}{k} = \binom{n}{n-k} \quad \text{and} \quad {}_nC_k = {}_nC_{n-k}$$

The number of subsets of size k of a set with n objects is the same as the number of subsets of size $n-k$. The number of combinations of n objects taken k at a time is the same as the number of combinations of n objects taken $n-k$ at a time.

Example 4 *Indiana Lottery.* Run by the state of Indiana, Hoosier Lotto is a twice-weekly lottery game with jackpots starting at $1 million. For a wager of $1, a player can choose 6 numbers from 1 through 48. If the numbers match those drawn by the state, the player wins the jackpot.

a) How many 6-number combinations are there?

$${}_{48}C_6 = \binom{48}{6}$$

$$= \frac{48!}{\boxed{}(48-6)!} = \frac{48!}{6!\,\boxed{}}$$

$$= \frac{48\cdot47\cdot46\cdot45\cdot44\cdot43\cdot\boxed{}}{6\cdot5\cdot4\cdot3\cdot2\cdot1\cdot42!}$$

$$= \frac{48\cdot47\cdot46\cdot45\cdot44\cdot43}{6\cdot5\cdot\boxed{}\cdot3\cdot2\cdot1}$$

$$= \boxed{}$$

b) Suppose it takes you 10 min to pick your numbers and buy a game ticket. How many tickets can you buy in 4 days?

$$4 \text{ days} = 4 \text{ days} \cdot \frac{\boxed{}}{1 \text{ day}} \cdot \frac{\boxed{}}{1 \text{ hr}}$$

$$= \boxed{} \text{ min}$$

$$\frac{5760}{10} = \boxed{} \text{ tickets}$$

c) How many people would you have to hire for 4 days to buy tickets with all the possible combinations and ensure that you win?

$$\frac{12,271,512}{\boxed{}} = \boxed{} \text{ people}$$

Example 5   How many committees can be formed from a group of 5 governors and 7 senators if each committee consists of 3 governors and 4 senators?

$$_nC_k = \frac{n!}{k!(n-k)!}$$

$$_5C_3 \cdot _7C_4 = \frac{5!}{\boxed{}(5-3)!} \cdot \frac{7!}{\boxed{}(7-4)!}$$

$$= \frac{5 \cdot 4 \cdot \boxed{}}{3! \cdot 2 \cdot 1} \cdot \frac{7 \cdot 6 \cdot 5 \cdot \boxed{}}{4! \cdot 3 \cdot 2 \cdot 1}$$

$$= \frac{5 \cdot 2 \cdot 2}{\boxed{}} \cdot \frac{7 \cdot 6 \cdot 5}{\boxed{}}$$

$$= 5 \cdot 2 \cdot 7 \cdot 5 = 10 \cdot \boxed{} = 350$$

There are $\boxed{}$ possible committees.

Section 12.7   The Binomial Theorem

Pascal's Triangle

$$1$$
$$1 \qquad 1$$
$$1 \qquad 2 \qquad 1$$
$$1 \qquad 3 \qquad 3 \qquad 1$$
$$1 \qquad 4 \qquad 6 \qquad 4 \qquad 1$$
$$1 \qquad 5 \qquad 10 \qquad 10 \qquad 5 \qquad 1$$
$$1 \qquad 6 \qquad 15 \qquad 20 \qquad 15 \qquad 6 \qquad 1$$

This pattern continues. Note that there are always 1's on the outside, and each remaining number is the sum of the two numbers above it.

The Binomial Theorem Using Pascal's Triangle

For any binomial $a + b$ and any natural number n,

$$(a+b)^n = c_0 a^n b^0 + c_1 a^{n-1} b^1 + c_2 a^{n-2} b^2 + \cdots + c_{n-1} a^1 b^{n-1} + c_n a^0 b^n,$$

where the numbers $c_0, c_1, c_2, ..., c_{n-1}, c_n$ are from the $(n+1)$st row of Pascal's triangle.

Example 1   Expand: $(u-v)^5$.

$$(a+b)^n \qquad a = u \qquad b = -v \qquad n = \boxed{}$$

$$(u-v)^5 = \left[u + \left(\boxed{} \right) \right]^5$$

We use the $\boxed{}$ row of Pascal's triangle.

$$1 \quad 5 \quad 10 \quad \boxed{} \quad 5 \quad 1$$

$$(u-v)^5 = 1(u)^5 + \boxed{}(u)^4(-v)^1 + 10(u)^{\boxed{}}(-v)^2$$

$$+ \boxed{}(u)^2(-v)^3 + 5(u)\left(\boxed{}\right)^4 + 1(-v)^{\boxed{}}$$

$$= u^5 \boxed{} 5u^4 v \boxed{} 10u^3 v^2 - 10u^2 v^3 + 5uv^4 - v^5$$

Example 2   Expand: $\left(2t + \dfrac{3}{t}\right)^4$.

$$(a+b)^n \quad a = 2t \quad b = \dfrac{3}{t} \quad n = 4$$

We use the $\boxed{}$ row of Pascal's triangle.

$$1 \quad \boxed{} \quad 6 \quad 4 \quad 1$$

$$\left(2t + \dfrac{3}{t}\right)^4 = 1(2t)^4 + 4(2t)^{\boxed{}}\left(\dfrac{3}{t}\right)^1 + \boxed{}(2t)^2\left(\dfrac{3}{t}\right)^2$$

$$+ 4(2t)^1\left(\boxed{}\right)^3 + 1\left(\dfrac{3}{t}\right)^4$$

$$= 1\left(\boxed{}\right) + 4(8t^3)\left(\dfrac{3}{t}\right) + 6(4t^2)\left(\boxed{}\right) + 4\left(\boxed{}\right)\left(\dfrac{27}{t^3}\right) + 1\left(\dfrac{81}{t^4}\right)$$

$$= 16t^4 + \boxed{} \cdot t^2 + \boxed{} + 216t^{\boxed{}} + 81t^{-4}$$

The Binomial Theorem Using Combination Notation

For any binomial $a + b$ and any natural number n,

$$(a+b)^n = \binom{n}{0}a^n b^0 + \binom{n}{1}a^{n-1}b^1 + \binom{n}{2}a^{n-2}b^2 + \cdots + \binom{n}{n-1}a^1 b^{n-1} + \binom{n}{n}a^0 b^n$$

$$= \sum_{k=0}^{n}\binom{n}{k}a^{n-k}b^k.$$

Example 3   Expand: $\left(x^2-2y\right)^5$.

$$(a+b)^n \quad a=x^2 \quad b=\boxed{} \quad n=5$$

$$\left(x^2-2y\right)^5=\left(\frac{5}{\boxed{}}\right)\left(x^2\right)^5+\binom{5}{1}\left(x^2\right)^{\boxed{}}(-2y)^1$$

$$+\binom{5}{2}\left(\boxed{}\right)^3(-2y)^2+\left(\frac{5}{\boxed{}}\right)\left(x^2\right)^2(-2y)^{\boxed{}}$$

$$+\binom{5}{4}\left(x^2\right)^1(-2y)^4+\left(\frac{5}{\boxed{}}\right)(-2y)^5$$

$$=\frac{5!}{0!5!}x^{\boxed{}}+\frac{5!}{1!\boxed{}}x^8(-2y)+\frac{5!}{\boxed{}3!}x^6\left(4y^2\right)$$

$$+\frac{5!}{3!2!}x^4\left(-8y^3\right)+\frac{5!}{4!1!}\boxed{}\left(16y^4\right)+\frac{5!}{5!0!}\left(\boxed{}\right)$$

$$=1x^{10}+\boxed{}\cdot x^8(-2y)+10\cdot x^6\left(4y^2\right)$$

$$+\boxed{}\cdot x^4\left(-8y^3\right)+5x^2\left(16y^4\right)+1\left(-32y^5\right)$$

$$=x^{10}\boxed{}10x^8y\boxed{}40x^6y^2-80x^4y^3+80x^2y^4\boxed{}32y^5$$

Example 4   Expand: $\left(\dfrac{2}{x}+3\sqrt{x}\right)^4$.

$$(a+b)^n \quad a=\frac{2}{x} \quad b=3\sqrt{x} \quad n=4$$

$$\left(\frac{2}{x}+3\sqrt{x}\right)^4=\binom{4}{0}\left(\frac{2}{x}\right)^4+\binom{4}{1}\left(\frac{2}{x}\right)^{\boxed{}}(3\sqrt{x})$$

$$+\left(\frac{4}{\boxed{}}\right)\left(\frac{2}{x}\right)^2(3\sqrt{x})^2+\binom{4}{3}\left(\frac{2}{x}\right)\left(\boxed{}\right)^3+\binom{4}{4}(3\sqrt{x})^4$$

$$=\frac{4!}{0!4!}\left(\frac{\boxed{}}{x^4}\right)+\frac{4!}{1!\boxed{}}\left(\frac{8}{x^3}\right)\left(3x^{\boxed{}}\right)$$

$$+\frac{4!}{2!2!}\left(\frac{4}{\boxed{}}\right)(9x)+\frac{4!}{3!1!}\left(\frac{2}{x}\right)\left(27x^{\boxed{}}\right)+\frac{4!}{4!0!}\left(\boxed{}x^2\right)$$

$$=\frac{16}{x^4}+\frac{96}{x^{\boxed{}}}+\frac{216}{\boxed{}}+216x^{\boxed{}}+81x^2$$

Finding the $(k+1)$st Term

The $(k+1)$st term of $(a+b)^n$ is $\binom{n}{k}a^{n-k}b^k$.

Example 5   Find the 5th term in the expansion of $(2x-5y)^6$.

$$\binom{n}{k}a^{n-k}b^k \qquad a=2x \qquad b=-5y$$

$$n=\boxed{}$$

$$5=4+1$$

$$k=\boxed{}$$

5th term is $\quad \binom{6}{4}\left(\boxed{}\right)^{6-4}(-5y)^{\boxed{}}$

$$= \frac{6!}{4!\,\boxed{}}(2x)^2(-5y)^4 = \boxed{}\,x^2y^4$$

Example 6   Find the 8th term in the expansion of $(3x-2)^{10}$.

$$\binom{n}{k}a^{n-k}b^k \qquad a=3x \qquad b=\boxed{}$$

$$n=10$$

$$8=7+1$$

$$k=\boxed{}$$

8th term is $\quad \binom{10}{7}(3x)^{10-\boxed{}}\left(\boxed{}\right)^7$

$$= \frac{10!}{7!\,\boxed{}}(3x)^3(-2)^7 = \boxed{}\,x^3$$

Total Number of Subsets

The total number of subsets of a set with n elements is 2^n.

Example 7   The set $\{A, B, C, D, E\}$ has how many subsets?

Number of subsets is $2^{\boxed{}}$, or $\boxed{}$.

Example 8   Wendy's, a national restaurant chain, offers the following toppings for its hamburgers:

{catsup, mustard, mayonnaise, tomato, lettuce, onions, pickle}.

In how many different ways can Wendy's serve hamburgers, excluding size of hamburger or number of patties?

$$\binom{7}{0} + \binom{7}{1} + \binom{7}{2} + \ldots + \binom{7}{7} = 2^{\square} = \boxed{}$$

Wendy's serves hamburgers in $\boxed{}$ different ways.

Section 12.8 Probability

Principle P (Experimental)

Given an experiment in which n observations are made, if a situation, or event, E occurs m times out of n observations, then we say that the **experimental probability** of the event, $P(E)$, is given by

$$P(E) = \frac{m}{n}.$$

Example 1 *Television Ratings.* There are an estimated 114,200,000 households in the United States that have at least one television. Each week, viewing information is collected and reported. One week, 28,510,000 households tuned in to the 2013 Grammy Awards ceremony on CBS, and 14,204,000 households tuned in to the action series "NCIS" on CBS (*Source:* Nielsen Media Research). What is the probability that a television household tuned in to the Grammy Awards ceremony during the given week? to "NCIS"?

$$P = \frac{28,510,000}{\boxed{}} \approx 0.2496 \approx \boxed{}\%$$

$$P = \frac{14,204,000}{\boxed{}} \approx 0.1244 \approx \boxed{}\%$$

Example 2 *Sociological Survey.* The authors of this text conducted an experimental survey to determine the number of people who are left-handed, right-handed, or both. The results are 82 right-handed, 17 left-handed, and 1 ambidextrous.

a) Determine the probability that a person is right-handed.

Right-handed	82
Left-handed	$\boxed{}$
Ambidextrous	1
Total	100

$$P(\text{Right-handed}) = \frac{\text{Number right-handed}}{\text{Total}} = \frac{\boxed{}}{100}, \text{ or } 0.82, \text{ or } 82\%$$

b) Determine the probability that a person is left-handed.

$$P(\text{Left-handed}) = \frac{\text{Number left-handed}}{\text{Total}} = \frac{17}{100}, \text{ or } 0.17, \text{ or } \boxed{}\%$$

c) Determine the probability that a person is ambidextrous (uses both hands with equal ability).

$$P(\text{Both}) = \frac{\text{Number ambidextrous}}{\text{Total}} = \frac{1}{100}, \text{ or } \boxed{}, \text{ or } 1\%$$

d) There are 120 bowlers in most tournaments held by the Professional Bowlers Association. On the basis of the data in this experiment, how many of the bowlers would you expect to be left-handed?

$$17\% \cdot 120 = \boxed{} \cdot 120 = 20.4$$

About $\boxed{}$ of the bowlers will be left-handed.

Example 3   *Dart Throwing.* Consider a dartboard divided into 3 equal sections: red, black, and white. Assume that the experiment is "throwing a dart" and that the dart hits the board. Find each of the following.

a) The outcomes

Hitting red (R)

Hitting black $(\boxed{})$

Hitting white (W)

b) The sample space

$$\left\{ R, \boxed{}, B \right\}$$

Example 4   *Die Rolling.* A die (pl., dice) is a cube, with six faces, each containing a number of dots from 1 to 6 on each side.

Suppose that a die is rolled. Find each of the following

a) The outcomes

$$1, 2, 3, \boxed{}, 5, 6$$

b) The sample space

$$\left\{ 1, \boxed{}, 3, 4, 5, 6 \right\}$$

Principle P (Theoretical)

If an event E can occur m ways out of n possible equally likely outcomes of a sample space S, then the **theoretical probability** of the event, $P(E)$, is given by

$$P(E) = \frac{m}{n}.$$

Example 5   Suppose that we select, without looking, one marble from a bag containing 3 red marbles and 4 green marbles. What is the probability of selecting a red marble?

$$P(\text{selecting a red marble}) = \frac{\boxed{}}{7}$$

Example 6   What is the probability of rolling an even number on a die?

1 2 3 4 5 6

$$P(\text{Rolling an even number}) = \frac{3}{\boxed{}} = \frac{1}{2}$$

Example 7   What is the probability of drawing an ace from a well-shuffled deck of cards?

A deck contains $\boxed{}$ cards.

There are $\boxed{}$ aces.

$$P(\text{drawing an ace}) = \frac{4}{52}, \text{ or } \frac{1}{\boxed{}}$$

Probability Properties

a) If an event E cannot occur, then $P(E) = 0$.

b) If an event E is certain to occur, then $P(E) = 1$.

c) The probability that an event E will occur is a number from 0 to 1: $0 \le P(E) \le 1$.

Example 8   Suppose that 2 cards are drawn from a well-shuffled deck of 52 cards. What is the probability that both of them are spades?

$$P(\text{drawing 2 spades}) = \frac{_{13}C_{\boxed{}}}{_{52}C_2}$$

$$= \frac{\boxed{}}{1326} = \frac{1}{\boxed{}}$$

Example 9   Suppose that 3 people are selected at random from a group that consists of 6 men and 4 women. What is the probability that 1 man and 2 women are selected?

$$P = \frac{_6C_1 \cdot _4C_2}{_{10}C_3} = \frac{3}{10}$$

Example 10   *Rolling Two Dice.* What is the probability of getting a total of 8 on a roll of a pair of dice?

There are $6 \cdot 6$, or 36 possible outcomes. Five of these give a total of 8.

$$P = \frac{\boxed{}}{36}$$

Answers

Chapter R

Section R.1

Your Turn: Set Notation and the Set of Real Numbers: 1. 6 **2.** $\{7, 8, 9, 10, 11, 12, 13, 14\}$
3. $\{x \mid x \text{ is a natural number greater than 6 and less than 11}\}$ **4. (a)** 125, 1; **(b)** 0, 125, 1;
(c) 0, -16, 125, 1; **(d)** 0.2, 0, -16, $-\dfrac{1}{9}$, 125, 1; **(e)** $\sqrt{7}$; **(f)** 0.2, 0, -16, $-\dfrac{1}{9}$, $\sqrt{7}$, 125, 1

Order for the Real Numbers: 1. $-11 < -6$ **2.** $1 > -15$ **3.** $\dfrac{2}{5} < \dfrac{4}{7}$ **4.** $10 > y$ **5.** True

Graphing Inequalities on the Number Line: 1.

2. **Absolute Value: 1.** 39 **2.** $\dfrac{7}{8}$

Practice Exercises: 1. False **2.** False **3.** True **4.** True **5. (a)** 4; **(b)** 0, 4; **(c)** $-1, 0, 4$;
(d) $-3.1, -1, 0, \dfrac{75}{2}, 4$; **(e)** $\sqrt{10}$; **(f)** $-3.1, -1, 0, \sqrt{10}, \dfrac{75}{2}, 4$ **6. (a)** 51; **(b)** 51; **(c)** 51;
(d) $-\dfrac{3}{8}, \dfrac{1}{2}, \dfrac{43}{2}, 51$; **(e)** $\sqrt{12}$; **(f)** $-\dfrac{3}{8}, \dfrac{1}{2}, \sqrt{12}, \dfrac{43}{2}, 51$ **7.** $\{h, i, s, t, o, r, y\}$
8. $\{5, 10, 15, 20, 25, \ldots\}$ **9.** $\{x \mid x \text{ is a multiple of 10 between 20 and 100}\}$
10. $\{x \mid x \text{ is an even number between 71 and 81}\}$ **11.** $>$ **12.** $<$ **13.** $>$ **14.** $-7 \le x$ **15.** True
16. True **17.** False **18.** **19.** **20.** 5
21. 3.8 **22.** 0

Section R.2

Your Turn: Addition: 1. 3 **2.** -17 **3.** 0 **4.** 8 **5.** $\dfrac{1}{10}$ **Opposites, or Additive Inverses:**
1. 15 **2.** $-7; 7$ **3.** $-\dfrac{2}{25}$ **Subtraction: 1.** -5 **2.** $\dfrac{3}{4}$ **Multiplication: 1.** -88 **2.** $\dfrac{4}{15}$
Division: 1. 7.4 **2.** 0 **3.** $-\dfrac{7}{4}$ **4.** $-\dfrac{3}{5}$

Practice Exercises: 1. add **2.** positive **3.** not defined **4.** multiply **5.** -23 **6.** -20.2
7. $\dfrac{3}{14}$ **8.** -13.6 **9.** $\dfrac{7}{2}$ **10.** 0 **11.** 14 **12.** 2.5 **13.** 32 **14.** -47.4 **15.** $-\dfrac{2}{15}$ **16.** -15
17. 48.8 **18.** $-\dfrac{3}{8}$ **19.** 2 **20.** -7 **21.** Not defined **22.** $-\dfrac{5}{7}$ **23.** $-\dfrac{1}{8}$ **24.** $\dfrac{8}{5}$ **25.** -4

Section R.3

Your Turn: Exponential Notation: 1. m^6 **2.** 49 **3.** $\dfrac{25}{64}$ **4.** 102.01 **5.** 16

Negative Integers as Exponents: 1. $\dfrac{1}{t^9}$ **2.** $\dfrac{1}{25}$ **3.** $\dfrac{125}{27}$ **4.** $(-9)^{-2}$

Order of Operations: 1. -8 **2.** -9 **3.** $\dfrac{1}{4}$

Practice Exercises: 1. division **2.** multiplication **3.** division **4.** subtraction

5. multiplication **6.** addition **7.** t^3 **8.** $\left(\dfrac{5}{11}\right)^5$ **9.** $(-54.9)^4$ **10.** 16 **11.** -125 **12.** 1.2

13. 1 **14.** $\dfrac{8}{27}$ **15.** y^5 **16.** $-\dfrac{1}{27}$ **17.** 5^{-3} **18.** x^{-5} **19.** $(-6)^{-3}$ **20.** 1 **21.** -43 **22.** -108

23. -7 **24.** $-\dfrac{1}{2}$ **25.** 6

Section R.4

Your Turn: Translating to Algebraic Expressions: 1. $s-5$ **2.** $\dfrac{1}{3}n-4$

3. $42\%\cdot w$, or $0.42w$ **4.** $2+\dfrac{a}{b}$, or $\dfrac{a}{b}+2$ **Evaluating Algebraic Expressions: 1.** -130

2. 12.6 ft^2 **3.** 57

Practice Exercises: 1. (c) **2.** (a) **3.** (b) **4.** (d) **5.** $3d$ **6.** $5+2x$, or $2x+5$

7. $a-b+6$, or $6+a-b$ **8.** $20\%y-7$, or $0.20y-7$ **9.** $10+m$, or $m+10$ **10.** $\dfrac{1}{5}(3q)$

11. $25y-1$ **12.** $r-34$ **13.** $P-15\%P$, or $P-0.15P$ **14.** \$8.65h **15.** 6 **16.** 54 **17.** 26
18. -33 **19.** -20 **20.** 14 **21.** 9.45 ft^2

Section R.5

Your Turn: Equivalent Expressions: 1. $19,-20$; $4,5$; $14.56,17$; The expressions are not

equivalent. **Equivalent Fraction Expressions: 1.** $\dfrac{12}{21t}$ **2.** $\dfrac{b}{5}$ **3.** $-\dfrac{7}{2}$

The Commutative Laws and the Associative Laws: 1. $-70;-70$ **2.** $15;15$

3. $(3\cdot y)\cdot x$; $(x\cdot 3)\cdot y$; $3\cdot(x\cdot y)$; answers may vary **The Distributive Laws: 1.** $5x+10$

2. $-10m+14n-18p$ **3.** $4(2a-3b+5)$ **4.** $m(x+y+z)$

Practice Exercises: 1. (c) **2.** (d) **3.** (b) **4.** (f) **5.** (a)
6. $12,-3,12$; $27.2,12.2,27.2$; $20,5,20$; $4(x+5)$ and $4x+20$ are equivalent expressions.

7. $\dfrac{3y}{7y}$ **8.** $\dfrac{6a}{15a}$ **9.** $-\dfrac{8}{3}$ **10.** $\dfrac{9a}{4}$ **11.** $n+6$ **12.** yx **13.** $u+(v+3)$ **14.** $5\cdot(c\cdot p)$

15. $y+(5+z)$, $(z+5)+y$, $(y+5)+z$; answers may vary

16. $(4\cdot x)\cdot y$, $y\cdot(x\cdot 4)$, $4\cdot(y\cdot x)$; answers may vary **17.** $3x+30$ **18.** $-21x+6y+3z$

19. $5x, -3y$ **20.** $-3a, -b, 18c$ **21.** $7(a+b)$ **22.** $2(2x-1)$ **23.** $3(2m+4n-5)$
24. $2 \cdot \pi \cdot r \cdot (h+r)$

Section R.6

Your Turn: Collecting Like Terms: 1. $13x$ **2.** $5x+6.9$ **3.** $\dfrac{1}{14}x - \dfrac{5}{24}y$

Multiplying by –1 and Removing Parentheses: 1. $19a+8b+12$ **2.** $10x-5y-6$
3. $-5d+10$

Practice Exercises: 1. equal **2.** opposite **3.** opposite **4.** equal **5.** $9m$ **6.** $-3n$ **7.** $13p$

8. $6a+13b$ **9.** $5.3s-7.6t$ **10.** $-\dfrac{7}{15}x + \dfrac{1}{10}y$ **11.** $5a$ **12.** $-d-14$ **13.** $-3x-2y+8$

14. $3f-4g-2h+j$ **15.** $x+\dfrac{1}{5}w+4.7y-15.4z$ **16.** $9b-7$ **17.** $8x-23$ **18.** $11y-57$

19. $-a-12$ **20.** $354+198x$ **21.** $6-34y$

Section R.7

Your Turn: Multiplication and Division: 1. $-24x^{6n}$ **2.** $-14a^{11}b$ **3.** $\dfrac{-3}{y^{4n}}$ **4.** $-9m^3n$

Raising Powers to Powers and Products and Quotients to Powers: 1. x^{-24p} **2.** $\dfrac{x^{28}y^8}{81z^{20}}$

3. $\dfrac{16}{x^4 y^{12}}$ **Scientific Notation: 1.** 7.01×10^{13} **2.** 0.004078 **3.** 7.92×10^3 **4.** 8.0×10^{10}

Practice Exercises: 1. add **2.** multiply **3.** scientific **4.** negative **5.** 3^{11} **6.** $32x^4$ **7.** $-\dfrac{18}{a^2}$

8. x^7 **9.** $\dfrac{5}{x^{20}}$ **10.** $-9x^3y^4$ **11.** $36x^2y^4$ **12.** $\dfrac{x^4}{4y^6}$ **13.** $\dfrac{27x^9}{64y^3}$ **14.** $\dfrac{81}{16x^{44}y^{24}}$

15. 4.7×10^{-2} **16.** 5.0179×10^{11} **17.** 1.9×10^5 **18.** $81{,}300{,}000$ **19.** 0.00204 **20.** $24{,}000$
21. 6×10^9 **22.** 1.305×10^{-1} **23.** 4.0×10^{-20} **24.** 2.5×10^{15}

Chapter 1

Section 1.1
Your Turn: Equations and Solutions: 1. True **2.** False **3.** Neither **4.** No **5.** Yes

The Addition Principle: 1. -24 **2.** -3.4 **3.** $-\dfrac{17}{12}$ **The Multiplication Principle: 1.** -6

2. 13 **3.** 2200 **4.** 32 **Using the Principles Together: 1.** $\dfrac{9}{2}$ **2.** 4.1 **3.** No solution

4. All real numbers are solutions.

A-4

Practice Exercises: 1. (f) **2.** (a) **3.** (d) **4.** (e) **5.** No **6.** Yes **7.** No **8.** 2 **9.** 13 **10.** 10.9 **11.** $-\dfrac{1}{12}$ **12.** 8 **13.** −17 **14.** 0.9 **15.** $-\dfrac{3}{2}$ **16.** 6 **17.** $-\dfrac{1}{2}$ **18.** 7.6 **19.** All real numbers **20.** No solution **21.** −6

Section 1.2
Your Turn: Evaluating and Solving Formulas: 1. 227.5 miles **2.** $y = 2A - x$ **3.** $n = \dfrac{B - 3m}{4}$ **4.** $b = \dfrac{2a}{ac + 1}$

Practice Exercises: 1. False **2.** True **3.** True **4.** False **5.** 10.75 meters per cycle **6.** $t = \dfrac{d}{60}$ **7.** $x = 26 - y$ **8.** $a = Tb$ **9.** $x = \dfrac{3y}{2}$ **10.** $x = \dfrac{y - b}{m}$ **11.** $q = 3A - p - r$ **12.** $w = \dfrac{P - 2l}{2}$ **13.** $r^2 = \dfrac{S}{4\pi}$ **14.** $r = \dfrac{d}{t}$ **15.** $y = \dfrac{x}{a^2 + z}$ **16.** 62 inches **17.** 12 ft

Section 1.3
Your Turn: Five Steps for Problem Solving: 1. 20,500 thousand metric tons **2.** −13, −12, −11 **Basic Motion Problems: 1.** $\dfrac{2}{3}$ hr, or 40 min

Practice Exercises: 1. Familiarize **2.** Translate **3.** Solve **4.** Check **5.** State **6.** $68.68 **7.** 6, 8 **8.** 215 units **9.** 42, 43 **10.** $12°, 60°, 108°$ **11.** 110 sec

Section 1.4
Your Turn: Inequalities: 1. No **2.** No **3.** Yes **Inequalities and Interval Notation: 1.** $(-4, 3]$ **2.** $(-\infty, -7)$ **3.** $(6, 11]$ **4.** $(8, \infty)$ **Solving Inequalities:**
1. $\{x \mid x > -6\}$, or $(-6, \infty)$; **2.** $\{x \mid x > 7\}$, or $(7, \infty)$;
3. $\{x \mid x \geq 5\}$, or $[5, \infty)$; **4.** $\{x \mid x \leq 9\}$, or $(-\infty, 9]$;
5. $\{x \mid x \leq 4\}$, or $(-\infty, 4]$; **Applications and Problem Solving:**
1. More than $1046.02 in sales; $\{S \mid S > 1046.02\}$ **2.** More than 25 guests; $\{g \mid g > 25\}$

Practice Exercises: 1. (d) **2.** (a) **3.** (e) **4.** (f) **5.** (c) **6.** (b) **7.** No **8.** Yes **9.** No **10.** $[-2, \infty)$ **11.** $[-3, 5]$ **12.** $(-\infty, 6)$ **13.** $(0, 5]$ **14.** $\{x \mid x > 4\}$, or $(4, \infty)$;
15. $\{x \mid x > -6\}$, or $(-6, \infty)$; **16.** $\{x \mid x \leq 1\}$, or $(-\infty, 1]$;
17. $\{x \mid x \geq -3\}$, or $[-3, \infty)$; **18.** $\left\{x \mid x < \dfrac{5}{3}\right\}$, or $\left(-\infty, \dfrac{5}{3}\right)$;

19. Times more than 170 hours; $\{t \mid t > 170\}$

20. Scores greater than or equal to 92; $\{S \mid S \geq 92\}$

21. Times less than 40 hours; $\{t \mid t < 40\}$

Section 1.5

Your Turn: Intersections of Sets and Conjunctions of Inequalities: 1. $\{0, 1, 8\}$

2. $\{y \mid -1 \le y < 4\}$, or $[-1, 4)$; **3.** $\{x \mid x > 4\}$, or $(4, \infty)$;

Unions of Sets and Disjunctions of Inequalities: 1. $\{0, 1, 3, 9\}$ **2.** $\{x \mid x < 2 \ or \ x > 4\}$,

or $(-\infty, 2) \cup (4, \infty)$; **3.** $\{x \mid x \le 8\}$, or $(-\infty, 8]$;

Applications and Problem Solving: 1. Temperatures from $167°F$ to $176°F$; $\{F \mid 167° \le F \le 176°\}$

Practice Exercises: 1. False **2.** True **3.** False **4.** False **5.** True **6.** True **7.** $\varnothing$ **8.** $\{a, c\}$

9. $\{x \mid 3 \le x < 7\}$, or $[3, 7)$; **10.** $\{x \mid -3 < x < 2\}$, or $(-3, 2)$;

11. $\{x \mid x > -3\}$, or $(-3, \infty)$; **12.** $\varnothing$ **13.** $\{1, 2, 3, a, b, c\}$ **14.** $\{1, 2, 3\}$

15. $\{x \mid x \le 0 \ or \ x \ge 3\}$, or $(-\infty, 0] \cup [3, \infty)$; **16.** $\{x \mid x < -1 \ or \ x \ge 4\}$, or

$(-\infty, -1) \cup [4, \infty)$; **17.** $\{x \mid x \ge 2\}$, or $[2, \infty)$;

18. $\{x \mid x \text{ is a real number}\}$, or $(-\infty, \infty)$;

19. $-56.2° \le F \le 44.6°$ **20.** Scores greater than or equal to 86; $\{x \mid x \ge 86\}$

Section 1.6

Your Turn: Properties of Absolute Value: 1. $5x^4$ **2.** $4|y|$ **3.** $\dfrac{x^2}{3}$ **Distance on the**

Number Line: 1. 21 **2.** 11 **Equations with Absolute Value: 1.** $\varnothing$ **2.** $\{-10, 10\}$

3. $\{-2, 4\}$ **Equations with Two Absolute-Value Expressions: 1.** $\{-1, 1\}$ **2.** $\{0\}$

Inequalities with Absolute Value: 1. $\{x \mid -2 < x < 2\}$, or $(-2, 2)$;

2. $\{y \mid y \le -1 \ or \ y \ge 1\}$, or $(-\infty, -1] \cup [1, \infty)$;

3. $\{x \mid -1 < x < 5\}$, or $(-1, 5)$;

4. $\{x \mid x \le 1 \ or \ x \ge 7\}$, or $(-\infty, 1] \cup [7, \infty)$;

Practice Exercises: 1. True **2.** False **3.** False **4.** True **5.** True **6.** $3|x|$ **7.** $6x^2$ **8.** 27

9. 5 **10.** $\{-16, 16\}$ **11.** $\{-6, 10\}$ **12.** $\left\{-\dfrac{2}{5}, 0\right\}$ **13.** $\left\{-\dfrac{2}{7}\right\}$ **14.** $\left\{-1, -\dfrac{1}{7}\right\}$ **15.** $\{-4\}$

16. $\left\{-4, -\dfrac{4}{7}\right\}$ **17.** $\{x \mid x \text{ is a real number}\}$

18. $\{x \mid -4 \le x \le 4\}$, or $[-4, 4]$;

19. $\{x \mid -7 < x < -1\}$, or $(-7, -1)$;

20. $\{x \mid x < -3 \ or \ x > 3\}$, or $(-\infty, -3) \cup (3, \infty)$;

21. $\{x \mid x \le -3 \ or \ x \ge -1\}$, or $(-\infty, -3] \cup [-1, \infty)$;

Chapter 2

Section 2.1

Your Turn: Plotting Ordered Pairs: 1.

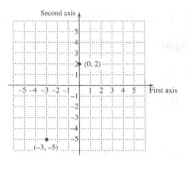

2. $(-3,-1)$: III

$(5,-2)$: IV

$(6,0)$: On an axis,
        not in a quadrant

$(-1,4)$: II

Solutions of Equations:

1. $(0,-2)$ is a solution; $(5,1)$ is not a solution.

2.

$$\begin{array}{c|c}
y = 2x - 5 \\
\hline
-1\ ?\ 2(2) - 5 \\
\quad\ \ 4 - 5 \\
\quad\ \ -1 \qquad\qquad \text{TRUE}
\end{array}$$

$$\begin{array}{c|c}
y = 2x - 5 \\
\hline
-5\ ?\ 2(0) - 5 \\
\quad\ \ 0 - 5 \\
\quad\ \ -5 \qquad\qquad \text{TRUE}
\end{array}$$

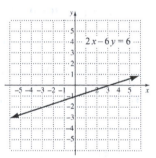

$(3,1)$; answers may vary

Graphs of Linear Equations: 1.

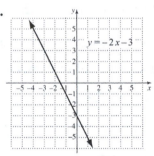

2.

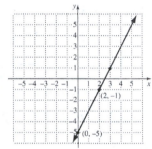

Graphs of Nonlinear Equations: 1.

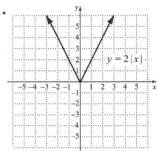

2.

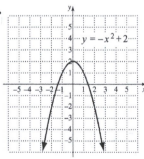

Practice Exercises: **1.** True **2.** False **3.** True **4.** True **5.** False

6.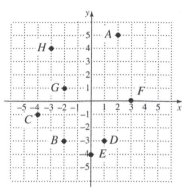

7. Yes **8.** Yes **9.** No

$$y = 1 - 2x$$

$$5 \; ? \; 1 - 2(-2)$$

$\begin{vmatrix} 1 + 4 \\ 5 \end{vmatrix}$ TRUE

$(-2, 5)$ is a solution

$$y = 1 - 2x$$

$$-1 \; ? \; 1 - 2(1)$$

$\begin{vmatrix} 1 - 2 \\ -1 \end{vmatrix}$ TRUE

$(1, -1)$ is a solution

10.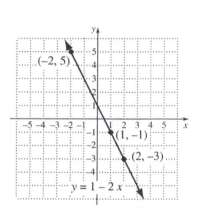

$(2, -3)$; answers may vary

11.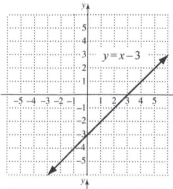

12.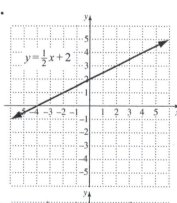

13.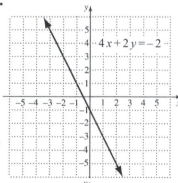

14.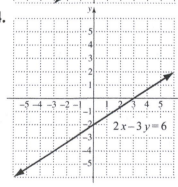

15.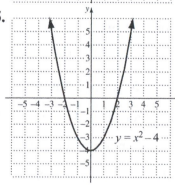

16.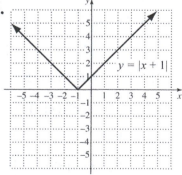

Section 2.2

Your Turn: Identifying Functions: 1. Not a function **2.** A function **3.** Not a function
4. A function **Finding Function Values: 1.** $g(0)=3$, $g(-1)=7$, $g(a+2)=-4a-5$

2. 7 **3.** $h(1)=-2$, $h(-1)=4$ **4.** 36π cm$^2 \approx 113.04$ cm^2 **Graphs of Functions:**

1.
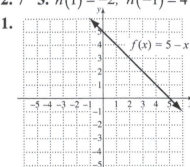

2.
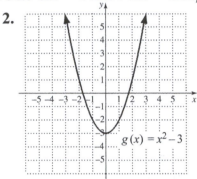

The Vertical-Line Test: 1. Yes **Applications of Functions and Their Graphs:**
1. Approximately 1000 for-profit hospitals

Practice Exercises:
1. function; domain; range **2.** vertical; function **3.** output **4.** A function
5. Not a function **6.** 14 **7.** 14 **8.** $4a-11$ **9.** 5 **10.** 150 cm^2

11.
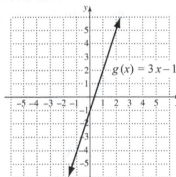

12.
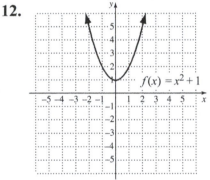

13. A function **14.** Not a function **15.** Approximately 248 hospitals

Section 2.3

Your Turn: Finding Domain and Range: 1. Domain: $\{-4,-2,2,4\}$; range: $\{-2,0,1,3\}$

2. Domain: $\{x|-5\le x\le 4\}$; range: $\{y|-5\le y\le 1\}$ **3. a)** 1; **b)** $\{x|x\ge -3\}$; **c)** 1; **d)** $\{y|y\ge 0\}$

4. $\{x|x$ is a real number *and* $x\ne 4\}$ **5.** All real numbers

Practice Exercises: 1. function **2.** relation **3.** domain **4.** range **5. a)** 3; **b)** $\{-3,0,2,4\}$;

c) -3; **d)** $\{-3,-2,1,3\}$ **6. a)** 1; **b)** all real numbers; **c)** 2, 4 **d)** $\{y|y\le 2\}$

7. a) 4; **b)** $\{x|-4\le x\le 5\}$; **c)** 0; **d)** $\{y|-2\le y\le 4\}$ **8.** All real numbers

9. $\{x|x$ is a real number $and\ x \ne 5\}$ **10.** $\{x|x$ is a real number $and\ x \ne -\dfrac{3}{2}\}$

11. All real numbers   **12.** All real numbers

Section 2.4

Your Turn: The Sum, Difference, Product, or Quotient of Two Functions:

1. $3x^2 - x + 4$ **2.** -8 **3.** $2t^2 + 7t - 4$ **4.** $-\dfrac{7}{2}$ **Determining Domain:**

1. $\{x|x$ is a real number $and\ x \ne 0\}$ **2.** $\{x|x$ is a real number $and\ x \ne 2\ and\ x \ne 7\}$

Practice Exercises: 1. (c)   **2.** (d)   **3.** (a)   **4.** (b)   **5.** 6   **6.** 2   **7.** $-x^3 + 3x^2 - x + 3$ **8.** $\dfrac{3}{5}$

9. 18   **10.** $9 - 6x + x^2$ **11.** $\dfrac{a^2 + 1}{3 - a}, \ a \ne 3$ **12.** -14 **13.** $\{x|x$ is a real number$\}$

14. $\{x|x$ is a real number$\}$ **15.** $\{x|x$ is a real number $and\ x \ne 0\}$

16. $\{x|x$ is a real number $and\ x \ne 1\}$ **17.** $\{x|x$ is a real number $and\ x \ne -2\}$

18. $\{x|x$ is a real number $and\ x \ne 3\}$ **19.** $\{x|x$ is a real number $and\ x \ne -4\ and\ x \ne 8\}$

20. $\{x|x$ is a real number $and\ x \ne 1\ and\ x \ne 3\}$

Section 2.5

Your Turn: The Constant b: The y-Intercept: 1. $(0, 13)$ **2.** $(0, -9.5)$ **3.** $(0, 3)$

The Constant m: Slope: 1. $\dfrac{2}{5}$ **2.** 9   **3.** $-\dfrac{1}{2}$ **Applications: 1.** 15 min/page

2. 1.5 cups/pie

Practice Exercises: 1. y-intercept   **2.** slope   **3.** down   **4.** slope-intercept form

5. $\left(0, -\dfrac{1}{2}\right)$ **6.** $(0, -1)$ **7.** 3   **8.** 1   **9.** -5 **10.** Slope: $\dfrac{3}{8}$; y-intercept: $(0, -6)$

11. Slope: $\dfrac{3}{2}$; y-intercept: $(0, 3)$ **12.** Slope: 1; y-intercept: $(0, -9)$

13. Slope: $-\dfrac{1}{3}$; y-intercept: $\left(0, \dfrac{2}{3}\right)$ **14.** $1\dfrac{1}{6}$ miles per minute   **15.** $\dfrac{1}{20}$ mile per minute

Section 2.6
Your Turn: Graphing Using Intercepts:

1.

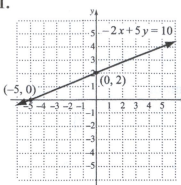

$-2x + 5y = 10$

$(-5, 0)$ $(0, 2)$

2.

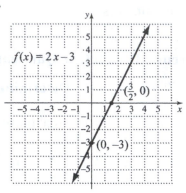

$f(x) = 2x - 3$

$\left(\frac{3}{2}, 0\right)$

$(0, -3)$

Graphing Using the Slope and the y-Intercept:

1.

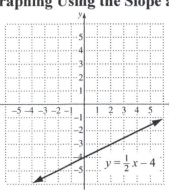

$y = \frac{1}{2}x - 4$

2.

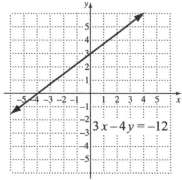

$3x - 4y = -12$

Horizontal Lines and Vertical Lines:

1.

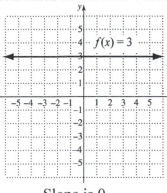

$f(x) = 3$

Slope is 0.

2.

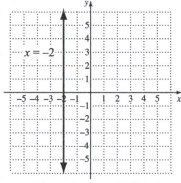

$x = -2$

Slope is not defined.

3.

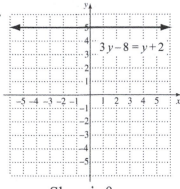

$3y - 8 = y + 2$

Slope is 0.

Parallel Lines and Perpendicular Lines: 1. Not parallel **2.** Perpendicular

Practice Exercises: 1. vertical **2.** parallel **3.** −1 **4.** rise; run

5.
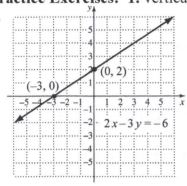

6.
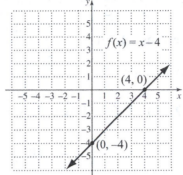

7.
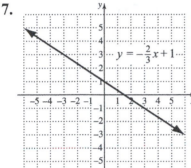

8.
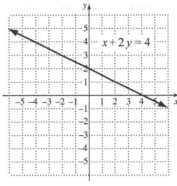

9.
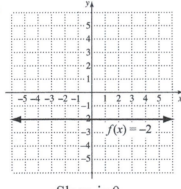

Slope is 0.

10.
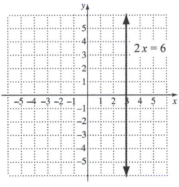

Slope is not defined.

11. Parallel **12.** Not parallel **13.** Perpendicular **14.** Perpendicular

Section 2.7

Your Turn: Finding an Equation of a Line When the Slope and the *y*-Intercept Are Given: 1. $y = -7x + 5$ **2.** $f(x) = 4x - 10$ **Finding an Equation of a Line When the Slope and a Point Are Given: 1.** $y = \dfrac{2}{3}x - \dfrac{35}{3}$ **2.** $y = -x + 7$ **Finding an Equation of a Line When Two Points Are Given: 1.** $y = -\dfrac{1}{2}x + \dfrac{1}{2}$ **2.** $y = -x + 5$

Finding an Equation of a Line Parallel or Perpendicular to a Given Line Through a Point Not on the Line: 1. $y = -\frac{2}{5}x + \frac{33}{5}$ 2. $y = -2x - 9$

Applications of Linear Functions: 1. $e(t) = -\frac{1}{2}t + \frac{49}{2}$; $17,500

2. $h(p) = -15p + 250$; 25 headbands

Practice Exercises: 1. (d) 2. (b) 3. (a) 4. (c) 5. $y = -4x + 8$ 6. $f(x) = \frac{1}{2}x - 1$

7. $y = 6x - 18$ 8. $y = -\frac{1}{2}x - 7$ 9. $y = \frac{1}{3}x + \frac{13}{3}$ 10. $y = 2x - 5$ 11. $y = x - 2$

12. $y = \frac{4}{5}x - 9$ 13. a) $C(t) = 22.95t + 150$; b) $517.20 14. a) $N(x) = 34x + 150$;
b) 524 students

Chapter 3

Section 3.1
Your Turn: Solving Systems of Equations Graphically:

1.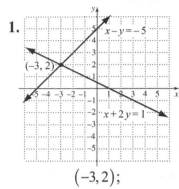

$(-3, 2)$;
consistent; independent

2.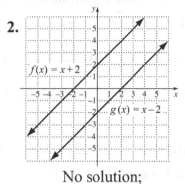

No solution;
inconsistent; independent

3.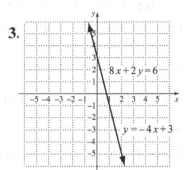

Infinitely many solutions;
consistent; dependent

Practice Exercises: 1. (a), (c), (d) **2.** (b) **3.** (d) **4.** (a), (b), (c) **5.** (b) **6.** (c)

7.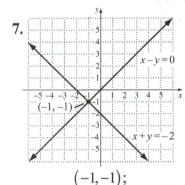

$(-1, -1)$;
consistent; independent

8.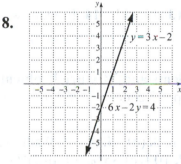

Infinitely many solutions;
consistent; dependent

9.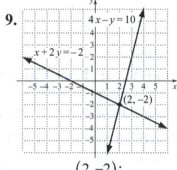

$(2, -2)$;
consistent; independent

10.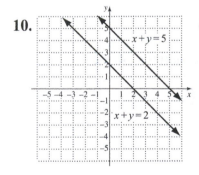

No solution;
inconsistent; independent

11.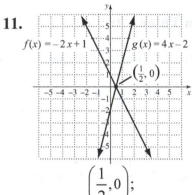

$\left(\dfrac{1}{2}, 0\right)$;

No solution;
inconsistent; independent

consistent; independent

12.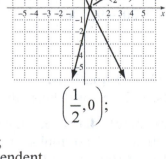

consistent; independent

Section 3.2

Your Turn: The Substitution Method: 1. $(2, -2)$ **2.** No solution

3. Infinitely many solutions

Solving Applied Problems Involving Two Equations: 1. Length: 35 ft ; width: 20 ft

Practice Exercises: 1. False **2.** False **3.** True **4.** False **5.** $(3, 1)$ **6.** $(2, 5)$ **7.** $(-5, -3)$

8. No solution **9.** Infinitely many solutions **10.** $\left(\dfrac{3}{4}, -\dfrac{1}{4}\right)$ **11.** $(-4, 3)$

12. Infinitely many solutions **13.** Length 9 in.; width 7 in. **14.** $11°, 79°$

Section 3.3

Your Turn: The Elimination Method: 1. $(1, -2)$ **2.** $(4, -1)$ **3.** $(3, -3)$ **4.** No solution

Solving Applied Problems Using Elimination: 1. Two-point shots: 24; three-point shots: 9

Practice Exercises: 1. True **2.** False **3.** True **4.** True **5.** $(4, 2)$ **6.** $(-3, 1)$

7. Infinitely many solutions **8.** $(-2, -3)$ **9.** No solution **10.** Infinitely many solutions

11. $\left(\dfrac{2}{5}, -\dfrac{7}{5}\right)$ **12.** No solution **13.** $15, -11$ **14.** Balcony seats: 80; floor seats: 295

Section 3.4

Your Turn: Total-Value Problems and Mixture Problems:
1. One-scoop: 8 cones; two-scoop: 17 cones **2.** 96 L of 60%; 24 L of 20%
Motion Problems: 1. 30 mi

Practice Exercises: 1. (b) **2.** (a) **3.** (d) **4.** (c) **5.** 9 adults; 18 children
6. $3.89: 14 gal; $3.49: 11 gal **7.** 10 L of 35%; 5 L of 20%
8. Fruit punch: 24.5 L; sparkling water: 10.5 L **9.** 15 km/h **10.** 650 mi

Section 3.5

Your Turn: Solving Systems in Three Variables: 1. $(-2, 4, 1)$

Practice Exercises: 1. three; three **2.** linear **3.** triple; all three equations **4.** $(2, -1, 3)$

5. $(-2, 0, -1)$ **6.** $\left(\dfrac{1}{2}, 3, 2\right)$ **7.** $(12, 20, -9)$ **8.** $\left(4, \dfrac{1}{2}, -\dfrac{1}{2}\right)$ **9.** $(18, -11, 3)$

Section 3.6

Your Turn: Using Systems of Three Equations:
1. Adult tickets: 296; child tickets: 102; senior tickets: 46

Practice Exercises: 1. (b) **2.** (d) **3.** (c) **4.** (a) **5.** 24, 48, 72
6. 2-point field goals: 24; 3-point field goals: 6; foul shots: 18
7. $20 gift cards: 6; $50 gift cards: 3; $100 gift cards: 1 **8.** 6, 24, 10

Section 3.7
Your Turn: Graphing Inequalities in Two Variables:

1.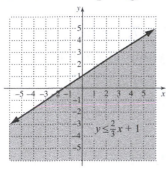

2.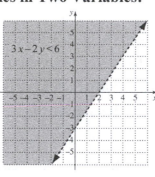

3.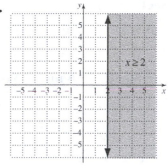

Systems of Linear Inequalities: 1.

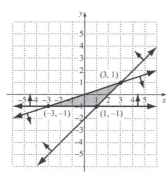

Applications: Linear Programming: 1. Maximum profit of $11,500 occurs when 50 tuxedos and 150 dresses are made.

Practice Exercises: 1. is not 2. dashed 3. same 4. a plane 5. intersection

6.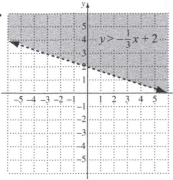

7.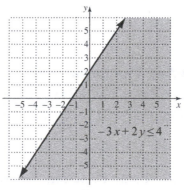

8.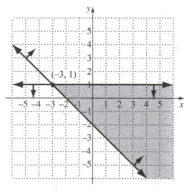

9.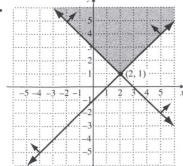

10.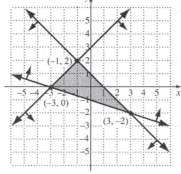

11. Maximum profit of $1300 occurs when 3 princess playhouses and 2 pirate playhouses are made and sold.

Chapter 4

Section 4.1

Your Turn: Polynomial Expressions: 1. $12y^4, -7y^7, -6y^3, -8$; $4, 7, 3, 0$; 7; $-7y^7$; -7; -8
2. $-7y^7 + 12y^4 - 6y^3 - 8$; $-8 - 6y^3 + 12y^4 - 7y^7$ **Evaluating Polynomial Functions:**
1. -70 **2.** 30.6 mpg   **Adding Polynomials: 1.** $-12x^2 - 7x + 5$ **2.** $17x^2y^3 - 7x^2y - 3x^4y$
3. $12y^4 + 2y^3 - 3y + 5$ **Subtracting Polynomials: 1.** $-\left(-6xy^2 + 5xy - 16y + 20\right)$;
$6xy^2 - 5xy + 16y - 20$ **2.** $-2x^2 + 8x^3y^3 - 11$ **3.** $x^2 - x + 1$

Practice Exercises: 1. binomial; 4, 0   **2.** 4; 5; 4   **3.** ascending   **4.** like terms

5. $-6x$, x^3, 8, $3x^4$; 1, 3, 0, 4; 4; $3x^4$; 3; 8

6. 5, $-12x^4$, $-6x^7$, $3x^3$, $-2x$; 0, 4, 7, 3, 1; 7; $-6x^7$, -6; 5

7. $-3xy$, $-6xyz$, $2x^2$, -7; 2, 3, 2, 0; 3; $-6xyz$; -6; -7 **8.** $-3x^4 + 2x^2 - x + 15$

9. $12x^6 - 8x^3 - 3x^2 + 4x - 10$ **10.** $9 + 2y^2 - 3y^3 + y^4$ **11.** $-2x^3 + 3x^2y + 6y^2 - 5xy^4$

12. 220; 4   **13.** 554; -4 **14.** \$15,822.40   **15.** $2x^3 + x - 4$ **16.** $-13x^2y^2 - 10x^3$

17. $6xyz - 3x^2y + 6xy$ **18.** $2x^2 + 2x - 1$ **19.** $-\left(9x^4 - 3x^3 + 2x^2 - 5\right)$, $-9x^4 + 3x^3 - 2x^2 + 5$

20. $-\left(-12xy - 7x^2y^2 + xyz - 6\right)$, $12xy + 7x^2y^2 - xyz + 6$ **21.** $2x^2 + 9x - 11$

22. $10a^3 - 3a^2 - 12$ **23.** $8x^7yz - x^2y^2z^2 - x$ **24.** $-15x^5 - 7xy + 3$

Section 4.2

Your Turn: Multiplication of Any Two Polynomials: 1. $-48a^7b^6c^2$ **2.** $36a^4 - 72a$
3. $3x^9 - 3x^5 - 6x^4 + 6$ **4.** $-10x^5 + 2x^4 + 22x^3 - 34x^2 + 2x + 6$ **Product of Two Binomials Using the FOIL Method: 1.** $18x^2 - 42x + 12$ **2.** $m^3 + 9m^2 - 12m - 160$ **Squares of Binomials: 1.** $9y^2 - 30y + 25$ **2.** $\dfrac{1}{4}x^2 - 5x + 25$ **Products of Sums and Differences:**
1. $25x^4y^2 - 1$ **2.** $0.36x^2 - 1.44y^2$ **Using Function Notation: 1. (a)** $a^2 + 5a - 4$;
(b) $a^2 - 3a - 4$ **2.** $16x^4y^{12} - 24x^2y^6 + 9$

Practice Exercises: 1. True   **2.** False   **3.** False   **4.** True   **5.** $8x^8$ **6.** $15x^8y^9$ **7.** $-27a^3b^2c^5$
8. $24 - 12x$ **9.** $-12a^3 - 28a^2$ **10.** $-35x^3y^4 + 30x^5y^2$ **11.** $10x^2 + 13x - 3$
12. $x^3 + 6x^2 + 3x - 10$ **13.** $3x^3 + 13x^2 - x - 20$ **14.** $x^2 - \dfrac{2}{3}x + \dfrac{1}{12}$ **15.** $12x^2 - 2x - 2$
16. $3.12x^2 + 0.7x - 15$ **17.** $x^2 - 12x + 36$ **18.** $4x^2 + 12xy + 9y^2$ **19.** $x^6y^2 - 4x^3y + 4$
20. $a^2 - 100$ **21.** $9 - 36x^2$ **22.** $-16a^2 + b^4$ **23.** $8x^3 - 26x^2 + 17x - 5$ **24.** $t^2 - 5t - 5$
25. $2a^2 + 4ah + 2h^2 + 3$ **26.** $-a^2 + 4a + 13$

Section 4.3

Your Turn: Terms with Common Factors: 1. $2xy(4y+2x^4-x^2)$ **2.** $-7x(x^2-2x+3)$
3. $2(lw+lh+wh)$ **Factoring by Grouping: 1.** $(3x-2)(2x+3)$ **2.** $(x-1)(2x^2+3)$
3. $(a+4)(2+b)$ **4.** $(x-y)(2x-3y)$

Practice Exercises: 1. $25x^5y^3z^2$ **2.** prime **3.** $x-2$ **4.** $-1(y-x)$ **5.** $6(2x^2-3x+4)$
6. $3a^3(2a^3+3a^2-4)$ **7.** $4x^4y(4y-2xy^2+x^2)$ **8.** $a^3(4a^5-a^2-1)$ **9.** $-9(x+5)$
10. $-3(x^2-3x+4)$ **11.** $-1(x^3+5x^2+6x-7)$ **12.** $-6a(-a+3b)$ **13.** $(r+2)(t+s)$
14. $(3a^2+2)(4a^3+1)$ **15.** $(4x^2+3)(x+2)$ **16.** $(x-y)(5x-6y)$ **17.** $P(x)=x(x-4)$
18. $R(x)=-0.5x(-700+x)$

Section 4.4

Your Turn: Factoring Trinomials: x^2+bx+c: **1.** $(x-9)(x+1)$ **2.** $3(x+1)(x+4)$

Practice Exercises: 1. (a) **2.** (d) **3.** (b) **4.** (c) **5.** $(x+1)(x+6)$ **6.** $(t+1)(t-12)$
7. $(x-3)(x-5)$ **8.** $(x+6)(x-2)$ **9.** Not factorable **10.** $3(x+2)(x-8)$
11. $\left(y-\frac{1}{4}\right)\left(y-\frac{1}{4}\right)$ **12.** $x^5(x+1)(x+27)$ **13.** $-1(x+4)(x-9)$,
or $(-x-4)(x-9)$, or $(x+4)(-x+9)$ **14.** $(a+b)(a+5b)$ **15.** $(x^2-8)(x^2+1)$
16. $(t+0.3)(t+0.3)$ **17.** $-x(x-1)(x-2)$

Section 4.5

Your Turn: The FOIL Method: 1. $(2x-3)(x+4)$ **2.** $-1(5a+6)(5a-3)$, or
$(-5a-6)(5a-3)$, or $(5a+6)(-5a+3)$ **3.** $a^2(4a+3b)(2a+3b)$ **The ac-Method:**
1. $(2x-1)(4x+3)$ **2.** $a^3(5a-2)(2a-3)$

Practice Exercises: 1. common **2.** last **3.** positive **4.** first **5.** $(2x-1)(5x+3)$
6. $(2a-5)(3a+2)$ **7.** $a(3a+2)(5a+1)$ **8.** $(2t-1)(t+1)$ **9.** $-5(4x-3)(x+1)$
10. $2(2p-3w)(3p-5w)$ **11.** $(5x-4)(x+1)$ **12.** $-2z(3z+1)(3z-5)$
13. $(4r-7)(r-7)$ **14.** $(2t-5)(5t-2)$

Section 4.6

Your Turn: Trinomial Squares: 1. $(y+6)^2$ **2.** $(4y+7)^2$ **3.** $(x-3y)^2$ **4.** $3x^2(2x+3)^2$
Differences of Squares: 1. $(x+2)(x-2)$ **2.** $(7t+3r)(7t-3r)$ **3.** $2(x+9y)(x-9y)$
More Factoring by Grouping: 1. $(a-1)(a+7)(a-7)$ **2.** $(4x+5+y)(4x+5-y)$
Sums or Differences of Cubes: 1. $(t^5-3)(t^{10}+3t^5+9)$ **2.** $(3x+1)(9x^2-3x+1)$
3. $2(4x^4+3y^3)(16x^8-12x^4y^3+9y^6)$ **4.** $(t+1)(t^2-t+1)(t-1)(t^2+t+1)$

Practice Exercises: 1. (b) **2.** (e) **3.** (c) **4.** (d) **5.** (a) **6.** $(x+3)^2$ **7.** $(t-9)^2$
8. $-y(y+5)^2$ **9.** $(3m+2n)^2$ **10.** $(5x-2)^2$ **11.** $3(x+6)^2$ **12.** $(x+12)(x-12)$
13. $(5a+8)(5a-8)$ **14.** $4(5m+1)(5m-1)$ **15.** $x^2(3x+4)(3x-4)$

16. $8x^2(y^2+z^2)(y+z)(y-z)$ **17.** $\left(\dfrac{1}{5}+x\right)\left(\dfrac{1}{5}-x\right)$ **18.** $(c+d+6)(c+d-6)$

19. $(y+1)(y+2)(y-2)$ **20.** $(12z+x-y)(12z-x+y)$ **21.** $(x+3)(x^2-3x+9)$

22. $(p-2y)(p^2+2py+4y^2)$ **23.** $(1-5a)(1+5a+25a^2)$ **24.** $y^3(xy-1)(x^2y^2+xy+1)$

25. $\left(a-\dfrac{1}{2}\right)\left(a^2+\dfrac{1}{2}a+\dfrac{1}{4}\right)$ **26.** $(a+1)(a^2-a+1)(a-1)(a^2+a+1)$

Section 4.7

Your Turn: A General Factoring Strategy: 1. $10(x+2)(x-6)$
2. $(m+3+5n)(m+3-5n)$

Practice Exercises: 1. True **2.** False **3.** False **4.** True **5.** $-7x(x+2)(x-4)$
6. $2y(3x+1)(3x-1)$ **7.** $-2(y+3)^2$ **8.** $(m+2)(3m+5)(3m-5)$
9. $4x(x-2)(x^2+2x+4)$ **10.** $a^2(3a-b)(9a^2+3ab+b^2)$ **11.** $(x-9+6y)(x-9-6y)$
12. $2p^3(2p+7)(2p-7)$ **13.** $-4t^2(t^3+3)$ **14.** $24(x+2y)(x^2-2xy+4y^2)$

Section 4.8

Your Turn: The Principle of Zero Products: 1. $-6,-\dfrac{5}{3}$ **2.** $-4,2$

3. $\{x\,|\,x$ is a real number *and* $x\neq 0$ *and* $x\neq 5\}$ **Applications and Problem Solving: 1.** 4 in.

Practice Exercises: 1. False **2.** True **3.** False **4.** False **5.** $0,\dfrac{1}{3}$ **6.** $-12,2$ **7.** $0,\dfrac{5}{2}$

8. $-4,4$ **9.** $-\dfrac{4}{3},\dfrac{2}{5}$ **10.** $-1,0,3$ **11.** $-9,-5$ **12.** $-7,-2,2$

13. $\{x\,|\,x$ is a real number *and* $x\neq -4$ *and* $x\neq 4\}$

14. $\{x\,|\,x$ is a real number *and* $x\neq 0$ *and* $x\neq 2$ *and* $x\neq 3\}$ **15.** 5 hundred televisions
16. 15 m

Chapter 5

Section 5.1

Your Turn: Rational Expressions and Functions: 1. -3
2. $\{x \mid x \text{ is a real number } and \ x \neq -4 \ and \ x \neq 1\}$, or $(-\infty, -4) \cup (-4, 1) \cup (1, \infty)$ **Finding**

Equivalent Rational Expressions: 1. $\dfrac{3t(t-1)}{3t(2t+5)}$ **Simplifying Rational Expressions:**

1. $\dfrac{3}{y}$ **2.** $\dfrac{x+4}{5}$ **3.** $\dfrac{x+2}{x-5}$ **4.** $\dfrac{1}{y-3}$ **5.** -1 **Multiplying and Simplifying: 1.** $\dfrac{2(x+3)}{3}$

2. $\dfrac{(x-8)(x+1)}{(x+3)(x-5)}$ **Dividing and Simplifying: 1.** $\dfrac{2(x+2)}{5x}$ **2.** $\dfrac{(x+3)^2}{x+2}$ **3.** $\dfrac{x-y}{x^2 y^2}$

Practice Exercises: 1. True **2.** False **3.** False **4.** True **5.** $-\dfrac{2}{3}$

6. $\{x \mid x \text{ is a real number } and \ x \neq -1 \ and \ x \neq 11\}$, or $(-\infty, -1) \cup (-1, 11) \cup (11, \infty)$

7. $\dfrac{(a+3)(a-1)}{(a+3)(2a+5)}$ **8.** $\dfrac{3a^2}{5}$ **9.** $\dfrac{x-4}{3}$ **10.** $\dfrac{5x}{2}$ **11.** $\dfrac{t-6}{t+3}$ **12.** $\dfrac{2x+1}{5xy}$ **13.** $\dfrac{x(x-1)}{(x+4)(x+1)}$

14. $-\dfrac{3}{2}$ **15.** $y^2 + 2y + 4$ **16.** $\dfrac{4a}{(a+3)(a+6)}$ **17.** $\dfrac{-5(2x-5y)(x-2y)}{x+5y}$

Section 5.2

Your Turn: Finding LCMs by Factoring: 1. 60 **2.** $45x^2 yz$ **3.** $(2x+1)(x-5)(x+5)$
4. $t(t+1)(t-1)$ **5.** $2y^2(y^2+1)(y+1)^2(y-1)$ **Adding and Subtracting Rational**

Expressions: 1. $\dfrac{9+2a}{a}$ **2.** $\dfrac{2y^2+9}{15y^3}$ **3.** $\dfrac{x^2+2xy-y^2}{(x+y)(x+2y)}$ **4.** $\dfrac{5m+1}{m-5}$ **5.** $\dfrac{x}{(x-1)(x+2)}$

Combined Additions and Subtractions: 1. $\dfrac{2(2x+1)}{(x-2)(x+2)}$

Practice Exercises: 1. True **2.** False **3.** False **4.** True **5.** 120 **6.** $30x^4 y$
7. $y(y+5)(y-5)$ **8.** $(x+1)(x-1)(x-7)$ **9.** $12a^2(a-2)^2$ **10.** $18x(x+3)^2(x-3)$

11. $\dfrac{3}{k}$ **12.** $\dfrac{y-13}{(y-4)(y+1)}$ **13.** $\dfrac{3x^2+5x-39}{(x+5)(x-6)}$ **14.** $\dfrac{x^2-8x+3}{(x-1)^2(x-5)}$ **15.** $\dfrac{-c^2-cd-d^2}{(c+d)(c-d)}$

16. $\dfrac{2y+9}{(y-3)(y+2)}$ **17.** $\dfrac{5(2m+1)}{(m-2)(m+3)}$

Section 5.3

Your Turn: Divisor a Monomial: 1. $-12x^3 + 4x - \dfrac{2}{x}$ **2.** $-4x^2y^3 + 6x^3y - 10y^4$ **Divisor**

Not a Monomial: 1. $x - 1 + \dfrac{6}{x+5}$ **2.** $y + 3 + \dfrac{-y-7}{y^2+1}$ **Synthetic Division:**

1. $x^2 - 8x + 18 + \dfrac{-41}{x+2}$ **2.** $5x^2 + 12x + 24 + \dfrac{47}{x-2}$

Practice Exercises: 1. False **2.** True **3.** False **4.** True **5.** $6x^3 + 2x - 5$

6. $8mn^5 - 5n + 6m$ **7.** $-4y^7 + y - 2$ **8.** $\dfrac{a^3b^2}{2} + a - \dfrac{5}{b}$ **9.** $x + 5$ **10.** $z - 11 + \dfrac{46}{z+6}$

11. $2x^2 + 5x + 1 + \dfrac{6}{2x-1}$ **12.** $y^2 - 2y + 5 + \dfrac{-15}{y+2}$ **13.** $x^2 - 2x - 4 + \dfrac{-13}{x-3}$

14. $4x^2 - 13x + 47 + \dfrac{-191}{x+4}$ **15.** $y^2 - y - 1 + \dfrac{9}{y+1}$ **16.** $n^4 - 2n^3 + 4n^2 - 8n + 16 + \dfrac{-64}{n+2}$

17. $9x^3 + 6x^2 - 21x + 6$

Section 5.4

Your Turn: Simplifying Complex Rational Expressions: 1. $\dfrac{12a + m^5}{4am^2\left(3 + m^3\right)}$

2. $\dfrac{k(1+k)(1-k)}{2(6-k)}$ **3.** $\dfrac{n^2 + nm + m^2}{mn(n+m)}$ **Difference Quotients: 1.** 4 **2.** $-2x - h$

Practice Exercises: 1. complex rational expression **2.** LCM **3.** numerator, denominator

4. numerator, denominator **5.** $-\dfrac{2}{5}$ **6.** $\dfrac{x}{3-x}$ **7.** $\dfrac{2(x-6)}{3(x+5)}$ **8.** $\dfrac{x-5}{(x-1)(x^2-x+3)}$

9. $\dfrac{cd(d-c)}{d^2 - cd + c^2}$ **10.** $\dfrac{(x+8)(x-2)}{x(x+1)}$ **11.** 4 **12.** $6x + 3h$ **13.** $\dfrac{-1}{6x(x+h)}$, or $-\dfrac{1}{6x(x+h)}$

Section 5.5

Your Turn: Rational Equations: 1. $-1, \dfrac{1}{3}$ **2.** No solution **3.** 0

Practice Exercises: 1. False **2.** True **3.** False **4.** True **5.** $\dfrac{15}{2}$ **6.** -32 **7.** $\dfrac{15}{16}$ **8.** $\dfrac{5}{9}, 3$

9. No solution **10.** $-\dfrac{2}{11}$ **11.** No solution **12.** 1, 8 **13.** -2 **14.** 3 **15.** $-3, 6$

Section 5.6

Your Turn: Work Problems: 1. $1\frac{5}{7}$ hr **2.** Gavin: 15 hr; Jeff: 10 hr **Applications**

Involving Proportions: 1. 10 gal **2.** 30 whales **Motion Problems: 1.** 4.5 ft/sec
2. Highway: 70 mph; country roads: 40 mph

Practice Exercises: 1. rate; time, or time; rate **2.** distance; time **3.** distance; rate **4.** 1

5. $5\frac{5}{11}$ hr **6.** Robert: 12 days; Georgia: 15 days **7.** 18 hr **8.** 54 hits **9.** 135 deer

10. 30 ft/sec **11.** 5 km/h

Section 5.7

Your Turn: Formulas: 1. $y = \dfrac{k}{3k-2}$ **2.** $t_2 = \dfrac{mSt_1 - H}{mS}$

Practice Exercises: 1. $yx = k$ **2.** $b = ac + c - b$ **3.** $w + p = t$ **4.** $b = \dfrac{ay}{x}$ **5.** $a = \dfrac{wc}{w-c}$

6. $c = \dfrac{ab-bx}{ax}$ **7.** $x_2 = \dfrac{2A - x_1 w}{w}$ **8.** $g = \dfrac{AG}{a+A}$ **9.** $t = \dfrac{1}{y_1 + y_2}$ **10.** $P = \dfrac{A}{rt+1}$

Section 5.8

Your Turn: Equations of Direct Variation: 1. 6; $y = 6x$ **Applications of Direct**

Variation: 1. \$13 **Equations of Inverse Variation: 1.** 7.5; $y = \dfrac{7.5}{x}$ **2.** 288; $y = \dfrac{288}{x}$

Applications of Inverse Variation: 1. 5 workers **Other Kinds of Variation: 1.** $y = 4x^2$

2. $W = \dfrac{2}{d^2}$ **3.** $y = 6xz$ **Other Applications of Variation: 1.** 12.8 W/m^2

Practice Exercises: 1. (d) **2.** (a) **3.** (b) **4.** (c) **5.** 5; $y = 5x$ **6.** 8; $y = 8x$ **7.** $\dfrac{2}{7}$; $y = \dfrac{2}{7}x$

8. $\dfrac{1}{2}$; $y = \dfrac{1}{2}x$ **9.** 48 mi **10.** 35; $y = \dfrac{35}{x}$ **11.** 9; $y = \dfrac{9}{x}$ **12.** 10; $y = \dfrac{10}{x}$ **13.** 330; $y = \dfrac{330}{x}$

14. 3 hr **15.** $y = \dfrac{0.64}{x^2}$ **16.** $y = \dfrac{8xz}{w}$ **17.** 1200 gal

Chapter 6

Section 6.1

Your Turn: Square Roots and Square-Root Functions: 1. 3 **2.** $\frac{5}{9}$ **3.** -11 **4.** 0.05

5. 9.110 **6.** $y+8$ **7.** $\sqrt{14}$ **8.** $\left\{x \mid x \ge -\frac{5}{2}\right\}$, or $\left[-\frac{5}{2}, \infty\right)$ **9.**

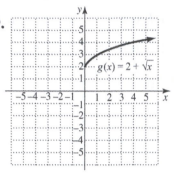

Finding $\sqrt{a^2}$: 1. $9|y|$ **2.** $|x+2|$ **3.** $x+2$ **Cube Roots: 1.** -2 **2.** $2x$ **3.** 3
Odd and Even kth Roots: 1. 10 **2.** -2 **3.** Does not exist as a real number **4.** $2x+1$
5. $-|3x|$, or $-3|x|$

Practice Exercises: 1. True **2.** True **3.** False **4.** True **5.** $10, -10$ **6.** $16, -16$ **7.** $-\frac{10}{9}$
8. 0.5 **9.** 12.369 **10.** $3x-7$ **11.** 3 **12.** Does not exist as a real number **13.** 1
14. $|3x|$, or $3|x|$ **15.** $|-5x|$, or $5|x|$ **16.** $|y-6|$ **17.** 5 **18.** 1 **19.** $-6x$ **20.** -1 **21.** $|x|$
22. $-\frac{1}{2}$ **23.** -1 **24.** a **25.** $|y+8|$ **26.** $2x+5$

Section 6.2

Your Turn: Rational Exponents: 1. 7 **2.** $\left(-3x^2y\right)^{1/5}$ **3.** 729 **4.** $\left(9xy\right)^{5/4}$

Negative Rational Exponents: 1. $\frac{1}{5}$ **2.** $\frac{1}{8}$ **3.** $\frac{1}{\left(11xyz\right)^{2/3}}$ **4.** $\frac{7y^{4/5}}{x^{3/2}}$ **5.** $\left(\frac{5ab}{2xy}\right)^{3/5}$

6. $a^{2/7}$ **Laws of Exponents: 1.** $x^{9/10}$ **2.** $a^{-1/2}$, or $\frac{1}{a^{1/2}}$ **3.** $3.4^{1/2}$ **4.** $x^{-2/15}y^{1/10}$, or $\frac{y^{1/10}}{x^{2/15}}$

Simplifying Radical Expressions: 1. t^3 **2.** $\sqrt[5]{x^2}$ **3.** $\sqrt{3}$ **4.** $\sqrt[24]{x^7}$ **5.** $x^{25}y^5z^{10}$ **6.** $\sqrt[15]{x}$

Practice Exercises: 1. (d) **2.** (a) **3.** (c) **4.** (b) **5.** $\sqrt[3]{x^2}$, or $\left(\sqrt[3]{x}\right)^2$ **6.** 9 **7.** $\sqrt[3]{x^2y^2}$

8. 32 **9.** $5^{1/3}$ **10.** $m^{3/4}$ **11.** $\left(4xy\right)^{5/2}$ **12.** $\left(2x^2y\right)^{8/3}$ **13.** $\frac{1}{9}$ **14.** $\frac{x^2z^4}{y^{1/3}}$ **15.** $12ac^{4/7}$ **16.** 9

17. $\frac{5c^{3/4}}{a^{3/2}b^{1/5}}$ **18.** $\frac{1}{\left(6xyz\right)^{3/5}}$ **19.** $7^{1/2}$ **20.** $12^{3/8}$ **21.** $2.1^{4/3}$ **22.** $x^{13/15}$ **23.** $\frac{b^{1/28}}{a^{1/6}}$ **24.** $\frac{1}{5}$

25. $\sqrt[5]{x}$ **26.** a^2b^2 **27.** $\sqrt{3}$ **28.** $\sqrt[12]{ab}$ **29.** x^8y^{12} **30.** $\sqrt[4]{8}$

Section 6.3

Your Turn: Multiplying and Simplifying Radical Expressions: 1. $\sqrt{65}$ **2.** $\sqrt[3]{20x^2y}$
3. $\sqrt[6]{200}$ **4.** $5\sqrt{5}$ **5.** $4y^2\sqrt{3x}$ **6.** $2a^3b^2\sqrt[3]{4ab}$ **7.** $2\sqrt[3]{9}$ **8.** $-24\sqrt{10}$ **9.** $3x^2\sqrt[4]{x^3y^2}$

Dividing and Simplifying Radical Expressions: 1. 5 **2.** 16 **3.** $\dfrac{2}{3}$ **4.** $\dfrac{x^4\sqrt{3x}}{5y^3}$

5. $\sqrt[10]{a^9b}$, or $a\sqrt[10]{\dfrac{b}{a}}$

Practice Exercises: 1. $81x^4$ **2.** simplify **3.** perfect cube **4.** the same **5.** $5\sqrt{2}$ **6.** $2\sqrt[3]{5}$
7. $2x^2\sqrt{30x}$ **8.** $-2x^3\sqrt[3]{3}$ **9.** $5b^3\sqrt{2ab}$ **10.** $2x^3y\sqrt[4]{2y}$ **11.** $2\sqrt{15}$ **12.** $7a^4\sqrt{2}$
13. $a\sqrt[3]{18}$ **14.** $x^2y^4\sqrt{x^2y^3}$ **15.** $2xy\sqrt[4]{180x^3y^3}$ **16.** 5 **17.** $x^2\sqrt{2}$ **18.** $5y^2\sqrt[3]{x^2}$ **19.** $\sqrt[12]{x^5y^2}$
20. $\dfrac{4}{5}$ **21.** $\dfrac{7}{x^2}$ **22.** $\dfrac{xy^2\sqrt[4]{3}}{z}$ **23.** $\dfrac{5\sqrt[3]{a}}{2b}$ **24.** $\dfrac{3x\sqrt[3]{3x^2}}{y^2}$ **25.** $\dfrac{ac\sqrt[5]{a^4b^2c}}{z^3}$

Section 6.4

Your Turn: Addition and Subtraction: 1. $10\sqrt{3}$ **2.** Cannot be simplified
3. $9\sqrt[4]{2xy}-5\sqrt[3]{2xy}$ **4.** $20\sqrt{2}$ **5.** $(y^3+1)\sqrt[3]{5x}$ **More Multiplication: 1.** $\sqrt{15}+y\sqrt{5}$
2. $5y\sqrt[3]{4}$ **3.** $16-6\sqrt{7}$ **4.** $20-14\sqrt{3}$ **5.** 23

Practice Exercises: 1. True **2.** False **3.** False **4.** True **5.** $6\sqrt{3}$ **6.** $3\sqrt[3]{5}$
7. $4\sqrt{6}+4\sqrt[4]{7}$ **8.** $-4\sqrt{5}$ **9.** $(7x+5)\sqrt{x}$ **10.** $12\sqrt[3]{2}$ **11.** $2\sqrt{x+4}$ **12.** $5x\sqrt[3]{x^2}$
13. $6\sqrt{5}-15$ **14.** $3y-3\sqrt[3]{2y^2}$ **15.** $3+2x\sqrt{3}+x^2$ **16.** $a-5\sqrt{a}-36$ **17.** $17-\sqrt{5}$
18. $2\sqrt[3]{25}-16\sqrt[3]{35}+24\sqrt[3]{49}$ **19.** 33 **20.** $x-w$

Section 6.5

Your Turn: Rationalizing Denominators: 1. $\dfrac{\sqrt{2}}{2}$ **2.** $\dfrac{\sqrt[3]{36}}{2}$ **3.** $\dfrac{\sqrt{5x}}{2x}$ **4.** $\dfrac{\sqrt[3]{4x^2y^2z^2}}{4xyz}$

Rationalizing When There are Two Terms: 1. $\dfrac{4x-4\sqrt{3}}{x^2-3}$ **2.** $\dfrac{4\sqrt{2}-4\sqrt{11}-\sqrt{22}+11}{-9}$

Practice Exercises: 1. True **2.** False **3.** False **4.** True **5.** $\dfrac{\sqrt{3}}{3}$ **6.** $\dfrac{\sqrt{42}}{7}$ **7.** $\dfrac{\sqrt{30}}{9}$
8. $\dfrac{\sqrt[3]{3x^2y}}{xy}$ **9.** $\dfrac{\sqrt{21x}}{6}$ **10.** $\dfrac{\sqrt[4]{x^2y^3}}{2yz}$ **11.** $\dfrac{5-\sqrt{3}}{11}$ **12.** $\dfrac{8\sqrt{y}+4y}{4-y}$ **13.** $\dfrac{2+\sqrt{14}+\sqrt{6}+\sqrt{21}}{-5}$
14. $\dfrac{6-13\sqrt{x}+6x}{4-9x}$

Section 6.6

Your Turn: The Principles of Powers: 1. 25 **2.** 2 **3.** −12 **Equations with Two Radical Terms: 1.** 5 **2.** 1 **3.** 2 **Applications: 1.** 45 ft

Practice Exercises: 1. isolate **2.** powers **3.** radical **4.** solve **5.** original **6.** 121 **7.** −16 **8.** 3 **9.** 2, 5 **10.** 1 **11.** No solution **12.** $\dfrac{1}{2}$ **13.** No solution **14.** 5 **15.** 3 **16.** −20 **17.** 10 **18.** 16 m

Section 6.7

Your Turn: Applications: 1. $\sqrt{2}$; 1.414 **2.** $\sqrt{409}$ ft; 20.224 ft

Practice Exercises: 1. True **2.** False **3.** True **4.** False **5.** $\sqrt{85}$; 9.220 **6.** 8 **7.** $\sqrt{82}$ cm; 9.055 cm **8.** $\sqrt{2}$ m; 1.414 m **9.** $\sqrt{160,900}$ ft, or $10\sqrt{1609}$ ft; 401.123 ft **10.** $\sqrt{4525}$ m, or $5\sqrt{181}$ m; 67.268 m

Section 6.8

Your Turn: Increasing, Decreasing, and Constant Functions: 1. (a) $(-2, 2)$; **(b)** $(-5, -2)$; **(c)** $(2, \infty)$ **Relative Maximum and Minimum Values: 1.** Relative maximum: 5.500 at $x = -1.500$; relative minimum: 0.870 at $x = 0.167$; increasing: $(-\infty, -1.500)$, $(0.167, \infty)$; decreasing: $(-1.500, 0.167)$ **Applications of Functions: 1. (a)** $A(x) = x(36 - x)$, or $36x - x^2$; **(b)** $(0, 36)$; **(c)** 18 ft by 18 ft **Functions Defined Piecewise: 1.** $f(-5) = -7$, $f(-2) = 1$, $f(0) = 1$, $f(3) = 9$

2.
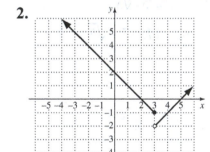

3.
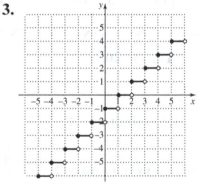

Practice Exercises: 1. False **2.** True **3.** True **4.** False **5. (a)** $(-\infty, -3)$, $(-1, 2)$; **(b)** $(-3, -1)$; **(c)** $(2, \infty)$ **6.** Relative maximum: 4.902 at $x = 0.975$; relative minimum: -4.766 at $x = -1.709$; increasing: $(-1.709, 0.975)$; decreasing: $(-\infty, -1.709)$, $(0.975, \infty)$ **7.** $A(b) = b^2 + 3b$ **8. (a)** $A(x) = x(360 - 3x)$, or $360x - 3x^2$; **(b)** $(0, 120)$; **(c)** 180 yd by 60 yd, where the side opposite the river is 180 yd **9.** $f(-3) = -9$, $f(-1) = 1$, $f(0) = 2$, $f(4) = 18$

10.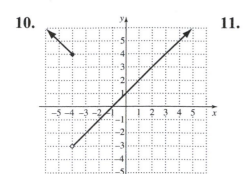

11.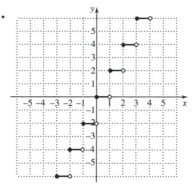